New Leadership in Strategy and Communication

Nicole Pfeffermann
Editor

New Leadership in Strategy and Communication

Shifting Perspective on Innovation, Leadership, and System Design

 Springer

Editor
Nicole Pfeffermann
Paris, France

ISBN 978-3-030-19683-7   ISBN 978-3-030-19681-3   (eBook)
https://doi.org/10.1007/978-3-030-19681-3

This Springer imprint is published by the registered company Springer Nature Switzerland AG
The registered company address is: Gewerbestrasse 11, 6330 Cham, Switzerland

Foreword

In the second edition of Strategies for Business Model Innovation: Challenges and Visual Solutions for Strategic Business Model Innovation, the cognitive challenges identified were based on the individual level. They stated three main themes: challenges based on the complexity of the task, the existing dominant logic, and the knowledge required. If we focus on the knowledge required, it stems from communications and the human abilities, which include intuition/strength, inner clarity, wisdom, etc.

Vision, leadership, and execution all ride on the back of communication. Does the leader communicate a benefit mind-set? Everyday leaders who seek to be well and do good, focus purposefully on why they do what they do and believe in developing their strengths and meaningfully contribute to a future of greater possibilities. Does the leader positively contribute to the organizational strategy?

The organizational strategy historically did not include both information technology (IT) and communication departments. Communication and IT were in a support role. Organizations strategic business model creation should include a seat at the table for both communications and information technology.

A communication strategy, first, should not be externally focused but an internal model. Most organizations neglect or spend very little time on the internal model.

A communication strategy is critical to success because it is the vehicle that supports vision, leadership, and execution. It will enhance brand value and mission/alignment for employees and is essential to business success.

Think of it as if you are planning a big trip to a foreign land. You have completed your research on the region and determined where you want to go. You made sure to have money exchanged, so you would get there with the proper currency in your pocket. You have secured your desired lodging. You have contracted several guided tours. You have made sure your clothing is appropriate for the weather or region. You have gathered books and viewed multiple Web sites on the destination. You have secured time off work and made backup plans if something is needed. You made sure the mail and newspaper delivery is stopped for that period, or e-mail has the appropriate out-of-office message.

However—if you did not secure transportation to the destination—you will not get there. A communication strategy is the transportation. It is the vehicle for your vision, goals, and corporate culture.

Executives will often believe failed agendas are because the goals were too big, or the evaluation of the business climate was off, or the target market was over-estimated and while it could be those things more than likely, it is because they **lacked a communication strategy**.

The Board of Directors and C-Suite usually construct a vision and strategy for an organization. Yet, in most organizations, if you ask them to outline the vision or strategy, they are unable to articulate it. A 2013 McKinsey study of 772 Board Directors reported that only 34% thought their Board understood their strategy.

It is often the lack of a communication strategy that results in the inability of employees (and even Board members) to articulate the goals of the company (and preferable their role in reaching those goals).

A communication strategy determines the hurdles facing new information to be shared. It is organized, clear, and concise and determines the vehicle of sharing. It resolves the potential hurdles such as:

- What will the resistance be?
- What might employees think/fear about it? (values, beliefs, fear, and assumptions)
- How to identify issues and create dialogue?
- What are the best means of creating this dialogue?

The how and the where are often confused. How you communicate is the way you share the tone and impact of the message. It also understands the appropriate channel for your communication, evaluating the strengths and weaknesses of alternative channels for message delivery.

For example, the most elementary form of infographics, which have been in existence for centuries, takes the form of maps, graphs, or illustrations. Starting in the 1970s, they could often be found in newspapers such as the Sunday New York Times and USA Today. However, advances in digital and social media technology in the twenty-first century have brought about a new type of infographic that propagates the Internet. The latest generation of infographics emphasize enhanced visual design, layout, and visual imagery that are now feasible with Adobe Flash, HTML5, and CSS3.

In the research conducted Young, Amy and Hinesly, Mary, Infographics as a Business Communication Tool: An Empirical Investigation of User Preference, Comprehension & Efficiency (January 12, 2014). SSRN: https://ssrn.com/abstract= 2548559 or http://dx.doi.org/10.2139/ssrn.2548559, it was found millennials were more likely to say they would read the article (information) and recommend it to others if they received the infographic version than if they received the text version.

However, baby boomers attained better comprehension when they received the text version compared to the infographic version. Findings indicate that infographics and traditional text approaches are needed to reach a multigenerational workforce, although infographics will likely grow in popularity in the next decade as more millennials enter the workforce and baby boomers continue to exit it.

This is a simple example of the need for evolution and integration of communications and information technology as a necessary inclusion in the strategy models going forward. The evolution of technology use and the successful integration within the organization will provide advantages.

Psychologist Abraham H. Maslow contended that human needs are structured in a hierarchy; as each level of needs is satisfied, the next higher level of unfulfilled needs becomes predominant. See Abraham H. Maslow, "A theory of human motivation," *Psychological Review*, 1943, Volume 50, Number 4, pp. 370–96; and Abraham H. Maslow, *Motivation and Personality*, first edition, New York, NY: Harper & Brothers, 1954.

Strategies for Business Model Innovation: Challenges and Visual Solutions for Strategic Business Model Innovation shares the need to begin at the proper level in which to build upon based on Maslow's theory.

Additionally, IT in most organizations is not involved in strategy or communication strategy and yet it should be because communicating through technology has only increased. Currently, in organizations when IT is discussed, it is as a procurement or support center. **Information Technology or CIO's should be at the strategy table alongside the Chief Communications Officer**.

Approximately 45 projects completed with Fortune 500 companies being studied on the use of how social/digital media tools are used internally hints at the need for knowledge sharing and creation stems in **the ability to communicate across silos in positive organizational or departmental cultures. Simplicity of resource retrieval or connection was important**. There is much more work to be done in this area to better understand how the knowledge portion fits into the cognitive challenges based on the complexity of the task, the existing dominant logic, and the knowledge required as identified in Strategies for Business Model Innovation: Challenges and Visual Solutions for Strategic Business Model Innovation.

This new contributed volume focuses on this linkage between strategy and communication from a leadership perspective to assist in shifting to the new leadership paradigm with an enhanced understanding about strategy communication, communication strategy and the important role of information technology (IT) and interaction with human individuals within and across organizations.

For example, engagement is essential in collaborative innovation, transformation, and business innovation but usually something is missing when individuals are engaged in open innovation. The connection is far more than engagement. Human connection as part of human leadership in business practice requires leadership communication skills for:

- Being in sync and relating to human individuals for sharing social, emotional moments
- Exchanging ideas and thoughts with a responsive partner who mutually participates
- Connecting with dialog partners in a present, meaningful, and enriching manner

The book structure helps readers to discover new approaches, gain advanced knowledge, learn about cutting-edge topics, and also dive deeper into the integrated perspective in four sections:

(1) New Leadership in Strategy—Focus on business model innovation, strategy, and transformation
(2) New Leadership in Communication—Focus on communication skills, tools (e.g., infographics, canvas, games, music, and multisensory), strategic communication, and strategy communication
(3) New Leadership in Education—Focus on leadership development in business practice and educational sector
(4) New Leadership in Practice—Focus on case examples, sharing practical experience, industry insights, and fresh ideas from a practical perspective.

Dr. Mary D. Hinesly
Ross School of Business
Ann Arbor MI, USA
mhinesly@umich.edu

Dr. Mary D. Hinesly earned her doctorate at Grenoble École de Management Technology and Innovation in France. She has over 20 years in the private sector. Dr. Hinesly has served as the Director of Research for Thompson Associates, and the Director of Educational Content and Research for the National Retail Federation. Dr. Hinesly has been with the Ross School of Business at the University of Michigan for over 14 years, teaching business history, business communications, and co-created new curriculum in the area of social/digital business systems, processes and communications. She has been involved with over 100 companies including Google, Cisco, Amazon, Jive, Microsoft, PwC, Unilever, Wacker Chemical, GE, LVMH, Saudi Arabian Monetary Authority (SAMA), STC, The New York Times, Metalsa, and more.

Acknowledgements

The collection of works, published in this edition, aims to make a valuable contribution to the area of strategic innovation management and communication management from a leadership perspective, covering recent and future developments in new leadership, system design, transformation, and empowerment in the advanced co-creation age.

A number of people have contributed to making the new edition of this book possible. First of all, I would like to thank Jan-Philip Schmidt, current Editor, Physical Sciences and Engineering, and his Book Coordination Manager Mrs. Petra Jantzen who saw the potential of this new project and gave their commitment to the concept of this book as well as the support that brought this edition to fruition.

Writing chapters is especially challenging when submission deadlines compete directly with other academic, professional, and social tasks—especially in the digitalized information age. I would like to express my strong gratitude to all contributors for taking the time to write a chapter and help to co-define and co-create the value of this book and have a great impact on shifting to the new leadership world.

Finally, my thanks go to the readers and reviewers of my previous editions, who have supported me preparing this volume. Last but not least, I sincerely hope that researchers, students, colleagues, business managers, and (advanced) co-creators will enjoy reading this book and be inspired by multiple perspectives and managerial implications provided by the insightful chapters.

Paris, France
August 2017

Nicole Pfeffermann
nicole.pfeffermann@googlemail.com

Contents

Editor and Contributors

About the Editor

Nicole Pfeffermann is a Associate Advisor and advanced pioneer specialized in new leadership, innovation communication, and leaderart communication. She is dedicated to enhancing leadership skills and updating humanity to truly empower (young) individuals, shift to new realities and worlds with art leaders, embrace artistic talent and (young) artists, and create new (neuronal) pathways for personal, business, and system growth that lasts. She has more than 15 years of professional experience in strategy and innovation, research transfer, and management consulting and is editor/author and senior lecturer in leaderart communication. She taught digital business, methodology, and information and knowledge management. She did her Ph.D. in engineering (robotics, Deutsche Post DHL Scholarship) and studied business economics.

Contributors

Bernardo Alayza Pontifical Catholic University of Peru (PUCP), Lima, Peru

Beatrice Bauer SDA Bocconi School of Management, Milan, Italy

Bettina Sophie Blasini University of Cambridge, Cambridge, UK

Hanane Bouzidi Deutsche Telekom AG, Bonn, Germany

Andrew Breen New York, NY, USA

Noel Capon Columbia Business School, New York, USA

Leonardo Caporarello SDA Bocconi School of Management, Milan, Italy

Phillip Cartwright PSB Paris School of Business, Paris, France

Rani J. Dang University of Cambridge, Cambridge, UK

Martin de Bree Hellevoetsluis, The Netherlands

Jean-Philippe Deschamps IMD, Lausanne, Switzerland

Joshua Entsminger IE University, Segovia, Spain;
Ecole des Ponts Business School, Paris, France

Mark Esposito Hult International Business School, Cambridge, USA;
Thunderbird Global School of Management at Arizona State University, Phoenix,
USA

Lene Foss The School of Business and Economics, UiT-The Arctic University of
Norway, Tromsø, Norway

Kai Gausmann Capgemini Invent, Berlin, Germany

Babak Ghassim The School of Business and Economics, UiT-The Arctic
University of Norway, Tromsø, Norway

Domingo Gonzalez Pontifical Catholic University of Peru (PUCP), Lima, Peru

Stefan Hofrichter Allianz Global Investors, Frankfurt, Germany

William Howe San Diego, CA, USA

Gysele Lima Ricci Bundeswehr Universität, München, Germany

Peter Malone UWA Business School, University of Western Australia, Crawley,
WA, Australia

Beatrice Manzoni SDA Bocconi School of Management, Milan, Italy

Tim Mazzarol UWA Business School, University of Western Australia, Crawley,
WA, Australia;
Burgundy School of Business, Dijon, France

Cindy McCauley Centre for Creative Leadership, Greensboro, USA

Sandy McLean Centre for Creative Leadership, Greensboro, USA

Tim Minshall University of Cambridge, Cambridge, UK

Letizia Mortara University of Cambridge, Cambridge, UK

Winfried Müller BMW Bank GmbH, Munich, Germany

Nadine Page Hult International Business School, Berkhamsted, Hertfordshire, UK

Robert Pearl Stanford University, Palo Alto, USA

Nicole Pfeffermann Paris, France

Joseph Press Centre for Creative Leadership, Greensboro, USA

Sophie Reboud Burgundy School of Business, Dijon, France

Joseph C. Santora Paris, France

Christoph Senn St. Gallen University, St. Gallen, Switzerland

Samah Shaffakat INSEAD, Fontainebleau, France

Jialu Shan IMD Business School, Lausanne, Switzerland

Thomas Sullivan Hult International Business School, Cambridge, MA, USA

Kadeshah Swearing University College of the Cayman Islands, George Town, Grand Cayman, Cayman Islands

Leigh Thompson Kellogg School of Management, Northwestern University, Evanston, USA

Henrik Totterman Cambridge, MA, USA

Ian C. Woodward INSEAD, Fontainebleau, France

Lisa Xiong EMLyon Business School, Lyon, France;
Judge Business School at University of Cambridge, Cambridge, UK

Howard Yu IMD Business School, Lausanne, Switzerland

Chapter 1
Introduction

The Paradigm Shift: Human Self-leadership #Authenticity

Nicole Pfeffermann

> *To be yourself in a world that is constantly trying to make you something else is the greatest accomplishment.*
> Ralph Waldo Emerson

It is the 'too good to be true' new age of high-tech innovation, cozy co-working spaces and innovative playgrounds worldwide which allow to easily start a business venture and create new avenues for personal and business success. Perfect, isn't it?!

But why leads open innovation and digital transformation only to survival, growth, and prosperity for some organizations and systems? Why do systems become more and more split? What is the missing piece that when applied can truly help a firm and people to be empowered and benefit from new digital concepts, shared disruptive ideas and new business ventures?

Many articles and trainings focus on strategy and new leadership communication skills, for instance, how to clearly communicate visions, engage employees in strategic programs, be present and ask the right questions that will foster productive and intelligent communication, and actively listening to other people (e.g. Brent Gleeson, 5 Tips for Improving Leadership Communication, Forbes, May 2016). But is it the missing piece?

A key theme of this new contributed volume is the provision of an integrated perspective on new leadership in strategy and communication which allow (senior) leaders, managing directors, project managers, and individuals to (1) better link strategic business innovation and leadership and (2) shift to the human self-leadership paradigm and in particularly leadership advances that consider ideas from multiple disciplines and transgenerational views. That includes a new understanding about knowledge, learning and change and how leaders re-discover and develop their human abilities, which include intuition, balance and clarity, reflection and wisdom.

N. Pfeffermann (✉)
Paris, France

© Springer Nature Switzerland AG 2020
N. Pfeffermann (ed.), *New Leadership in Strategy and Communication*,
https://doi.org/10.1007/978-3-030-19681-3_1

1.1   The Role of Strategy and Misconception

For so many years we have built businesses and learned about value creation based on Porter's value chain (1985): Innovation was understood as a closed, primary production process from idea generation all the way to commercialization—meaning from inbound logistics all the way to customer service—supported by activities such as human resource management, procurement, and technology/information management.

After coming to terms with the extreme period of growth 1.0 with all the fast-growing Internet platforms and dot-com bubbles, a supplementary e-value chain was introduced to explain value creation in the so-called net economy (interchangeable with digital business): Information is thereby key to succeed in digital business, for instance, a platform creates value by gathering, selecting, and offering new information via mobile devices (e.g. UBER, Airbnb.com). In this context, the following three main added values for users are:

- Structuring value: Online offer provides an overview of a large amount of information
- Communication value: Online offer allows different consumers to communicate
- Selection value: Online offer allows consumers to locate the desired info/service.

Linking the two value chains is the ultimate formula for digital business success. Many business consultants and startups are involved in developing new IT-driven operating models and strategic programs to perform open innovation and expand into new (e-)markets. The firm infrastructure represents thereby still the foundation of digital business and the design of innovation-information processes.

What are the main challenges?

Linking value chains: Many digital transformation projects do not deliver the promised growth results and returns due to a missing knowledge base of how to best link the two value chains and successfully design open innovation-information processes.

Firm infrastructure: Many businesses have problems and misunderstand how to (1) build capabilities as a foundation for innovation in the digital age, (2) establish an open innovation culture, and/or (3) invest in real co-creation activities (not design methods).

Fragmentation of systems: It is per se a split system design (support and primary activities) and 'IT/Information Management', 'Communication' and 'Innovation' are separate units in the value creation process leading to silo thinking, budgeting issues.

Corporate innovation methods: Corporate-startup partnerships are a new means for exploitation and oftentimes it is not the right engagement model to effectively interact and collaborate with startups and use external value drivers in innovation management.

Resentment, fear, and emotional age: We experience a tech-driven imbalance in corporate systems and leaders who are involved in power games due to changing structures instead of focusing on expanding their leadership communication skills.

In fact, it is not complex, it is a paradigm shift and the 'old' familiar world is not working any longer. We have become digital and focus all our human energy and manpower on (a) ego-based power-control games to win the digital race and (b) new IT-driven operating models to desperately find ways to re-create the growth 1.0 pattern and conceptions. What is the outcome? We build enmeshed[2] (engagement level) and enmeshed[3] (co-dependent level) systems. And the new bubble is already in the making…

It is our collective, learned belief system—our navigation system and world map—how we think businesses are built and individuals relate, engage and connect with other individuals in systems to successfully innovate and achieve goals. We have learned how to communicate, innovate, and strategize to get a job done, achieve greater impact, and make progress in the world. If we want to change it, we must understand how we can modify our navigation system and how we communicate at a system level, that involves starting with leaders who can have a deep impact on systems and make a difference in the world.

1.2 The Role of Communication and Blueprints

Long before the economic meltdown Brooksley Born tried to warn about the threat to the financial system and was shut down by leaders of the Clinton administration. Why? In fact, nobody wanted to listen and the 'man behind the curtain', Alan Greenspan, could communicate in a way that "no one understood what he said, but he said it in such a way that everybody bought it" Arthur Levitt, SEC Chairman, 1993–2001, Frontline 'The Warning'.

The recent example of Hollywood, published by the journalist Ronan Farrow, reveals a very similar story of one 'man behind the curtain', Harvey Weinstein, and his enablers, the 'army of spies', who helped to track targets and shut down their attempt to speak up and speak out what needed to be heard for making better decisions and taking the initiative in leading necessary change to the system including strategy, structure, operations, and culture.

Both examples demonstrate the importance of knowing communication blueprints and its effects on financial, political and societal systems as well as biological, psychological-emotional systems of human beings. What is a communication blueprint? A communication blueprint is a navigation system, so-called knowledge scheme, which helps us to interact, act and reflect on situations and events in our life. We have learned in childhood how to relate, engage and interact with other individuals and, hence, it is the knowledge scheme 'how to communicate' which is deeply anchored in our subconscious. It is so basic that oftentimes the way how we communicate is given.

The Example 'There and Not There'

Maybe Brooksley Born and the abused women and men could improve their communication skills (body language, voice and negotiation skills) to obtain different results? Certainly, we can learn business communication, negotiation techniques and how we can use our voice, actively listen to others and ask the right questions in many courses at universities, business schools and educational institutions. But it would have made no difference.

A better question is: How can we communicate with a person who is there and not there, can't we? A communication blueprint is different for every individual based on how the upbringing looked like and how an individual has experienced relationships with caregivers and other individuals. If a person has experienced emotional abuse in childhood or had a caregiver who was emotionally and/or physically unavailable, this person has learned to relate to and interact with a caregiver who was 'there and not there' resulting in a similar communication blueprint in adulthood.

In sum, understanding communication blueprints is key to explain system dynamics resulting in the need for a better understanding of communication in leadership positions, for instance, learning about the diversion tactics and how new leadership communication helps to better relate, connect and engage in transition phases, politics and new business development.

1.3   The Role of Education and New Instruments

We live in an achievement-oriented society and companies focus on innovation, profit growth, performance and process efficiency. Human talent is supposed to micro-manage the defined project and job tasks and social media has become the ultimate psychological e-friend for breaks at work to get distracted, judge others or push harder to achieve goals and high-ranked profiles compared to others.

This work-lifestyle of 'being there and not there' and driving force of being not good enough—not fast enough, innovative enough, and high-ranked enough—is leading to stress reactions (PTSD) and the pressing need for gaining deeper insights into system resilience and growth at different levels rather than solely focusing on generic themes in management practice. The three different levels are as follows:

1. Management level: Micro-management focusing on innovation (process) design
2. Individual level: Builder 'Creating Ideas' focusing on information-interaction design
3. System level: Gardener 'Growing Ideas' focusing on system (re-)design.

From ideas and visions to innovation success, it is important to understand all three design levels and the main difference between growth 1.0 and 2.0, which means the difference between builders (=change communicators) and gardeners (=innovation communicators) relating to nourishment, foundation, and culture for personal, business and system growth that lasts.

Managers cannot invest time and energy in ideas/visions, getting things implemented and executed and eventually take a little bit of time for self-reflection and personal development/**leadership development**, and never go to the bottom (system level). Yes, it works for a short period of time, for short-term goals, but not in the long-run and with KPIs measuring the long-term goals and business performance

(#longevity). The management level is only the visible surface with max. 5–10% impact on system growth in the long-run. 90% is about the individual and system/collective level and how individuals interact, connect and learn in relationships and systems (relating and system patterns).

It is very important to gain a better and deeper understanding about system patterns because it gets worse over time. It never stops because it gives energy only in one direction for those who desperately needs energy supply from the energetic exchange system of human beings. A good example for enmeshment is innovation management and new business development: After building on internal ideas submitted by employees (idea management), now startups, students, the crowd, and creative people are an energetic source of ideas in the so-called open innovation economy. A misinterpretation or intentional misuse of this concept in dysfunctional systems causes more and more mental-emotional health problems including coming up with new concepts, such as design thinking, business model innovation, and especially ambidexterity for more exploitation and dysfunctional system patterns and cycles (dark age movement). As for diplomacy, there is a simple answer for why it doesn't work for those systems and it disappears: You cannot use logic and reasoning with mentally-emotionally disordered people. A realistic term would be 'depletioncy'.

Taking the leap, re-thinking and focusing on new educational (leadership) programs and instruments for empowerment and healthier relationship and system patterns is key to have a deep impact and build a new legacy. There is a pressing need for the development of new instruments in strategy and communication and new leadership development programs.

1.4 Outlook on Human Self-leadership

Individual leaders can surely make a difference in the world, but I think it is difficult when a solid foundation is absent and being 'there and not there' has become the status quo in relating networks. It is exhausting and makes it difficult to build impactful businesses and create growth opportunities. How can we fix it?

A first step in this process is to be prepared and committed to better understand how financial, educational and corporate systems were built. It is not complex because it points to similar problems and key themes:

Assumptions and expectations of how relating networks function based on communication blueprints. It is a collective, learned belief system that needs change and, hence, it needs amendments and refinements in educational and corporate systems.

For example, Porter's value chain (1985) is still the basis for organizational functions and job descriptions and corporate tax systems define how to build a firm's resource set and assets. In digital business, however, we can see a change from primary to secondary activities in order to develop new business models and successful community platforms (e.g. information and knowledge management is not a secondary activity in true digital innovation, it is at the heart of value creation). Another example is even bigger than the Enron case and Arthur Andersen LLP & Co., it is the so-called culture of 'gamesmanship' (Arthur Levitt) and deceptive reporting of financial results and turnarounds. In this context, communication is key to build trustful relationships with partners, investors and the public and a communication design can help to better avoid wrongdoings and misuse.

From Politics to Capability: A New Capability-Based View

A second step in this process is to re-think systems and commit to building new solid foundations. It presents, however, tremendous challenges for executives, digital/business transformation managers and strategy & innovation units. Why? Because the true cause is often ignored in outsourced project management with short-term goals and in particular in dysfunctional systems it is excluded from the 'ingredients list' for transition and growth.

One new approach is the dynamic capability-based view which ultimately leads to a shift from politics to capability and from political communication to leadership communication. That includes relational communication management and new leadership communication programs to empower individuals in building mature systems in the long-run #longevity.

We lost ourselves in the 'doing' world and micro-manage our entire day, weeks, and years focusing on the digital and material world. Relating has become the new status instrument—how many followers, likes, engagement activities, and 'connections' on social media channels—for instance, app users pay in restaurants and cafés with social sharing activities depending on their social status or invitations to special events and visible sessions. The growth space is defined as an opportunity to become a better micro-manager, not missing out a chance to better act, engage, and manage visible relating activities.

Leaders have become the facilitators at all three levels: They envision and shape the future of business with new ideas and visions; plan and execute strategy and innovation activities; and lead and motivate business units, teams, and individuals to perform open innovation. The growth spaces are limited to management and leadership development in terms of developing new concepts & methods and leadership skill sets to better DO something, which allows managers and leaders to successfully develop the digital company and manage digital innovation.

In sum, the **masculine-driven, old leadership world** is designed for **human doings who focus on open innovation activities and engagement based on information exchange**. The hashtags #läuft #doer #created are the mantras of smart innovators. It is the market-based, competitive, outside-in approach, functional and image-driven, to successfully innovate and develop (new) digital businesses and business models.

Summary: Main Characteristics of the Old Leadership World

- **[ab]USE Mindset in strategy and leadership**: Mainly focusing on innovation management with strategic activities, information exchange, and open engagement to feed the top
- **Unhealthy relating and interaction patterns**: (Attachment) trauma and toxicity in interaction styles leading to co-dependent relationships and abusive behavioral patterns and cycles
- **Growth—Exploitation and experimentation**: Digital is the new target for the company 2.0 in terms of USING digital technologies and USERS to grow new digital businesses.

The new leadership world is designed for human self-leaders who focus on open dialogs—from innovation design to system design—and human connection based on interaction. It is the inside-out, dynamic capability approach to seize and transform, create and co-create, and successfully govern ordinary capabilities and resources. Inspiration and solutions as well as clarity and methodology play a vital role for advanced co-creators to better strengthen the gardener capacity and grow their ideas and visions at a system level. The new hashtags are #weareWE #authenticity #real and the new mantra is 'Be the inspiration!'

Summary: Main Characteristics of the Human Self-leadership World

- **WE-Mindset in strategy and leadership**: Mainly focusing on innovation communication with dialog activities, interaction design, collaboration and human connection to empower human beings
- **Healthy relating and interaction patterns**: Communication design helps individuals and organizations to balance out information (energy) exchanges and master the art of relating
- **Growth—Authentic co-creating and learning**: Co-defining and co-creating value with prosumers and human self-leaders who self-express their own personality #authenticity.

What is the missing piece that when applied can truly help a firm and people to be empowered and benefit from new digital concepts, shared disruptive ideas and new business ventures?

It's simple. Communication is the answer

According to Gardner ('5 Minds for the Future') human beings need to develop five minds: Disciplined mind, creating mind, respectful mind, ethical mind, and synthesizing mind. I would like to add the ***communicating mind*** to the ensemble of the minds.

As human beings we have a well-organized system of ideas which helps us to understand what is going on in the world around us and inside (self-talk). It is a specific perceptual lens—a navigation system—that works for us in terms of how we communicate and feel comfortable with events and situations in life. What happens when our navigation system is challenged? To understand (new) ideas and how we react in unfamiliar situations or when we experience disruptive events, we use our knowledge schemes, our templates to organize information and understand interactions with others. These schemes (information-interaction designs) can be activated and modified every time we seek information, reflect, act, and interact with others based on our ability to communicate with others and ourselves.

Hence, our information-interaction design is the critical impact factor to help us understand how we can bring forward a new idea, be understood in back-and-forth exchanges with responsive partners, and (co-)create new worlds or transform systems. The communicating mind helps to systematically develop information-interaction designs over periods of time and, hence, our navigation system in different phases and relating stages in life (e.g. attachment and interaction, learning and transition phases) which automatically affects new leadership in strategy and communication.

At a Glance: The Communicating Mind: Listening carefully and active, willingness to reflect and understand before sending out messages, managing information transmission considering openness, time/timing, and interrelation, being aware of the impact and effects of information-interaction designs including instruments, activities, recipients, and channels.

Examples (formal education and place of work): Effectively presenting ideas in a kick-off meeting or conference session, inviting students or co-workers from abroad to co-create a new publication or organizational workflows, mastering critical reflection and mental strength (inner dialog) in conflicts and talks.

Period of development: Starting in childhood, continues as lifelong learning, increasing awareness and competence building over time for specific situations, activities, and events.

Pseudoforms: Pretending to be involved in interactions and understand other perspectives, thoughts, information, strategies or new ideas, only focusing on external conditions with a lack of authenticity and meaningful conversations, inappropriate organizing of information, documentation, and interactions (avoiding learning from mistakes), non-mutual conversation with reactive partners.

Appendix: 10 Most Common Diversion Tactics in Dysfunctional Systems—The Communication View

It is not a complete list of diversion tactics. This list provides an overview of common tactics in dysfunctional systems independent of the abuse type.

1. Toxic triangulation
It is a transaction model with an abuser, controlled target, and passive enabler(s). The triangle makes it very hard to detect the true cause of conflict, drama, and stress. It is a power game to create tensions between controlled target and enabler(s) leading to distraction from the abuse. Communication seems to be ineffective, not fruitful or progressive and cannot lead to (re)solutions because the abuser does not want it. Normally it is defensive communication: The controlled target and enabler(s) want to avoid power plays, anger, negative consequences, and violent outbursts of the abuser at all costs. The enabler(s) are passive and tolerating to the point where it becomes an extremely toxic, fear-based communication climate and negative, unhealthy system
2. Projection (Blaming)
It is a subtle or direct one-directional transaction between the abuser and target to dump wrongdoing and avoid responsibility. This defensive mechanism helps the abuser to blame the target and avoid being accountable for any actions and words. The abuser is not able to self-differentiate which means to understand the concept of self and other individuals. Projection helps thereby the abuser in a (passive-)aggressive way to violate boundaries and project his/her negative feelings and core beliefs onto the target. The target feels disorientated and stressed out because the feelings are not 'real'. On the other hand, the target also projects onto the abuser his/her good qualities and, hence, oversees the abuse
3. Gaslighting/cognitive dissonance
Another name for gaslighting is crazy-making or brain-washing. It is the most dangerous and subtle manipulation tactic out there. The abuser misdirects, reframes and twists conversations with the aim to disorient, confuse and frustrate the target. Communication with the abuser never leads to any concrete results or goes on and on in a circle with changing topics, argumentation lines and confusing, senseless remarks. Over a period of time the target starts self-gaslighting, dissociation, losing his/her own reality and feeling guilty for being a human individual with needs, thoughts, opinions, values, and moral standards—life-fading is the worst-case scenario which can lead to helplessness, depression, disempowerment and regression. The target loses his/her own voice
4. Word salad/mindreading/drama/lies
Word salad is a conversation tactic to deliberately overwhelm a target with buzzwords, images and circular reasoning to finally say and commit to nothing. It helps the abuser to confuse the target and prevents getting closure in a cycle of abuse: (1) with holding closure is power and control over the target, (2) avoidance of committing to wrongdoing and taking responsibility, and (3) it means unfinished business and, hence, a possibility to start a new cycle of mental and emotional abuse because the target wants to solve the problem and understand what was said and happened. It is a subconscious programming which leads to believing all the lies and emotionally-charged, negative messages of the abuser over and over again. Non-closure often leads to more drama and pain in those cycles of abuse

(continued)

(continued)

5. Ignoring/witholding/secretiveness
Communication is used in a cat-and-mouse game. The abuser intentionally withholds information, emotions, attention, intimacy and opinions to keep the target stuck in uncertainty and reactivity. In the silent treatment the target feels confused, worried, anxious, and doubtful, for instance, the abuser disappears or is present but does not answer questions, suddenly leaves the room, checks e-mails or starts a phone call while the target is talking about a specific topic. It is very effective in generating fuel from the target and makes the abuser powerful. Being punished with the silent treatment means feeling invisible, devalued, rejected and not enough in the presence of the abuser. Non-communication is thereby a way to deny appreciation, validation, approval, and love
6. Entitlement/superiority/grandiosity
This tactic is often used interchangeable with the term 'narcissist'. Engaging with those types of abusers can re-traumatize targets because the abuser diminish achievements, manufacture stress and fear-based, catastrophizing situations, sets impossible standards and expectations, and generalize with blanket statements. The abuser feels threaten by the achievements and talents of others and must win the game. The inferior target most often feels being controlled, out of balance, and shameful for not being worthy. The abuser wants to be the center of attention and is completely self-absorb in conversations including a hyper-focus on flaws, 'I am busy' interactions, and micro-management of PR and other activities to control events and outcomes. The abuser feels entitled and shows no remorse
7. Promises/fantasy land/enmeshment
In an abusive relationship the abuser expects from the target to take care, fix and heal, and make the abuser feel better. The target burdens him/herself with his/her emotional baggage and the abuser's destructive, negative emotions. Oftentimes the target grew up in a dysfunctional system and learned to take care of the parents and other siblings (the parentified child). As a result, the reactive target over-owns his/her responsibility and wants to fix problems and make things work. The abuser knows it and can easily play with promises, future faking and 'tangle the carrot': One day we are happy, you get this job, I recognize your competence, etc. It is a false core belief and the enmeshment leads to an undeveloped self, depression, self-doubt, obsessive thinking, and addiction
8. Play victim/hoovering/love bombing
The abuser plays the victim at the beginning or in the hoovering stage of an abusive relationship. This subtle tactic supports opening up to the abuser and believing 'It is me.' because the abuser is so kind, innocent and friendly. In conversations with the target the abuser effectively mirrors the target, gives great compliments, pays attention to details and is very engaging and available. After building trust and letting the guard down, the abuser starts to devalue and constantly violate emotional, physical and psychological boundaries or feels boredom and disengage/discard. It is a toxic cycle of abuse which leads to trauma bonding and internalized toxic shame resulting in going back to the abuser to find closure, get love, feel safe again, understand what happened and solve the problem

(continued)

(continued)

9. Smear campaign/stalking/isolation
The abuser goes further and not only controls the target but oftentimes the whole system which leads to smear campaigns, stalking and isolation. Typical communication forms are gossip, storytelling, insults, and re-framing to provoke reactivity, silence and self-sabotage. The target should behave 'crazy' so that the abuser can destroy the target's reputation and support network. The abuser craves the position of power and control and therefore knows very well that it is also important to deliberately build a community of supporters, so-called flying monkeys, who keeps the abuser in his/her own power strategy. The addictive flying monkeys/enablers and the abuser constantly trigger, provoke and make the target feel less than and 'not belonging' to the group. That builds a toxic 'powerhouse' and web of abuse
10. Bullying & ambient abuse/indirect insults
Most overt abusers like to bully, attack and violate boundaries. It is obvious and everyone knows about bully behavior from schools, workplaces or public situations. Contrary to the typical harsh comments and 'Oh, you are too sensitive' remarks of overts, ambient abuse takes places on an 'invisible' level to silence and put the target down. For example, a seemingly great compliment is painful and threaten the target in a specific context and 'in-between the lines' this insult opens a wound. Nobody else can understand the target's pain and it is very difficult to explain others, even present witnesses, that the seemingly great compliment is a threat. The abuser is sadistic and enjoys seeing the target in pain, fear, obligation, and guilt (FOG) due to the abuser's indirect and/or nonverbal communications

Part I
New Leadership in Strategy

Chapter 2
To Mars on a Bike—Images of Regulation

Martin de Bree

Abstract In this chapter, we argue that the current paradigms of regulation are not productive for the growing complexity in society. First we explain why these traditional paradigms leads to problems and what these problems look like. Then we demonstrate how regulators try to cope and struggle on. New approaches are being explored by several scholars and practitioners, but encounter substantial challenges which seem to stem from the traditional black and white thinking of traditionalists. We conclude by arguing that new approaches, although hard to implement, are promising and may have significant social benefits.

2.1   Introduction

It is fascinating to see how the interface between legal systems and entrepreneurship has developed. The Dutch waste industry for example has experienced a very fast growing amount of rules in the 1970 and 1980s. After the necessary scandals in which waste was cheated (Uniser, EMK, Zegwaard, TCR) strict rules were drawn up by the government. Not only many rules were imposed on the industry. The violation of most of the rules could lead to penalties under criminal law so that a violation could have immediate and massive consequences. At that time it was also said that as a managing director of a waste company you were always standing with one leg in prison. Not a very attractive position to be in while building a business.

Studying this dilemma of an industry which was badly needed to solve an obvious problem was under such a tight regulation regime, led to deepen the understanding of the relationship between regulation and entrepreneurship (Bree 2006). At first glance one would presume that the waste sector is very specific because of the questionable reputation of the companies, and that therefore it was no more than logical that tight regulation was necessary. However, the complaints about the pressure of too many detailed rules have been expressed in many other industries, from health care

M. de Bree (✉)
Marshalllaan 91, 3223 HC Hellevoetsluis, Netherlands
e-mail: mbree@rsm.nl

© Springer Nature Switzerland AG 2020
N. Pfeffermann (ed.), *New Leadership in Strategy and Communication*,
https://doi.org/10.1007/978-3-030-19681-3_2

to education and from chemistry to finance. One could say that the illusion of the engineered society has left its marks in many more sectors than just the waste sector.

In more and more industries the balancing of rules and room for entrepreneurship and innovation is discussed. Not only are the rules often experienced as an obstacle for business, but they are also putting an enormous pressure on inspection agencies and regulators to enforce all those rules. A somewhat dated inventory shows that in the Netherlands we have 1,883 formal laws, 2,307 royal decrees and 5,378 ministerial regulations, so a total of about 10,000. That is still without the municipal and provincial rules and regulations. That is where the European rules come in again and these are currently as much as 34,132. Many of them are regulations, directives and decisions. To know all these rules is already impossible, let alone understand the content and then convert it into actions.

If the number of rules is so large, it is not surprising that the number of violations is also large even to the point that the rules have little distinctive character. If everyone breaks the rules, where is the distinction between good and bad companies? If there is an incident and the rules appear to have been violated, the media and politics often respond strongly. Hastily generalizing does the rest: 'If there are so much violations, it is clear that the whole industry is malicious!' Several industries have been incriminated this way.

In recent years, in addition to the waste sector, construction, the chemical industry, the financial world, housing corporations, education and health care, all kinds of incidents have become public. The standard reaction from the political and administrative apparatus often consists of setting more rules. Obviously, this reaction does not solve the problem but only makes it bigger. The result is not much less than a Kafkian madhouse.

In this chapter I will discuss the problems with regulation and supervision. I argue that the problems are caused by an outdated discourse about rulemaking and therefore require drastic changes. I show examples of an innovative approach and conclude with some perspectives for the future.

2.1.1 Trying to Travel to Mars on a Bike

How did we end up in this situation? In this section I will attempt to find an answer to this question. It seems to me legitimate to say that thinking about regulation and supervision has for a long time been dominated by a legal-economic discourse. Regulation and supervision are the domain of lawyers, economists and criminologists, both in the academic world and in practice. The most well-known scholars in this area often are either criminologist or economists. The legal-economic discourse has brought us a lot of benefits, but also quite a few limitations. The main problem here is that the legal-economic discourse is based on assumptions and fictions whereof the validity is at present debatable. Although we tend to be quite satisfied with our legal systems, discomfort about its suitability to deal with future challenges like climate change and immigration has been expressed (Ruhl 1997).

Firstly, regulation and supervision are dominantly based on step-by-step processes. The legislator establishes the law on the assumption that by implementing that law the intended goal is realized. Citizens and companies comply with the law and if they do not, the regulator will ensure enforcement. This step-by-step process falls short in a number of respects. Despite attempts to include third parties through consultation rounds and the like, it is implicitly assumed that the legislator has all the knowledge in-house to formulate effective rules. It is also assumed that the rules are effective and that their implementation is feasible. However, these assumptions are not at all a guarantee that the process of lawmaking results in the desired outcomes. That a top-down linear sequential design process does not result in a suitable product in a dynamic and complex context, has long been well understood by companies. That is the reason why innovative organizations do not create new products sequentially, but simultaneously. In other words, the design process is organized in such a way that all knowledge carriers contribute simultaneously and equally to the design.

It seems that the design process of rules is far from optimal and this has consequences for the quality of the rules in terms of effectiveness. There is an ingrained process of producing rules that is triggered in the event of incidents and that runs according to outdated design principles. No wonder we are facing too many rules that are also of questionable quality.

Secondly, this legal-economic discourse does not take into account the possibility that citizens and businesses behave ethically. The law is often designed to deal with the baddest of all guys and then applied on everybody. In this deterrence paradigm, we assume that citizens and businesses must be kept in fear of punishment and that the threat of sanctions is the means to achieve this. Deterrence is still the dominant principle in regulation, neglecting the fact that there are people with other motives than just avoiding punishment. Numerous studies (e.g. Wingerde 2012; Hodges 2015) show that punishment itself is not effective in many cases, certainly not if citizens and companies are of good will. Many citizens and companies that are of good will experience this as inappropriate, frustrating and sometimes insulting. Another effect is that citizens and businesses quickly become lazy when detailed rules are imposed and no longer think about the purpose of the rules themselves. This is not a plea to just end all penalties, sometimes penalties are badly needed, but the we need to better understand that penalties are not panacea for all undesirable events.

It seems that the dominant deterrence paradigm, as one of the pillars of regulation which has been precipitated and solidified in our institutional organization, is outdated. Of course there are all kinds of input and input from third parties such as in consultation rounds, but these are sometimes more window dressing (Kreveld 2016) than real contributions to the quality and the basic process is still step by step and it is the government that has a dominant voice in designing rules. Innovative ideas about a different design process of rules still seem very far from applicable. The mechanism of deterrence is neatly and bureaucratically organized. The criminal law is there to punish and has the freedom to work autonomously according to its own processes. The question of whether the criminal law is effective in a particular situation is hardly or not at issue. We just assume that in some magical way people will start to behave if they have been punished or fear to be punished. But most of the time what really happens, is that they get smarter and smarter in avoiding punishment instead of behaving in a responsible way. A little bit more wildly reasoned, one could posit that punishing in inappropriate situations may create crooks who become professionals in not being caught.

If the regulation system in its very basic design flaws, it looks like we try to travel to march on a bike.

2.1.2 Struggling On

Many of the problems that arise as a result of the aforementioned incident-rule reflex and over reliance on deterrence is a guiding principle, arrive on the plate of the regulator. The task of the regulator becomes even more difficult because he is sent out with a double assignment. On the one hand, his mandate is limited to maintaining compliance with the letter of the law. Maintaining the letter of the law becomes increasingly difficult with a proliferation of rules. If there are incidents in a given industry, media and politicians tend to primarily blame the regulator, whether the problems are due to non-compliance are not. On the other hand, if the regulator tries to focus on the goal of the law instead of the letter of the law, like Sparrow has suggested (2012), this boomerangs back because it is outside his formal mandate.

Either way, incidents will continue to happen, for no system generates 100% safety. This is what we might call the crystal house in which the regulator has to operate. Obviously, for the regulator this is not a very attractive position to be in. So it comes down to the problem that the regulator is being caught in the inconsistency of social goal (e.g. public safety) and the rules meant to help achieve this goal which are increasingly ineffective. Most regulators have a hard time being blamed over and over again.

Should the regulator just accept this and carry on? The Dutch Scientific Council for Government Policy (WRR) suggests a more reflective attitude (WRR 2013). The WRR calls regulators not only to supervise compliance with the law, but also to provide feedback to the legislator about any risks that have not been regulated. But how this can be implemented in the reality of daily practice is still far from clear.

Ayres and Braithwaite (1992) have argued that there is an alternative for deterrence based regulation. They advocate a model based on cooperation known as responsive regulation. For some time a discussion has been raging about what the best approach is, deterrence or cooperation. However, deterrence appears to be unworkable in a large number of cases and, where it works, it works mainly indirectly. Collaboration sometimes seems to work, but free rider behavior and capture are lurking. Regulators have been too confident in the past when an industry promised to self-regulate. The concept of responsive regulation which, although well-meant, seems to be a kind of trial-and-error method: the regulator should escalate its interventions until the desired effect is achieved.

The overall picture of the issue is that regulators in both approaches perceive the sector or organizations they supervise as a black box. The effects of regulators' interventions on the future behavior of the target group is largely unclear. Not knowing and understanding what is going on inside the black box renders every regulation strategy automatically into a reactive one. In this approach, incidents cannot be anticipated and are always a surprise. The obvious conclusion is that regulators need to understand much more about how the supervised organizations behave and which factors influence this. Then they can use this knowledge to come up with more effective strategies. This concerns in particular knowledge about how a regulator can assess whether desirable behavior is assured by an organization and what a regulator can do to promote desired behavior. Some regulators have already developed strategies just based on this principle: proactively looking at the very root causes within regulated organizations. The next part dives into this new proactive regulation strategy considering its possibilities and limitations.

2.1.3 New Roads

What options are there for improvements? A combination of two directions is obvious. On one hand, it seems important to make maximum use of the self-regulating capacity that comes from governance, that is from non-public forces that can promote ethical behavior. This step seems to be a hard necessity, given the limited capacity of regulators to carry out reactive strategies. After all, you cannot put a regulator behind every citizen or employee of a corporation. On the other hand, it seems important that there is room form customization for every situation has its own specific conditions. At this moment, regulators are stuck to detailed rules that are generic and leave too little room to optimize for each situation. Regulators should be able to let the letter of the law be what it is in a situational way, provided that the underlying societal goal is achieved.

There is a lot of discussion about whether self-regulation can work. Opponents use examples of the financial crisis, oil spills, explosions, abuses at housing corporations, etc. to show that self-regulation does not work and business are not to be trusted. At the same time, we see that in many places where government supervision is not always present, it usually goes well. Buildings do not collapse in great numbers while construction supervision is minimal. Many chemical plants are still operating while the regulator is present only a fraction of the company's time. Surgery generally goes quite well while the inspection does not look at the surgeon's fingers. Apparently there is a potential for self-regulation that ensures that things go well without direct government intervention. A clear example that shows how self-regulation can work is commercial aviation. Since the Second World War, people in this sector have been working on a global system based on self-regulation. Commercial aviation is now 100 times safer than driving a car and is 10,000 times safer than motorcycling. Interestingly, in the aviation industry there *is* government intervention, but of a kind that leaves very much space for corporations to self-regulate. Sometimes maybe, this space is too big, but notwithstanding recent incidents, the overall long term safety performance based on self-regulative principles has been impressing and an example for many other industries. So it is not the question whether self-regulation can work, because that is very clear. The question is under which conditions self-regulation can contribute to the protection of public interests and what role should the regulator take in either situation with regard to more or less self-regulating industries and parties. Although to the answer to this question is not quite yet clear, there are interesting notions to take into account.

The discussion coming from the outdated discourse raises the question: how can we make use self-regulative potential of non-government parties and corporations to assure social values. Sparrow (2012) proposes that risk management can be transferred to the private sector in certain circumstances. However, this author does spend little attention to the question what this means for the role of the regulator and the conditions to go down this road.

The aviation case suggests, that using self-regulation of companies implies that the government regulator will have to monitor in a different way. The public authorities realize that the best way to manage the risks is by the companies themselves. They know best where things might go wrong and they know best what to do about it. So, it is very logical that they should be responsible for risk management. This implies that Government supervision shifts from a technical challenge ('are these bolts fixed properly?') to a social issue ('how has the company organized that these bolts are always well secured?'). The regulator assesses the quality of the self-regulation and checks whether and how the company identifies, analyzes and controls the risks. This includes assessing whether the required competencies are present, whether the process is managed adequately, whether the control measures are being implemented adequately and whether the company checks and corrects itself. The regulator in the aviation sector does hardly impose any penalties because the sector has outgrown this, it just makes no sense. A learning system has been created where information can flow freely precisely by leaving the deterrent mechanism and thus it becomes safer. It is also interesting that the number of regulators is very limited. The result is a high degree of safety with minimal effort from the government side because the regulator works in a different way and companies react positively to this.

2.2   Regulation that Makes Companies Assure Compliance

An example of how regulators may approach the problems in modern regulation as discussed according to the notions described above has been explored in a pilot project in Dutch health care. Of course, the circumstances in healthcare are different than in aviation and therefore also the specific risks. But because regulation drifts away from a technical issue to a social issue, roughly the same generic approach can be followed. At the end of the day, the risks in aviation may be different than in health care, but the methods to implement risk management can probably be quite similar.

In the pilot, the board of directors of a health care organization was asked how they guarantee quality and safety and was given the opportunity to explain this. The inspectorate provided feedback on how the institution had organized risk management and compliance assurance. The healthcare institutions reacted positively and set to work on structural improvements in the provision of responsible care. The traditional technical challenge (does the company comply?) is changing into a social challenge (how can we know that compliance is assured and risks are controlled?).

Part of the pilot was to study the impact of system based regulation (SBR) on the regulated organization. The findings indicate that SBR is an effective strategy to check whether the goal, management system, daily practice and final outcome of a company are consistent. Furthermore, by articulating the findings of an SBR inspection to the company, the company is encouraged to close any gaps between these four layers. In this way, compliance assurance is improved. As a result of the study, a model was constructed to visualize the gaps between the layers as the main focal points for the inspector (Fig. 2.1).

Obviously much more work has to be done to better understand several relevant aspects, but in a world that is becoming more complex and dynamic, new regulation strategies are very welcome. In this research it was demonstrated that SBR can be effective because of its leverage to increase compliance assurance and risk management instead of dealing with non-compliances case by case. It is therefore not surprising that SBR has demonstrated its value in several industries varying from health care to environment and safety. How effective SBR may be if carried out under beneficial circumstances, the implementation of SBR may be challenging. This is because it requires thorough understanding of the approach throughout the regulation agency

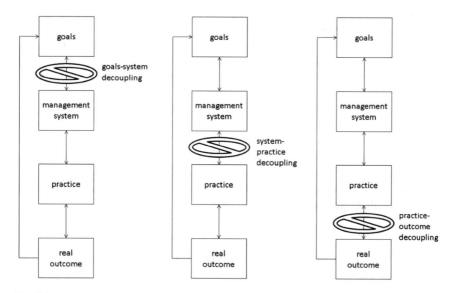

Fig. 2.1 Decoupling in regulated organizations (Bree and Stoopendaal 2018)

applying it. Furthermore, SBR requires additional competencies of inspectors for they will have to be able to assess the design and implementation of management systems, governance and corporate culture. If applied under the right conditions, SBR can be an interesting way to benefit from the vast potential for self-regulation in a world of decreasing agency budgets of regulators and growing complexity in society. Important conditions for success of this strategy are the feasibility both on the side of the regulated industry (can they actually organize self-regulation and are they really willing to behave responsibly?) and the regulator (do inspectors have the competences needed, is there enough support from the management of the agency and important stakeholders?).

Perspectives

Two things can be achieved by applying this approach. Firstly, the inspectorate can form an opinion on the quality of self-regulation. This judgment has predictive value because lack of control means that the organization may take too many risks that have not yet manifested itself as incidents. This offers the board of directors the opportunity to intervene before things really go wrong. Secondly, the inspectorate can stimulate to start an improvement process by linking findings to the organizations management.

In order to be able to do business and to run an organization in the harsh physical day-to-day reality, it is not helpful to let all kinds of rules get in the way. Risks differ per situation and therefore also require space that allows professionals to design the best solutions for each situation. If rules are too detailed on a general level, they stand in the way of optimal risk management. Both the regulator and the regulated organization must have room to maneuver in order to find the best solution for that situation, given the public and private interests. For more space, more trust is needed that the individual regulator in every situation chooses the right approach to optimally serve the interests of society. Without this space, the regulator can simply not do his job properly and may go for the easy way of ticking boxes. With regard to discretional space, the regulator can use his own professionalism and morality to best serve public interests.

The trust needed to make such an approach work will not come automatically. Trust must be earned by taking responsibility visibly. Regulators will have to demonstrate that this trust is justified individually at the organizational level. The management of regulatory organizations should claim the space from their political clients on the basis of their state-of-the-art policy. As the regulators job becomes more complicated and full of uncertainty, support and leadership from the management of the agency is badly needed to back their position.

Debottlenecking

For these new approaches to really contribute, important bottlenecks are yet to be solved. This part explores the work needed to make progress in the alignment of regulation strategies with today's complex and dynamic world.

New roads only are accessible through the adoption of other thinking. The main issue here is the difference between compliance and compliance assurance and the difference between incidents and risk management.

The first bottleneck is that many regulators, their administrators and political clients do not seem to realize that an individual incident does not automatically mean that a system or system is not right. An organization causing an incident may actually have a much more effective better risk management system than an organization without an incident. Effective self-regulation does not imply that mistakes are never made, but that any mistakes are noticed, corrected and used to learn from. The old paradigm would label Lionel Messi as a poor soccer player based on a couple of penalties the best player in the world has missed.

Representatives of the old discourse usually have great difficulty with this reasoning. But how unruly the old paradigms may be, there are hopeful signs that this risk-based view is starting to catch up with public regulators. In addition to the aforementioned example of supervision in aviation, there are various examples of regulators who realize that they sometimes do not have to respond directly to an incident. For example, in the US Federal Sentencing Guidelines, it is recognized that the occurrence of a single incident does not mean that the management system of a company is not functioning properly. Also, other hopeful signs give reason to believe that slowly things are changing in the discourse of the regulation community. The Province of North Brabant in the Netherlands is reluctant to intervene if the company takes adequate action itself in the event of an incident from a properly operating management system.

Reasoning the other way around, the absence of incidents does not mean that there is good risk management. It is possible that the organization takes great risks and that chances of a major incident are sky high. This too requires anti-intuitive response from the regulator, namely adequate intervention despite the absence of incidents, but on the basis of identified shortcomings in risk management. That this is not self-evident is demonstrated in numerous cases were the media had to dive into the case to force regulators to take action. This bottleneck manifests itself in the extension of the first one. Companies and institutions tend to promise a lot on paper, but not to translate this into concrete actions and results, also called decoupling as explained earlier in the model of Fig. 2.1. It usually has advantages to symbolically pursue a certain policy as a company, for example by endorsing a governance code in which all sorts of matters are regulated in terms of compliance and risk management. In doing so, their legitimacy among stakeholders is enforced. However, actually implementing that policy in daily practice is often a real challenge. As explained, this can lead to a decoupling between the formal policy that is outwardly propagated and the physical reality.

In deliberate decoupling we also speak of free rider behavior, window dressing or greenwashing. In this case, the organization chooses deliberately to benefit from the advantages of the policy, while not having the burden of implementation. Decoupling can also be unintentional if implementation is more difficult or more expensive than initially thought in spite of the sincere intention for implementation at the start of designing a policy. This phenomenon ensures that the regulator who is prepared to take account of self-regulation cannot rely on the paper reality as expressed in governance code, certification standards, code of conduct, policy statements and the like. The ease with which promises are expressed in these documents about compliance and risk management is striking and often not very realistic. For example, in the pilot project in health care it appeared that no institution had a good overview of the rules and that there was a need to improve in the area of risk management. The modern regulator's key task therefore is to collect information about whether the formal policy actually is effectuated in practice and generates desired results.

Supervision becomes more and more a social challenge and requires customization based on a good understanding of the target group and the organizations being a part of it. In an effective approach, both punishment, forms of cooperation and system-based strategies can be applied that do justice to the intentions and competences of the organizations under supervision.

2.3   Conclusion

The way we regulate is in urgent need for renewal. What used to work well in the past, no longer suffices because the circumstances have changed in terms of complexity, dynamics, innovation and globalization. Holding on to old habits will not help us any further and may even endanger the level of control over public values.

It is clear that the new roads require leadership. Leaders on both the public and private side are badly needed who have the courage to make clear that a risk-free society does not exist and that this illusion of full control actually stands in the way of improving the system. There is a better approach than the current one, even if it is not perfect. Several parties have already recognized this and are paving the way for bigger changes. They build upon trust rather than mistrust as the starting point and on this basis entering into cooperation with the business community. They are experimenting with system-based supervision and supervision on governance and have the courage to put trust and respect on the regulation agenda, even if it is not yet entirely clear what these conceptions look like in practice. These are important attempts and experiments needed to bring the original goal of regulation back into sight.

Regulators can learn from each other more than ever. Because supervision increasingly acquires a social component in addition to the traditionally technical, supervision on several industries becomes more alike. A risk management system of a chemical plant or an airline is essentially not very different from that of a hospital or a nuclear power plant. Knowledge about safety and quality assurance and the influence of this on culture and behavior is relatively industry-independent. This means that regulatory arrangements from one sector become more applicable in the other sector than before.

With the development of an adult relationship between regulators and companies and institutions, it is important to take steps from both sides in professionalism and morality. Leadership of regulators and leadership in the private industries may strengthen each other. Allowing the possibility of trust and having an open dialogue about what is needed to serve both public and private interests, is the way to go to create smarter solutions. This does not mean that the regulator looses its teeth but rather that the regulator realizes that using or showing its teeth should be reserved for the regulated organizations who do behave irresponsibly.

Bibliography

Ayres, I., & Braithwaite, J. (1992). *Responsive regulation: Transcending the deregulation debate.* Oxford University Press.

Bree, M. A. (2006). *Waste and innovation.* Berghauser Pont Publishing.

de Bree, M. A., & Stoopendaal, A. (2018). De- and recoupling and public regulation. *Organization Studies, on-line first.* https://doi.org/10.1177/0170840618800115.

Downer, J. (2010). Trust and technology: The social foundations of aviation regulation. *The British Journal of Sociology, 61*(1).

Hodges, C. (2015). Corporate behaviour: Enforcement, support or ethical culture? *Legal Research Paper Series,* Oxford paper nr. 19.

Perezts, M., & Picard, S. (2014). Compliance or comfort zone? The work of embedded ethics in performing regulation. *Journal of Business Ethics, 131*(4).

Ruhl, J. B. (1997). Thinking of environmental law as complex adaptive system. *Houston Law Review, 34*(4).

Sparrow, M. K. (2012) Crime reduction through a regulatory approach; Joining the Regulatory Fold. *Criminology & Public Policy. American Society of Criminology, 11*(2).

Stoopendaal, A., de Bree, M. A., & Robben, P. B. M. (2016). Reconceptualizing regulation: Formative evaluation of an experiment with system-based regulation in Dutch Health Care. *Evaluation, 22*(4).

van Kreveld, H. (2016). *Consultatie bij fiscale wetgeving.* Leiden University.

van Wingerde, K. (2012). *De afschrikking voorbij: Een empirische studie naar afschrikking, generale preventie en regelnaleving in de Nederlandse afvalbranche.* Wolf Legal Publishers.

WRR Toezien op Publieke Belangen rapport nr. 89 (2013).

Dr. Ing. Martin de Bree, M.B.A. is a researcher at Rotterdam School of Management Erasmus University in Rotterdam. He published in several international journals about innovative regulation and law enforcement. He and co-author Annemiek Stoopendaal have recently been awarded for the VIDE publication prize, a Dutch award for the best national publication about regulation. He is a member of the Expertforum Rechtmatigheid in een Nieuwe Tijd, a thinktank initiated by the Vrije Universiteit Amsterdam which strives to innovate regulation processes.

Chapter 3
Win-Win Negotiation in a Global Economy

Leigh Thompson

Abstract Negotiation is an essential skill for business leaders. However, the global, networked economy presents vexing challenges for negotiators who want to reach win-win outcomes. Negotiation involves a delicate balance of two motivations—cooperation that paves the way toward information sharing necessary to increase mutual gain, but also competition in that parties do not share similar incentive structures. We explore obstacles as well as solutions to negotiation on a global stage. Specifically, we examine how 6 factors, including: cultural differences, ethical concerns, political events, information technology, "big data," and networks affect the process and outcome of effective negotiation.

3.1 Introduction

The "win-win negotiation" concept has near universal appeal. Behavioral scientists and economists have identified the key theoretical concepts of win-win negotiation (typically referred to as integrative negotiation; e.g. Bazerman and Neale 1992; Lax and Sebenius 1986; Raiffa 1982; Thompson 2020; Walton and McKersie 1965). And, a corresponding education and training industry has developed to equip and train managers to create value at the negotiation table.

The best practices associated with integrative (value-creating) negotiation are rooted in a large body of social science research. To date, much of this research has been conducted in traditional laboratory and organizational settings. What do leaders need to know when facing complex business negotiations that involve cultural, communication, and socio-political challenges in a global economy?

This chapter briefly reviews the key concepts, or bedrock principles of effective, win-win negotiation. Then, important considerations that leaders need to wrestle with in the new global economy are identified. The insights and best practice suggestions reviewed in this chapter are based on laboratory and field research studies, as well as in-depth interviews with leaders negotiating in complex business and global

L. Thompson (✉)
Kellogg School of Management, Northwestern University, Evanston, USA
e-mail: leighthompson@kellogg.northwestern.edu

© Springer Nature Switzerland AG 2020
N. Pfeffermann (ed.), *New Leadership in Strategy and Communication*,
https://doi.org/10.1007/978-3-030-19681-3_3

environments. These best practices do not eclipse the large body of extant research, but rather, provide an important global context for optimally utilizing negotiation strategy.

3.1.1 Key Concepts of Effective Negotiation

Traditional research and theory on negotiation is divided into two key tasks or challenges of negotiation: distributive and integrative bargaining (Bazerman and Neale 1992; Raiffa 1982; Thompson 2020). The distributive aspect of negotiation refers to the fact that the involved parties need to divide or allocate resources between them, and most often, parties have opposing interests (e.g., the buyer wants a low price; the seller wants a high price). An analogous construct refers to negotiation as a mixed-motive task, involving both competition and cooperation, such that parties' interests are opposed, but they nevertheless are motivated to cooperate so as to identify a mutually-agreeable solution and avoid costly impasse (Lax and Sebenius 1986). Key questions center on exactly how much information should negotiators reveal so that they can capture all of the relevant mutual gains, but at the same time not risk the possibility that the counterparty claims most of the surplus value.

BATNA. Perhaps no other construct has quite as much universal appeal as BATNA (Best Alternative To a Negotiated Agreement; Fisher and Ury 1981). A negotiator's BATNA is colloquially referred to as a "plan B" (i.e., the course of action that the negotiator will take in the event of no agreement). Negotiators who have attractive BATNAs (e.g., multiple job offers, etc.) have more power and leverage than do negotiators who have unattractive BATNAs. From a practical perspective, negotiators are encouraged to improve their BATNA, but never reveal its exact value to the counterparty.

Reservation Price. Whereas reservation prices are based on BATNAs, they are not actually the same thing. A BATNA is the course of action a negotiator will take if they don't reach mutual agreement. A negotiator's reservation price is the monetary equivalent of their BATNA. Stated another way, it refers to a negotiator's "bottom line". A reservation price represents the very lowest (or highest) price a negotiator would agree to before walking away from the table.

ZOPA. The Zone of Possible Agreement represents the overlap (or lack thereof) between negotiators' reservation prices. In many situations, the zopa is positive; but in some situations the zopa is negative, which means that parties would be better off walking away from the table rather than reaching mutual settlement. Rationally speaking, negotiators should reach a settlement when a positive zopa exists (i.e., the very most a buyer is willing to pay is greater than the very least the seller is willing to accept); and walk away from the table when a negative zopa exists (e.g., the very most a buyer is willing to pay is less than the very least a seller is willing to accept).

Target Price. Negotiators should never initiate negotiations by asking for their reservation price; rather they should set a goal or target price. The risk in setting a too-aggressive target price is that the counterparty may be insulted and terminate negotiations. The risk in setting a too-conservative target price is that the other party may immediately accept the opening offer, resulting in the Winner's Curse (Thaler 1988).

The ideal opening offer should be on or near the counterparty's reservation price, thereby allowing the focal negotiator to claim most of the bargaining surplus. Practically speaking, negotiators will never (or rarely ever) know the counterparty's reservation price and so, determining the ideal opening offer is challenging.

Positions versus Interests. A position is a stated demand. An interest is an underlying need. The most fabled story illustrating the important difference between positions and interests is recounted by Mary Parker Follett who described a conflict between two sisters and a single orange (Follett 1994). After much contentious negotiation, the sisters agreed to compromise, thereby cutting the orange exactly in half. One sister used only the juice and the other only the rind; and thus, 50% of the orange was wasted. This story illustrates how compromise, or splitting things down the middle is often a suboptimal agreement. It also illustrates how negotiators who only state their position (e.g., "I want the orange") are less likely to reach an integrative agreement than negotiators who reveal their interests (e.g., "My recipe calls for orange zest"; Thompson 1991).

First Offers and Anchoring. A great deal of research has focused on whether negotiators are best advised to make the first offer or conversely, invite or cajole the counterparty to make the first offer. The optimal strategy depends on information symmetry. According to the anchoring information model (AIM), when negotiators possess roughly equivalent information about one another, a negotiator is best advised to make the opening offer so as to try to "anchor" the other party (Loschelder et al. 2016). However, when the information is asynchronous (i.e., such as in the case wherein Party A knows much more information relevant to the negotiation object than does Party B), it is advisable for Party A to encourage or cajole Party B to make the opening offer. Why? Party A is unlikely to be anchored because she knows a lot of information, but it is possible that Party B may not ask for enough, allowing Party A to reap a greater share of the zopa.

Concessions. Once negotiators have made their first offers, they begin the "dance of negotiation," wherein negotiators exchange offers and counteroffers (Raiffa 1982). It is advisable for negotiators to not engage in unilateral concession-making, in which they make more than one consecutive concession. Rather, negotiators should invite the counterparty to make concessions. Negotiators are also best served by not making large concessions. Rather, negotiators should match the concessions made by the other party and signal that they are getting near their reservation point by narrowing the magnitude of their concessions.

Tradeoffs (Logrolling). Thus far, our review of key negotiation principles has focused on "claiming", the competitive aspect of negotiation. Negotiations often contain potential for integrative agreement, but more often than not, negotiators fail to realize that potential. This is often because they hold a fixed-pie perception, or the belief that the counterparty's interests are diametrically opposed to their own interests (Thompson and Hastie 1990). Most negotiations contain potential for integrative agreement in which parties—like the sisters—can do better for themselves than simply dividing things down the middle. A key strategy for fashioning integrative agreement is the tradeoff or logrolling principle (Froman and Cohen 1970). When negotiators make tradeoffs, they gain on a high-priority issue and make a concession on a low-priority issue.

Consider how the CEO of a professional services company used the logrolling strategy in a sensitive HR situation. The CEO was having trouble with one of her managing directors (MD) in a regional office. The problem was that MD was not very "available" to her colleagues and her clients; indeed, her calendar was blocked out every day between 3 and 5 pm. One client bemoaned that they were not able to schedule a 30 min phone call for two months! This lack of accessibility had a very negative effect on internal colleagues and key clients. To further complicate matters, the organization was a ROWE environment (Results Only Work Environment) and so, employees did not feel compelled to be physically present. The MD's direct report grew increasingly frustrated trying to manage the mysterious calendar of her managing director and also felt that the task of managing someone's schedule was demeaning. With tensions rising, the CEO thought her only option was to ask the MD to leave or perhaps to work part-time.

However, the CEO decided to find out if there were some parts of the orange that she did not know about. During a meeting with the MD, the direct report, and herself it became clear that there were at least 3 parts of the orange: training sessions, client coaching, and calendar scheduling. Even more enlightening, the MD and direct report had very different preferences. The direct report wanted to be more involved in training clients—something that the MD strongly preferred not to do. Conversely, MD wanted to focus on coaching. And, she wanted to manage her own calendar. So, the CEO suggested that the direct report focus on training exclusively—no more calendar management and that the MD focus on coaching exclusively and manage her own calendar. Both parties received their most preferred parts of the orange.

MESOs. When negotiations contain more than one issue, negotiators face a choice. Namely, to either engage in sequential or simultaneous negotiation of the issues at hand. Negotiators are best advised to avoid the sequential discussion of negotiation issues because this can lead to a myopic focus wherein both parties compete for the best terms on each of the issues. By fashioning proposals (offers) that contain more than a single issue, negotiators can more likely create integrative agreements. In the MESO (multiple equivalent simultaneous offer) strategy, the proposer presents two or more multi-issue proposals of equal value to herself and invites the counterparty to choose. Given that the proposer is indifferent between the offers, if the counterparty finds one of them more attractive, by definition, this means that more mutual gain has been created. In addition, the proposer can gain insight into the counterparty's preferences by examining their reactions to the offers. Negotiators who use the MESO strategy realize several advantages, including the ability to use more aggressive anchoring and overcoming concession aversion (Medvec and Galinsky 2005).

Consider how the leader of a mid-market sales group used the MESO strategy in a SMS pricing deal with a new customer. The customer was a global procurement manager, and the negotiation centered on price, service, terms, and SLAs. The sales leader proposed two options, which he referred to as "Option A" (list price for a 10 store pilot) or "Option B" (10% of chain pilot: $2500 per bundle and $60 per device, per month for SMS + Insights + MI). In this scenario, the customer was intrigued and opted for "B".

Post-Settlement Settlements. Negotiators are often quick to settle for the first set of terms that exceeds their reservation price. However, if they settle too quickly, parties may suboptimize. These same parties may be reluctant to prolong negotiations because they fear that they will frustrate the other party. The post settlement-settlement (PSS) technique is an ideal strategy for the risk-averse negotiator. In this strategy, parties commit to a set of terms and then both parties then agree to brainstorm for a mutually-better deal with the understanding that the initial terms are binding unless both parties choose to revise the terms (Raiffa 1982). Thus, the only way a negotiator can improve her outcome is to generate more value for the other party. In this sense, the post settlement-settlement technique aligns both parties' interests.

For example, consider how a buyer (global manager) used the post-settlement settlement in a complex acquisition deal. The manager was buying a company in New Jersey, but the seller (an entrepreneur) was very hesitant. The buyer had provided clear financial details that both parties admitted would net the seller more money than by not selling. Eventually, the seller relented and agreed to the terms and a contract was drawn up and signed. The buyer then asked the seller (who still seemed remorseful), "what are the intangibles that you care about?" The seller sighed and said, "I have to figure out how to tell the manager in Florida that he does not have a job anymore." The buyer had a strict, no remote worker policy and so the seller did not even ask for an exception to this policy. This gave the buyer an idea: he proposed to transfer all the current clients (that he did not want) to the Florida-based project manager (who he did not want, either), allowing the seller to give the Florida manager a promotion, not a pink slip. This post-settlement settlement resulted in an even lower purchase price for the buyer than originally agreed to! The parties drew up another contract that both preferred over the original!

Pre-Settlement Settlements. Some business situations are so complex that negotiations may drag on for years while attorneys and finance people look over reports and fashion the appropriate language. Waiting to finalize terms can hurt both parties. The PreSS, or pre-settlement settlement technique is ideal for complex negotiations (Gillespie and Bazerman 1998). In the PreSS strategy, negotiators agree to a subset of the terms so that they can both avoid incurring immediate business losses and yet, have a clear understanding that major parts of the negotiation will be continued or even re-negotiated in the future. For example, in the negotiations between the United Kingdom broadcasters and independent producers of TV content, negotiations threatened to drag on indefinitely concerning who owned the rights to broadcast TV shows across broadband and mobile platforms (White 2018). Parties realized that even a year's delay could result in a lose-lose outcome. So, the parties agreed to a temporary agreement so as to generate revenue from new audiences, a shared interest for both partners.

Contingent Contracts. Negotiations are often stalemated because parties cannot agree on various future conditions. For example, in one negotiation, the buyer believed that resale home prices would decrease and property taxes would increase; whereas the seller believed the opposite, namely that resale home prices would increase and property taxes would decrease. Given that neither party can perfectly predict the future, negotiators can become entrenched and intransigent. In the contingent contract strategy, parties base their agreement on an as-yet-undetermined set of circumstances. For example, one business development manager was doing a complex negotiation with a VP of Operations concerning an upgrade and increased footprint for a digital marketing campaign. At loggerheads, the business development manager proposed, "If we can show you an ROI of between 2 and 5% in the first 4 months, will you agree to work with us in 10 of your stores?" The VP of Operations was intrigued and responded with, "If you can look after the cost of our upgrades, we will increase our footprint."

3.2 Successful Negotiation in the Global Economy

The constructs reviewed above are considered to be the foundation of effective negotiation skills. However, those constructs (and corresponding best practices) are not always sufficient to guide the leader and manager through the complexities of negotiating in a global economy. Our examination of over 30 global leaders and managers reveal 6 new considerations that must be wrestled with at the global negotiation table: ethics, cultural styles, communication medium, networks, information and big data, and social—political linkage effects.

Ethics. Negotiators are subject to different rules and mores when it comes to bargaining behavior. Whereas it might be acceptable to lie about a BATNA in some circumstances (i.e., claiming to have an alternative that in reality one does not have), in other business situations, this would constitute unethical and illegal behavior. Ethical understanding is made more complex because negotiators have more favorable views of their own behavior than they do of others. In short, negotiators look at themselves through rose-colored glasses, but are quick to judge others. In our study of global managers from 7 cultures (U.S., China, Hong Kong, Germany, Israel, Canada, and Latin America), negotiation ethics emerged as a key issue for business negotiations. One program director in a large government organization recounted an ethical dilemma, "My SVP disagreed about how to handle a situation and suggested that we ignore our boss, the Executive Vice President (EVP), arguing that 'the EVP will never know.'" Troubled, the program director said, "I need to think about this." Later than evening, she informed HR what the SVP had said and the situation was resolved.

Global and Cultural Styles. It is nearly impossible to conduct exclusively intra-cultural negotiations. In the global environment, leaders need to negotiate across cultural divides. One global executive lamented that culture affects information sharing and getting data from [sales] prospects is essential to demonstrate the value of their business solutions commenting that, "Latin and Middle Eastern [sales] prospects are much more transparent (or less defensive) than prospects from Europe or the USA."

A particularly contentious intercultural negotiation involves the use of domain names. In a global negotiation that spans several countries, US-based Amazon.com pushed up against the Amazon Treaty Organization regarding who owns the rights to domain names that end in ".amazon". Bezos argued that his company, Amazon.com (established 1994) has trademarks for the name. Conversely, the Amazon Treaty Organization argued that the use of the name infringes on their sovereign rights, noting that, "before 'Amazon' was a global e-commerce company, it was a rain forest in South America" (Fung 2015). In turn, Bezos accused Peru, Brazil and other groups of South American countries of making the issue "politicized", arguing they (Amazon.com) have every right to use ".amazon" in a non-geographic context. The dispute over ".amazon" has been going on for several years, but the negotiation is taking on greater urgency as the U.S. prepares to officially transition its Internet oversight authority away to a third party, the Internet Corporation for Assigned Names and Numbers (ICANN).

Communication Medium. The great majority of managers and executives in our survey reported that they conduct negotiations in a non-face-to-face fashion (e.g., email, phone, text, and video conference). In some situations, it may be wise to avoid face-to-face negotiation and opt for CMC (computer-mediated-communication). For example, in their communication orientation model, Swaab et al. (2012) propose that when negotiators do not have a favorable relationship, they are best advised to avoid face-to-face interaction. One business leader recounted a "failed" negotiation in which he pleaded to his board of directors to be present for the final round of negotiations with a Chinese counterpart. The board suggested that being on the phone would suffice. However, the negotiations failed because the Chinese counterpart felt disrespected in light of the absence of the other party.

Hidden Tables, Networks and Linkage Effects. The explosive growth of social media platforms has created a means by which people can learn detailed information about others. Most of the managers in our survey reported using Linked In, and that they use l Linked In to learn more about their customers, colleagues, and business associates. One manager in a large transportation company commented that, "We frequently need to negotiate with each other and with business leaders to balance multiple interests and arrive at the right end product. This isn't a typical one-side-vs-the-other negotiation, but it does require negotiations skills. Sometimes we need to balance our desire to communicate with employees with our desire to prevent negative media attention. Achieving the right end product requires discussion and debate with business leaders and the media team." Another business leader noted that because their industry is small, most senior leaders have worked for a competitor organization at some point in their career and that hidden tables and social linkages are quite common.

Information and (Big) Data. The era of "big data" has created an information overload for managers and executives. In the past, negotiators often suffered from a dearth of knowledge when seated at the bargaining table. At present, negotiators are challenged to discern the most relevant knowledge. Most typically, big data models are used to guide companies to understand markets and customer behavior. However, such information could also inform negotiators. However, "big data" does not always mean "clean data". One executive noted that, "analysts are good with data but lack the business and negotiation savvy concerning how to use the data." In addition, access to data is challenging. A high-level leader of a government organization attempted to create a win-win solution, but the other party was not forthcoming with key military information. Moreover, the information systems were not programmed to share certain information outside the organization.

Social-Political Events. Political identities can affect negotiators' ability to make decisions or even come to the table with someone who has a different view. One executive who works with nonprofit organizations on reproductive healthcare that includes abortion and birth control commented that political leaders and the news media dramatically shape the "landscape" of their negotiations. A particularly thorny socio-political negotiation centered on the National Football League (NFL). Player Colin Kaepernick's was released by the San Francisco 49ers after the 2016 season when he famously "took a knee" during the playing of the national anthem. Kaepernick accused NFL owners of collusion in a grievance, claiming his protest led teams not to sign him (Davis 2018). Kaerpernick's 2018 negotiation with the Denver Broncos grew even more complex when the NFL banned players from taking a knee when major sponsors, such as Nike, Ford, Hyundai, and Under Armour pressured powerful political groups to pull their support of the league (Meyer 2018).

3.3 Conclusion

We briefly reviewed the core skills of negotiation and outlined strategies to create integrative agreements. Our interviews with global business leaders reveal that the global economy presents special challenges for effective negotiation. When negotiations occur on a political, digital, or global stage, business leaders need to carefully examine how best to utilize the key value-claiming and value-creating skills of negotiation. What might be a straightforward strategy in one context could lead to unintended effects in a complex, global situation.

Bibliography

Bazerman, M. H., & Neale, M. A. (1992). *Negotiating rationally*. New York: Free Press.

Davis, S. (2018, August 16). 'He had his chance to be here': John Elway says the Broncos won't consider signing Colin Kaepernick because he turned down a contract 2 years ago. *Business Insider*. businessinsider.com

Fisher, R., & Ury, B. (1981). *Getting to yes: Negotiating agreement without giving in*. Boston: Houghton Mifflin.

Follett, M. (1994). In P. Graham (Ed.), *Mary Parker Follett: Prophet of management—A celebration of writings from the 1920s*. Boston: Harvard Business School Press.

Froman, L. A., & Cohen, M. D. (1970). Research reports. Compromise and logroll: Comparing the efficiency of two bargaining processes. *Behavioral Science, 15*(2), 180–183.

Fung, B. (2015, June 23). Amazon wants to control '.amazon.' but so does half of South America. The *Washington Post*. washingtonpost.com

Gillespie, J. J., & Bazerman, M. H. (1998). Pre-settlement settlement (PreSS): A simple technique for initiating complex negotiations. *Negotiation Journal, 14*(2), 149–159.

Lax, D. A., & Sebenius, J. K. (1986). *The manager as negotiator*. New York: Free Press.

Loschelder, D. D., Trötschel, R., Swaab, R. I., Friese, M., & Galinsky, A. D. (2016). The information-anchoring model of first offers: When moving first helps versus hurts negotiators. *Journal of Applied Psychology, 101*(7), 995–1012.

Medvec, V. H., & Galinsky, A. D. (2005). Putting more on the table: How making multiple offers can increase the final value of the deal. *HBS Negotiation Newsletter, 8*, 4–6.

Meyer, D. (2018, May 25). As the NFL bans kneeling during the national anthem, a new petition urges sponsors to boycott the league. *Fortune*. fortune.com

Raiffa, H. (1982). *The art and science of negotiation*. Cambridge, Mass: Belknap Press of Harvard University Press.

Swaab, R. I., Galinsky, A. D., Medvec, V., & Diermeier, D. A. (2012). The communication orientation model: Explaining the diverse effects of sight, sound, and synchronicity on negotiation and group decision-making outcomes. *Personality and Social Psychology Review, 16*(1), 25–53.

Thaler, R. H. (1988). Anomalies: The winner's curse. *Journal of Economic Perspectives, 2*(1), 191–202.

Thompson, L. L. (2020). *The mind and heart of the negotiator* (7th ed.). Upper Saddle River, NJ: Prentice Hall.

Thompson, L. L. (1991). Information exchange in negotiation. *Journal of Experimental Social Psychology, 27*(2), 161–179.

Thompson, L., & Hastie, R. (1990). Social perception in negotiation. *Organizational Behavior and Human Decision Processes, 47*(1), 98–123.

Walton, R. E., & McKersie, R. B. (1965). *A behavioral theory of labor relations*. New York: McGraw-Hill.

White, P. (2018, September 11). Viacom's channel 5 strikes terms of trade deal with UK producers. *Deadline*. deadline.com

Leigh Thompson is the J. Jay Gerber Distinguished Professor of Dispute Resolution and Organizations, Kellogg School of Management, Northwestern University, Evanston, IL. She is the author of 11 books including: *Creative Conspiracy: The New Rules of Breakthrough Collaboration, Making the Team: A Guide for Managers, The Mind and Heart of the Negotiator, The Truth about Negotiations*, and *Stop Spending, Start Managing*. Thompson is the director of the Leading High Impact Teams, High Performance Negotiation Skills, and Constructive Collaboration Executive Education programs. Her work on team creativity has appeared in *Fast Company* and *Business week*. She serves on the editorial boards of several academic journals and is a member of the Academy of Management.

Chapter 4
Do You Have the Right Profiles in Your C-Suite for an Effective Transformation?

Jean-Philippe Deschamps

Abstract This concurrence of several change factors can be summed up by the acronym VUCA (for **V**olatile, **U**ncertain, **C**omplex and **A**mbiguous). The speed at which globalization and digitalization are proceeding in most industries is adding an urgency dimension to these necessary mutations. Board directors are not just witnessing the acceleration of everything in the company they govern. They are in the midst of it, given their mission to support the company and its management throughout this ongoing metamorphosis. As a consequence, directors are bound to ask whether the CEO *and* the C-suite are on top of it *as a team*. Do they have the right leadership profile and the required multi-faceted and complementary competencies, individually as well as collectively?

4.1 Introduction

Many companies, even traditional ones, have entered a process of continued internal and market transformation through innovation. Think of the announced revolution in the automobile industry with electric and self-guided vehicles, or in banks with mobile banking applications. The acceleration of technological changes and the emergence of new business models often require a radical rethink of processes and even entire organizations, as well as a change in culture.

This concurrence of several change factors can be summed up by the acronym VUCA (for **V**olatile, **U**ncertain, **C**omplex and **A**mbiguous). The speed at which globalization and digitalization are proceeding in most industries is adding an urgency dimension to these necessary mutations.

Board directors are not just witnessing the acceleration of everything in the company they govern. They are in the midst of it, given their mission to support the company and its management throughout this ongoing metamorphosis. As a consequence, directors are bound to ask whether the CEO *and* the C-suite are on top of it

J.-P. Deschamps (✉)
IMD, Lausanne, Switzerland
e-mail: jean-philippe.deschamps@imd.org

as a team. Do they have the right leadership profile and the required multi-faceted and complementary competencies, individually as well as collectively?

This is a legitimate question that many boards may not feel fully equipped to answer. In the past, boards have always felt responsible for the selection, hiring and evaluation of the CEO, giving him/her a free hand to choose colleagues for the C-suite. Today, directors should ask whether the whole C-suite, not just the CEO, is prepared to deal with the necessary transformative challenge.

Indeed, today's technology and business environments exert huge pressure on the top team to contribute a variety of complementary skills, attitudes and leadership profiles. The CEO alone is unlikely to be able to cover all the required talents. He/she is expected to orchestrate a diverse team and to ensure that each member focuses on some of the tasks that, combined, will bring about the necessary transformation.

4.2   Is Your CEO a Fixer? Or a Grower?

Traditionally, the task of selecting and hiring a CEO involved responding to fairly straightforward questions. Does the company need to be streamlined, restructured, reorganized? Is its performance unsatisfactory compared to its peers? Or does it mainly need continued expansion into new market territories, either organically or through mergers and acquisitions (M&As)? This dilemma led boards to focus their search on what best-selling management author Robert Tomasko called either *fixers*, in the first case, or *growers* in the second.[1]

The profile of these two CEO types was relatively easy to define. They were generally hired on the basis of their track record, either within the company or with their previous employer. In a way, many CEOs could easily fit either of the two labels.

Since companies tend to experience performance waves, fixers were often required to come and "clean the house," or at least stabilize it after a period of fast growth. Similarly, growers were brought in to breathe new life into technologies and markets after a period of successful restructuring or performance improvement. This kind of changeover between fixers and growers has marked the history of many large companies. It is in this spirit that the board of GE chose Jeff Immelt—a presumed grower—after the departure of Jack Welch, a remarkable fixer.

To help boards go beyond these basic types—fixer or grower may be too rough as descriptors—a typology of leadership profiles could be useful. As with all typologies, it may appear overly simple and incomplete. However, it has proved to be intuitive and easily grasped by all kinds of leaders. It can be applied to the CEO as well as to C-suite members because they often complement the profile of the CEO. This typology may also be applied internally to the board itself and its various members, who may include both fixers and growers. This can be done during one of the board self-evaluation retreats which, besides being fun, can prove very instructive.

---

[1] Tomasko (2006).

4.3   A Typology of Leadership Profiles

The generic typology of leadership profiles proposed by Robert Tomasko was covered in my book on innovation governance.[2] It is summarized below.

Leaders, Tomasko advocates, need to develop four complementary abilities:

- To be aggressive, like a *warrior*
- To be conserving, like a *judge*
- To adapt to situations and people, like a *diplomat*
- To support people and ideas like a *mentor.*

The ideal leader would be that balanced individual who has these four abilities in equal proportions. But such leaders rarely exist! Real leaders tend to privilege one profile—their major—often with one (or two) complementary styles—their minors—depending on the tasks at hand and the company circumstances.

4.3.1   The Warrior Profile

Warrior leaders are eager to lead their troops into tough challenges, like entering new markets or pre-empting competitors in new technologies. In a nutshell, their philosophy is: "To be successful, let's make things happen, and be fast." They are at their best when they have the opportunity to take charge and show initiative. As dynamic and optimistic personalities, warriors tend to be practical, confident, persuasive, risk-taking and forceful. Some of these traits, if carried too far, can lead them to adopt utilitarian, arrogant, coercive, gambling and even domineering attitudes. In this sense they may be less good at delegating and winning cooperation from all in the company, but they create excitement.

4.3.2   The Judge Profile

Judge leaders like to work through processes, to rationalize and standardize. Their philosophy can often be captured as: "To be successful, let's base our decisions on facts." They are at their best when analyzing and weighing up options before acting. They care about always delivering what they promise. As concrete and rigorous leaders, they are factual, economical, tenacious, thorough, sometimes reserved, and always detail-oriented. But again, these traits sometimes comprise excesses. For example, to some people, they may appear unimaginative, stingy, rigid, perhaps somewhat dull, overelaborate and too data-bound. This explains why they may be

---

[2]Deschamps and Nelson (2014).

perceived as less prepared to react quickly to unplanned environmental or market changes and to improvise solutions. For judge leaders, the "analysis/paralysis" syndrome may be a reality.

4.3.3 The Diplomat Profile

Diplomat leaders believe that progress will be achieved only if consensus and win-win solutions prevail. Their philosophy is: "To be successful, let's ensure everyone gets a fair deal." This attitude makes them quite effective in pursuing trade-offs and handling cross-functional or departmental conflicts. Their profile helps them achieve their objective because they tend to be flexible, experimenting, tactful, patient, socially skilful and shrewd. But, once again, the excesses of their traits —they may be seen as inconsistent, speculative, overcautious, indecisive and even manipulative—reduce their ability to force a decision or stick their neck out in an open conflict.

4.3.4 The Mentor Profile

Mentor leaders believe in the power of staff development and the creation of a positive work environment. Their philosophy is: "To be successful, let's motivate and mobilize all our staff behind our objectives." Their "soft power" orientation makes them considerate, somewhat idealistic, supportive and responsive. They may even be seen as modest. They are therefore at their best when working with others to improve the social climate within the company and to create and develop teams. But their critics sometimes call them gullible, impractical, unimposing, paternalistic and even passive. Their internal focus may make them less prepared to look outside and engage in competitive battles.

Clearly, all these profiles are needed in a top management team because they represent complementary leadership facets. As Preston Bottger, an IMD faculty colleague, stated: "Leaders do or cause to be done all that needs to be done, and is not being done, to achieve what we say is important!" When they work together, these diverse profiles contribute to providing a sense of purpose, direction and focus, as well as to building alignment and obtaining commitment.

4.4 What Is the Leadership Profile of Your CEO?

Let's now revert to CEOs and characterize the leadership profiles of the two types —fixers and growers. Let's also propose that a third type of CEO is needed —*transformers*.

4.4.1   The CEO as a Fixer

Fixers tend to have a rather well-defined leadership style. Although there may be different variants of fixers, they can often be characterized as having several of Tomasko's *judge* personality traits. Because of their down-to-earth attitude and trustworthiness—they hate uncertainty and like to deliver what they promised—fixers are highly appreciated by boards who generally hate unexpected (bad) surprises. As judges, fixer-type CEOs tend to focus alternately, or in parallel, on two main fronts:

- Radical cost-cutting and/or restructuring, generally for performance improvement.
- Organizational change, as a consequence of their cost cutting or to better adapt the company to its globalization or market challenges.

Of course, leadership cannot be reduced to a single dimension. So, fixer CEOs may share some of the traits of the three other profiles, and this will generally reflect their preferred mode of operation. Some of them—let's call them judges/warriors— will go through their change campaigns in a fairly tough, not to say brutal, style. "Chainsaw Al" (Dunlap) at Scott Paper was one of the best known leaders in this group, hence the nickname that made him proud. Others may be more inclined to adopt softer more diplomatic or participative approaches; they may be characterized as judges/diplomats or judges/mentors. But in all cases, they remain intrinsically fixers, dedicated to streamlining their organization and/or their company performance.

4.4.2   The CEO as a Grower

Whereas fixer-type CEOs are primarily focused on the internal functioning of their company, grower CEOs are more interested in exploiting opportunities in the external world. They want to grow their market and expand the company's footprint. Their more externally aggressive style makes them closer to Tomasko's *warrior* profile.

As mentioned earlier, boards appoint grower CEOs when they are dissatisfied with the company's lack of growth. Whereas fixers are highly appreciated by bankers, financial analysts and shareholders, growers may create excitement in the media, and ultimately with shareholders, particularly if they adopt new technologies, open new markets or launch new product categories.

Growers may focus on the two complementary growth strategies to different degrees at different times:

- External growth, i.e. searching for the best opportunities available through complementary M&As.
- Internal growth, i.e. developing their company's creative capabilities and processes around home-grown innovations.

Warrior growers may have different modes of operation, reflecting their personality. The more careful ones may combine their warrior profile with strong judge overtones. Others will adopt a diplomat stance to advance their M&A agenda, or a mentor sub-profile to mobilize their organization behind their growth-through-innovation objective.

4.4.3 The CEO as a Transformer: An Emerging Phenomenon

CEOs who created entirely new industries have always existed—think of Edwin Land at Polaroid or Bill Gates at Microsoft. But the new-economy revolution has seen the multiplication of a completely different breed of entrepreneurial CEO.

Convinced of the need to move fast, pre-empt competitors and grab new market opportunities, they have capitalized on e-technologies to create entirely new industries or radically new business models. These strong and visionary personalities, from Amazon's Jeff Bezos to Tesla's Elon Musk, have captured the imagination of venture capitalists and analysts with their stated ambition to transform the world, and in their mind this is not an empty statement. But these new industry captains are not found only in the tech sector. Some, like IKEA's Ingvar Kamprad, have built empires in traditional industries. Others have dramatically changed their company's footprint by divesting entire lines of business and developing new ones, like Lou Gerstner at IBM, or by entering into bold alliances, like Carlos Ghosn at Renault-Nissan-Mitsubishi. All these CEOs can be characterized as transformers.

The arrival of this new type of leader brings us to review Tomasko's four leadership profiles and add a new one: the *visionary* profile/style. Visionary leaders see opportunities in all manner of external changes. Their philosophy is: "To be successful, let's create a totally new future." They tend to be curious, bold, unorthodox, passionate and, in all cases, determined. In short, they are at their best when they can challenge the industry, technology and market status quo. Their possible flaws—some may be overly inquisitive, forceful, impractical, emotional and even obstinate—may make them less at ease with the day-to-day handling of concrete implementation details. This explains why they are sometimes obliged to hire a strong and practical N°2, as Steve Jobs did with Tim Cook at Apple.

Transformer CEOs typically focus on different priorities depending on the situation and maturity of their industry. Their focus can take many forms, but at least two seem to prevail:

- The exploitation of new technologies—including digitalization but also new materials sciences and biosciences—to revolutionize their industry.
- The radical alteration of their corporate footprint by expanding their business and/or introducing totally new business models.

Like fixer and grower CEOs, transformers have complementary profile; their visionary attitude is often complemented by an aggressive warrior style. They more rarely adopt Tomasko's three other profiles, particularly the judge. This explains their need to hire strong COOs or CFOs. Similarly, because of the aggressive way they pursue their goals, they lean more toward the warrior than the diplomat. And their sometimes abrasive personality—think of the late Steve Jobs' reputation—doesn't induce them to behave like mentors!

4.5   The Board's Dilemma

In theory, all board directors should be aware of the type and profile of the CEO they have appointed to run the company and, above all, to acknowledge and understand why they have done so. This is in theory only, because personalities are complex and leadership profiles may present a number of gray areas. Actual CEOs may not fall easily into any one of the boxes in Fig. 4.1. Most CEOs—usually the most effective—develop complementary styles, straddling several boxes depending on circumstances and their business requirements.

Nevertheless, the board would be wise to reflect on the profile of their chosen CEO—possibly in his/her presence—and the extent to which this profile is effectively complemented by other C-suite members. This review can be achieved by raising the following questions:

- If our CEO is primarily a fixer, who are the growers in the management team? Do they understand the importance of their countervailing responsibilities?
- If our CEO is primarily a grower, who is going to ensure that the deals, alliances and projects launched make economic sense and are rigorously managed?
- If our CEO is not fundamentally a visionary transformer, how can the C-suite, as a team, launch a collective effort to introduce radical change and start a transformation process?
- In sum, do we have a balanced profile of leaders in our C-suite—balanced in terms of competencies, profiles and focal points—to support our CEO in his/her innovative transformation of the company?

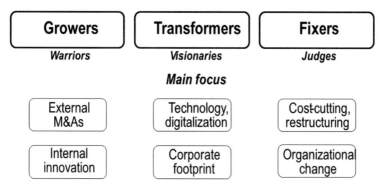

Fig. 4.1 Typical CEO leadership "default" profiles (not mutually exclusive)

The last question is essential because the board should be prepared to fight the tendency of some executives to appoint leaders with whom they feel comfortable. It is indeed quite natural to surround oneself with people who share the same beliefs and values. Warriors love to be supported by highly dynamic warrior-style colleagues, and judges by rigorous fact-oriented judge-profile subordinates. Diversity of profiles in a top management team does not always come naturally!

4.6   Radical Transformations Typically Require a Broad Range of Talents

Whatever their origin and purpose, radical transformations are likely to put a lot of pressure on the top team, who have to manage the transition while keeping the business viable during the change. This is why they often require ambidextrous leadership.[3] Former IMD dean, Derek Abell, characterized it as mastering the present while pre-empting the future. Since few leaders would claim to be equally inclined— or talented—to cover both objectives, in practice this means that the top team ought to include transformers, as well as growers and fixers.

But, underneath, there are quite a few roles to be played if all the changes are to proceed in parallel. Figure 4.2 summarizes these key roles and highlights the possible leadership profile of the various players.

- In the transformer category, technology (and digitalization) navigators should be available to propose a roadmap and lead the various implementation projects.

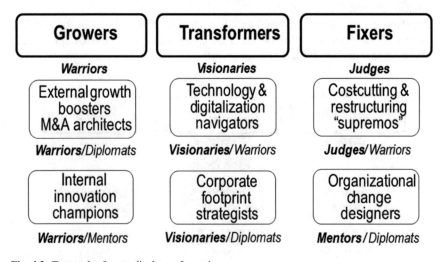

Fig. 4.2 Team roles for a radical transformation

---

[3] Abell (1993, 2010).

Similarly, corporate footprint strategists will need to devise how the new activities and new business models will compete in the market, and what to do with the old ones. These roles require a visionary attitude on both sides, complemented by practical considerations.

- In the grower category, marketing-oriented growth boosters and M&A specialists will be needed to define how the company can quickly gain access to the market, be it organically or through complementary supportive acquisitions. And simultaneously, all the creative energy of the organization will need to be harnessed by innovation champions to ensure the new products and services become winners. These growth boosters and innovation champions are likely to have strong warrior overtones.
- In the fixer category, it will be important to appoint a leader of the many cost-cutting and restructuring jobs that will inevitably accompany the change from the old world to the new. An organizational change designer will also need to propose and structure the new tasks and teams for implementation. These roles require a diversity of leadership profiles.

In summary, even though boards are unlikely to become involved in all the steps that will lead to the transformation, it would be wise if they made sure that all these roles are properly recognized by the CEO and C-suite, and adequately filled. The success of the transformation depends on it.

References

Abell, F. D. (1993, 2010) *Managing with dual strategies*. New York: The Free Press.
Deschamps, J. P., & Nelson, B. (2014). *Innovation governance: How top management organizes and mobilizes for innovation* (pp. 332–340). San Francisco: Jossey-Bass.
Tomasko, R. M. (2006). *Bigger isn't always better: The new mindset for real business growth*. New York: Amacom.

Jean-Philippe Deschamps is emeritus professor at IMD in Lausanne with 45 years of international experience as innovation practitioner/consultant and teacher. He focuses his research on the management and governance of innovation and the profile and focus of innovation leaders. He has (co-)authored three best-selling books: *Product Juggernauts: How Companies Mobilize to Generate Streams of Market Winners—Innovation Leaders: How Senior Executives Promote, Steer and Sustain Innovation*—and *Innovation Governance: How Top Management Organizes and Mobilizes for Innovation*. He has delivered lectures throughout the world, including twice in front of heads of states and entire governments, and at the prestigious Millenium Prize in Helsinki, the "Nobel" for technology. He was twice invited to speak at the World Economic Forum in Davos.

Chapter 5
Strategic Leadership: A Paradoxical Mindset of Value Creation

Kai Gausmann and Gysele Lima Ricci

Abstract Strategic leadership is a distinctive success factor for current and future competition. Strategic leadership requires that leaders have strategic thinking competencies, involving creative thinking and the ability to recognize emerging opportunities. This research describes the distinctive responsibilities for strategic leaders and develops a model for strategic leadership to facilitate the realization of strategic and innovation changes. This model called *Strategic Leadership Matrix* describes a new strategic leadership approach which not only contains capabilities of strategic leaders, but it focuses on their core responsibilities regarding three domains: corporate environment, value creation and people. As a result, we conclude that the ultimate goal of strategic leadership is to provide for innovation and the creation of new market opportunities.

5.1 Introduction

For the past 25 years the field of strategy has focused on how to build competitive advantage. Lately, the competitiveness in various sectors of the economy has imposed on companies an increase in the demand for innovative strategies and solutions. Thus, innovation becomes a distinctive success factor for current and future success.

Leaders throughout organizations know they need to change the way they work. As they seek to drive results, they are looking for new rules to have competitive advantage and encounter new industries, markets, products, and services. A brilliant strategy may put a leader at competitive edge but only solid execution will keep him there. In this way, a major concern for leaders everywhere is, how to achieve a sustained competitive advantage that ensures survival.

K. Gausmann (✉)
Capgemini Invent, Berlin, Germany
e-mail: kai.gausmann@gmail.com

G. Lima Ricci
Bundeswehr Universität, München, Germany
e-mail: gysele.lima@unibw.de

© Springer Nature Switzerland AG 2020   47
N. Pfeffermann (ed.), *New Leadership in Strategy and Communication*,
https://doi.org/10.1007/978-3-030-19681-3_5

Developing strategic leadership competencies involves changing the habit of managers. Given the inherent uncertainty of a VUCA economy (volatility, uncertainty, complexity, ambiguity) (Bennett and Lemoine 2014), they have to be able to learn and adapt. Leaders in today's digital world must be flexible to exploit discoveries, likewise, they need to deal with difficulties and failure as well as uncertainties in the market. More than ever, strategic leaders have to break the habit of silo thinking and only in their existing markets, they need to be able to think from a macro perspective and have a holistic view of the company as a whole (Goldman et al. 2015; Stigter and Cooper 2016).

Changing habits requires more than just organizing training about strategic tools. In our view, developing strategic leadership skills requires education and experience. This involves learning experiences that are capable of forming a mindset, transforming behavior and developing new competencies. Essentially, we translated our ideas and learning's of strategy and leadership over the last ten years of research and consulting services in various industries into this paper. More specifically, in this study we propose a strategic leadership model, and see how to develop strategic responsibilities for leaders to facilitate the realization of strategic and innovation changes.

5.2   Strategic Leadership

A brilliant strategy may put a leader on the competitive scenario, but only solid execution keeps him at competitive edge. Most companies struggle with differentiation from the market. However, strategy aims at the creation of a unique and valuable position, involving a different set of activities that requires you to make trade-offs to choose what not to do.

The term strategy is relatively complex, no consensus on the issue. Although there is no consensus, there are common elements, such as the elaboration of goals, decisions and the establishment of a course of actions to achieve predetermined objectives, aiming to increase the competitive advantage. Strategies are associated with the word "how": how to grow business, how to beat the competition, how to satisfy customers, how to respond to market variables, among others. They are specific to each company, adapted to the company's own situation and its performance objectives. According to Porter (1996, p. 68), strategy means the "creation of a unique and valuable position, involving a different set of activities". The idea of strategy is related to competition; it is a factor of differentiation of a company towards its rivals; it is the way that organizations must continue to survive (Ricci 2011; Kim and Mauborgne 2015).

Strategy is indispensably related to strategic leadership. Strategy defines the direction of the organizational journey (Mintzberg 2001) while leadership consists of influencing the attitudes and behaviors of individuals within and between groups to achieving goals (Bass 1990). Strategic leadership requires that leaders have strategic thinking competencies, involving creative thinking and to recognize emerging opportunities (Mintzberg 2001). Unlike leadership at the operational level which

Table 5.1 Strategic leadership responsibilities

Strategic leadership responsibilities	Activities	Concept	Author
Initiate	1. Establish sense of urgency	8 steps	Kotter (2010)
	2. Create a guiding coalition		
	3. Develop a vision and strategy		
	Awareness, Desire	ADKAR	Hiatt (2006)
	Unfreezing	–	Lewin (1947)
Guide	4. Communicate the change vision	8 steps	Kotter (2010)
	5. Empower broad based action		
	Knowledge ability	ADKAR	Hiatt (2006)
	Moving	–	Lewin (1947)
Reinforce	6. Generate short-term wins	8 steps	Kotter (2010)
	7. Build on the change		
	8. Anchor new approaches		
	Reinforcement	ADKAR	Hiatt (2006)
	Refreezing	–	Lewin (1947)

relies on technical and social skills, has a functional focus and is usually procedure-bound, strategic leaders need a different mindset. In other words, strategic leadership involves dealing with issues commonly addressed by a firm's top management team.

Strategic leadership also involves dealing with how to respond to changes in the external environment, how to deal with challenges and discontinuities that emerge from time to time (Goldman 2012; Stigter and Cooper 2016) and requires the development of insights and wisdom, the ability to think outside the box and must be able to connect and synthesize ideas (Goldman et al. 2015).

General leadership research refers to leaders at any level of the organization, whereas strategic leadership research refers to the study of people at the top of the organization as the guiding force behind organizational learning (Harrison 2018; Magsaysay and Hechanova 2017; Stock et al. 2016; Dinh et al. 2014; Parris and Peachey 2013; Hambrick and Pettigrew 2001; Lahteenmaki et al. 2001). In the sense of our proposal strategic leadership is more than the guiding force and not only refers to the top management. In our opinion and experience, strategic leaders that impactfully drive change initiatives are the *initiating, guiding* and *reinforcing* force behind organizational learning. This is particularly linked to the concept of Change Management as each change encompasses a journey of organizational learning, if done right. Strategic leaders are the sponsor for the activities that are described in Table 5.1. These are also consensus of all contemporary research and publications that involve leadership activity within change management research (Kotter 2010; Lewin 1947; Hiatt 2006).

From a strategic leadership perspective, we understand the leader as the "captain of a ship", who elaborates and implements strategic leadership, who guides the tasks formulation and implementation of strategic planning for the entire organization. The essence is that leaders lead, i.e. they carry out certain activities to direct or influence followers. The opinions and conclusions of leaders regarding the direction and focus of strategy constitute the strategic vision of the organization. Thus, they align the organization to a certain direction, indicating the strategic trajectory and shaping the organizational identity (Ricci 2011).

5.3   The Strategic Leadership Matrix

For us, strategic leadership in a VUCA world is obliged to focus on innovation and their market strategy in a sense of blue oceans (Kim and Mauborgne 2015). This is a direct result of increasing competition where everyone who owns a laptop, has an internet connection and the necessary business acumen can start to build a business from scratch that will change the rules of how traditional industries operate. This is true for Facebook for the communication services industry, for Uber for the transportation industry, for Spotify for the music industry, for Netflix for the entertainment industry, for Airbnb for the lodging industry and many more. This does not mean that any of the traditional industries will cease to exist. However, it does mean that the way those industries operate are already fundamentally changing. All the above-mentioned companies have one pattern in common: They created a new uncontested market space by innovating new business models. Innovation is, so to say, in their DNA. These two factors walk hand-in-hand for start-ups.

This paper shows, why it is important exactly for enterprises to adapt those patterns and what are the characteristics that strategic leaders need to adapt to create more impact within their area of responsibility. To understand the concept that we propose it is important to understand the roots of the idea and to look at the comprehensive effects of leadership from a systemic perspective that encompasses a variety of interdependencies. This is, what we will elaborate in the following.

The Strategic Leadership Matrix (Fig. 5.1) is characterized by four categories which describe different aspects of market and innovation strategy: *conservative leadership*, *innovative leadership*, *explorative leadership* and *entrepreneurial leadership*. Section 5.3.1 focuses on the characteristics of each of the four categories. The characteristics of strategic leadership create a context in which sustainable business growth happens. In this part we explain how to outperform your competitors by developing and implementing the characteristics and leadership capabilities of quadrant four: Entrepreneurial Leadership.

It is important to understand that the Strategic Leadership Matrix is rather a directing than a prescribing approach This means it is rather about the journey than about reaching a fixed status quo. It is a model to understand how corporate identity on a systemic level evolves into either a fixed or a growth mindset. In other words, the matrix applies for every company along the company life cycle (Sect. 5.4).

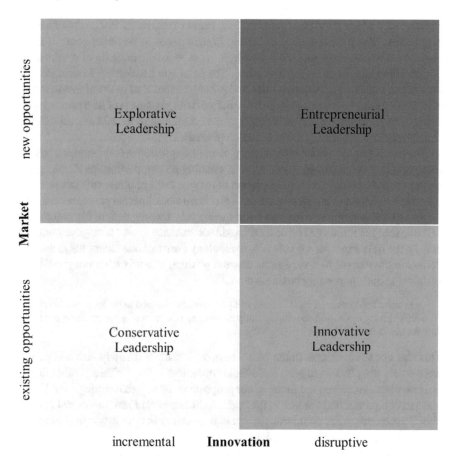

Fig. 5.1 The strategic leadership matrix

5.3.1   *Categories of the Strategic Leadership Matrix*

Conservative Leadership in the bottom left category is characterized by maintaining the status quo (case 1). Enterprises in this sector are most likely to become extinct over the course of the next ten years. As a matter of fact, those companies focus their resources at fighting their competition and reducing cost instead of finding new ways of creating profit. In the best case, they follow market developments as competitors prove to succeed with new ideas or products. Thus, they implement innovations incrementally instead of risking failure. Companies in this category refrain failure and risk, and implement new strategies with old mindsets, which is bound to fail.

However, there are companies that show a similar profile but do create innovative ideas. Those companies fall into the bottom right category *Innovative Leadership*. The case of Kodak for instance shows, they were even the innovators of digital photography when they presented the first digital camera in 1991. This example emphasizes how the Strategic Leadership Matrix needs to be understood: It is not about the obvious, it is about the invisible; it is not about the creation of new markets nor on innovation itself in the first place. The Strategic Leadership Matrix is about the underlying leadership capabilities and principles that lead to the aforementioned effects. Kodak did not want to jeopardize their existing business of film photography. Thus, despite innovative approaches they were missing all of the aforementioned key characteristics *initiation*, *guidance* and *reinforcement*.

Companies in the upper left category display *Explorative Leadership*. They are those that do even strive for new markets, creating new opportunities through partnering up and opening new collaboration networks. Nevertheless, still this does not guarantee sustainable business success. Apart from slack internal processes that may impede quick decision making and other factors that are prevalent in big companies, still one thing is missing. The case of Nokia for instance shows how your company may be the trailblazer for a whole new technology (smartphone technology) that will dominate the market in a very short amount of time, still this does not provide for sustainable and succeeding business models.

> Only when two factors go hand in hand will your business succeed in the long run: Continuously driving innovation and consequently conquering new market opportunities is key for sustainable business success.

This was not so much true in the past where other factors (namely size and power) were dominating factors to push aside the competition. In the digital age other factors fall into place. However, the focus of our proposal is the role that strategic leadership plays in companies that venture in the upper right category: *Entrepreneurial Leadership*. Those companies continuously invent themselves in new ways: they innovate, they experiment, they fail and quickly learn from their errors. They may also face crisis. It is important to note that the matrix is not about a fixed status quo but about the journey along the company lifecycle that every company undergoes. It is about the underlying strategic leadership principles and capabilities that are the motor of a company. Those principles are like a magnetic field. Sometimes traction is stronger, sometimes it is weaker. But always the center of traction is innovating, creating new opportunities and implementing them into the market.

5.4   Strategic Leadership as the Driving Force Along the Company Life Cycle

The Strategic Leadership Matrix suggests an approach of how to prepare a company for sustainable success based on strategic leadership responsibilities. The company life cycle of a company (Fig. 5.2) that is characterized by *Entrepreneurial Leadership* consists of the phases startup (1), growth (2), maturity (3), challenge (4), re-invention (5.1), termination (5.2) and growth in new markets (6). This model is particularly interesting for us as in phase (5) it distinguishes between termination

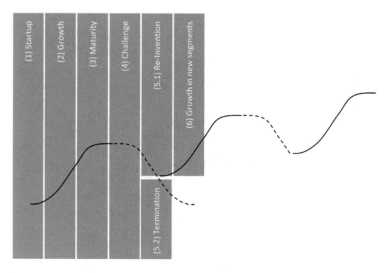

Fig. 5.2   The company life cycle. Adapted from Simon et al. (2010)

(5.2) and re-invention (5.1) of the company. Many other life cycle schemes only contain decline and/or termination of the company as the ultimate phase. Although our model focuses on re-invention other factors may fall into place, why a company goes extinct. For the sake of completeness, we kept this phase. However, we disagree with other approaches and think that how the company makes strategic decisions in the fifth phase is a strategic leadership choice. Nevertheless, it is not only one decision but a direct result of all strategic decisions that have been made until then. With growth comes control, with control comes rigidity and with rigidity comes a fixed company mindset (Ricci 2017). In this kind of state, innovation and novelty do not come naturally to the company anymore. This is the opposite of what the Strategic Leadership Matrix suggests. The matrix proposes a genuine continuous development process on all corporate levels. This means that in the sense of Kim and Mauborgne (2015) corporate development aims at:

– Eliminating all activities that do not add value to the customer;
– Reducing all activities that are still necessary but do not add value to the customer;
– Raising all activities that add value to the customer;
– Creating activities that are not yet in place but would add value to the customer.

In order to provide for this genuine continuous development process, the mindset of *Entrepreneurial Leadership* must be implemented at all corporate levels as all leaders to some extent have got responsibility for reinforcing and implementing the company strategy (Fig. 5.3).

On their journey through the life cycle the company undergoes changes regarding their market orientation (left axis of the matrix) as well as their innovation approach (bottom axis of the matrix).

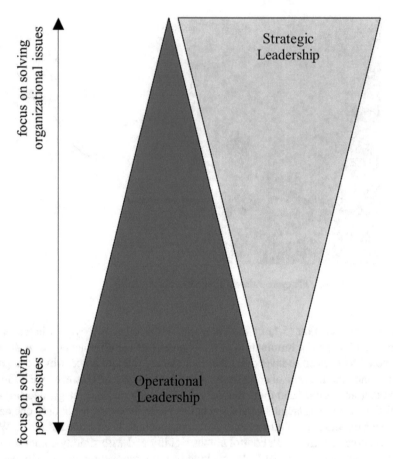

Fig. 5.3 Strategic leadership is part of a role, distributed over several hierarchy levels and not only bound to senior executives

A typical—even if not the only possible—journey is shown in Fig. 5.4. Whilst starting up (1) a company generally creates new market opportunities and at the same time uses innovative approaches. Thus, companies usually start with a style of *Entrepreneurial Leadership*. Within the growth phase (2) the company develops into a well-established mature business (3) that is being copied by other companies. The critical phase for further company development is the challenge phase (4) or rather the transition between maturity and challenge phase, when the company possibly loses market share and significance in the competition. At this point two scenarios are possible: re-invention (6) by disrupting innovation and creating new business opportunities again or termination (5) because of missing innovation and tough competition which does not leave any room for profitable business.

In our opinion the tipping point is inherently connected to strategic leadership decision making in the first place and empowering all leadership levels to act upon

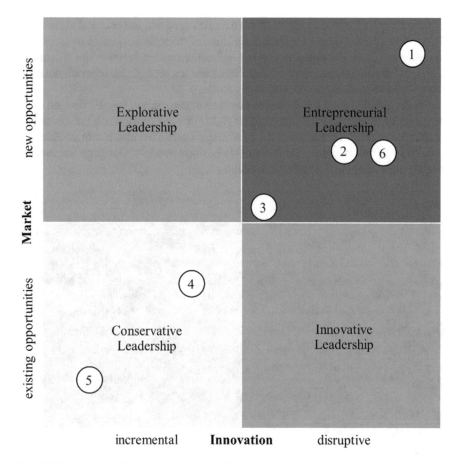

Fig. 5.4 Leadership styles along the business life cycle

it in the second place. As we show in the following, there are critical leadership capabilities for the leadership team that need to be in place. Having applied the leadership principles and capabilities of *Entrepreneurial Leadership* provides for new growth opportunities.

5.5   Areas of Responsibilities and Capabilities of Strategic Leadership

While companies undergo the life cycle, they experience severe challenges which top management's need to find strategic answers for. After research of leadership concepts, we recognize that regarding strategic leadership, research is still scarce. Therefore, we look at the most relevant leadership concepts and propose a comprehensive strategic leadership model that is based on our profound practical experience. This model focuses on the key responsibilities of strategic leadership to create and innovate sustainable business models in a replicable manner. Furthermore, this model highlights the differences and commonalities between strategic and operational leadership to provide direction for which one to use.

Leadership literature is comprehensive in characteristics of capabilities and competencies of leaders on the operational level. In our view an approach is missing that encompasses not only characteristics of strategic leaders but also responsibilities for leadership to create impact for sustainable business models. Our proposal focuses on the responsibilities for strategic leaders.

Case 1: Automotive Industry

Background: The automotive industry is facing fundamental challenges in the way they are operating. New business models are growing and the core of mobility itself is changing. Among others some major trends encompass "sharing instead of owning", moving from petrol to e-mobility, electrification, digitization of mobility and therefore the rise of completely new competitors (i.e. Tesla since 2003 and Google since 2009) that are natively digital.

Challenge: Facing severe financial problems an international German automotive company launches several initiatives to change the way they operate: They call on their employees to start using agile approaches, to transform their culture and change the way they lead. Apart from those cultural initiatives they have shortened their product development lifecycle from seven to five years and start new product initiatives.

Observed implementation gap: We work very closely at the intersection of all of these topics facilitating workshops on various hierarchical levels regarding leadership and collaboration. Observing the challenges all of the teams are facing, the company reveals an implementation gap of many of their initiatives. The reduction from seven to five years product development life cycle itself comprises several challenges (among others: provide for faster alignment, reduce non-productive processes, implement digital ways of working). This is a direct result of missing strategic leadership. Regarding cultural initiatives top management is relatively quick at initiating change but shows missing guidance and reinforcement. Our observations are based on over 50 workshops, with more than 1000 participants of various hierarchy levels over the course of half a year. The pattern the participants report is always the same: They lack top management support in a sense of guidance and reinforcement for instance for the implementation of agile approaches or digital process flows. Core workflows are still being executed in PowerPoint slides or Excel spreadsheets. Even workflows that have been digitized are deficient. Efficiency suffers from poor software execution as core functionality is not being covered.

Solution approach: Implementing new ways of working and changing a corporate culture is not a top-down approach and the enterprise needs to start with pilot projects. However, initiatives need to be created on a cross-functional level and they need to be driven by strategic leaders.

Those responsibilities derive from the main domains that lay at the heart of each corporation: the value proposition, the employees that deliver the value proposition and the corporate environment that creates the basis for your employees to deliver value.

All of these three domains have huge impact on the corporation's strategy. Thus, strategic leaders' responsibilities need to be aligned with these three main pillars (Fig. 5.5). In brief, strategic leaders need to:

– anticipate and envision future customer's value propositions;
– influence and persuade by role modelling and claiming behavioural change;
– create an environment that fosters the development of people and value.

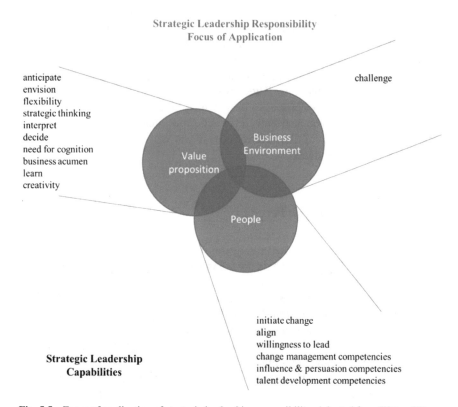

Fig. 5.5   Focus of application of strategic leadership responsibility. Adapted from Hitt and Duane (2002), Schoemaker et al. (2013)

5.5.1 Responsibility for Value Proposition

The responsibility of anticipating and envisioning future value proposition is focused on the strategy of creating new opportunities on the market and reinventing the company's value proposition. Therefore, to follow this responsibility the following capabilities are necessary as a strategic leader: anticipating, envisioning, maintaining flexibility, strategic thinking, interpreting, deciding, need for cognition, strategic thinking competencies, business acumen, learning, creativity.

In order to create new value propositions a strategic leader needs to anticipate the customer's needs and envision new services. However, without the necessary business acumen he will not succeed putting these ideas into practice. Strategic thinking is what links new value propositions to the existing business.

> Creating new opportunities or even markets is not about selling cookies when your business is in automotive. Therefore, you need to identify the very core of your business. In this case: mobility

It is about thinking beyond physical, technological and even social borders. As it was the case for Kodak, they would not look beyond the status quo to see digital photography beneficial for their existing business nor did they see it as a thread before they fell into decline. This part is about decision making. Nevertheless, it is not about retro focused decision making (based on experience and past truths) but about future focused decision making (based on experience and future opportunities). For future focused decision making there are important characteristics that make a strategic leader, that have seldom been included into strategic leadership capabilities: creativity and constant learning. Those two characteristics are key because they link generic leadership capabilities to those which lead to entrepreneurial business innovation. Only when you constantly learn and show interests even outside of your context of responsibility in the creative process you will be able to come up with genuine out-of-the-box ideas.

5.5.2 Responsibility for People

Especially long-lasting established enterprises display a strong resistance-to-change when the responsibility for people is not being taken charge of by top management (case 1). This is because over decades incentives have been put into place by executives that determine which values count in a corporation independent of which values are being proclaimed indeed. Strategic leaders are in charge for claiming behavioural change, influencing and persuading employees and other leaders and acting as a role model. Thus, you can match incentives to the desired culture, push forward structural changes and creating new opportunities. In this way, the following capabilities are key to take charge of people responsibility: initiate change, align, be willing to lead, change management competencies, influence and persuasion competencies, talent development competencies.

As argued before it is not enough to only initiate change, but strategic leaders additionally need to guide and reinforce change. This capability includes the willingness to lead because people will mistrust you as change discomforts them. The logical consequence is, that they will mistrust you unless you have the influence and persuasion skills to persuade them of the contrary. Even as cultural change is not a classical top-down activity, the command and control environment in traditional industries, requires top management to *initiate*, *guide* and *reinforce* cultural change. In the aforementioned case, cultural change will only happen when the top leaders start role modelling new behaviour and claiming to do so by their direct reports. They then need to cascade role-modelling down the corporate ladder. The method of role-modelling has been successfully implemented by other companies the authors have co-worked with, namely in the automotive and logistics industry. The time needed for cultural changes to become visible, in those companies that use role-modelling, is a fraction of those companies that do not use it. The reason is that change becomes visible the minute you start.

5.5.3 Responsibility for the Corporate Environment

The corporate environment in our understanding consists of processes, infrastructure (including the IT infrastructure) and the incentive-system. This is an area which in most cases is treated as the least impactful one by executives. The Airbus CIO Luc Henneken is an impressive example of how to take charge of the corporate environment seriously. In March 2018 Airbus announced to migrate their complete 130,000 workforce from Microsoft products to Google G-suite. Henneken announced this migration to be part of the digital transformation process which focuses on the collaboration of their workforce. One of the main reasons Henneken proclaims is not even about functionality it is about collaboration, as functionality has been enhanced in Office 360.

The underlying capability of taking charge of the corporate environment is challenging the status quo. However, none of the research around strategic leadership covers any other capabilities that fall into this category which is worth further investigation. Processes, infrastructure and incentive-system are the key-characteristics that influence how people work together. They are strategic functions, thus need to be taken care of deliberately. A metaphor easily explains how these three play together: While the infrastructure is the road that you let your people travel on, the processes are the rules that apply of how to use that road. You may not overtake on the right side and you must follow the signs that indicate the flow. Finally, to make sure people use the highway for driving at high speed and not the minor roads and to make sure they obey to the rules like not using the wrong lane, incentives (positive and negative) apply. *If you are slack on any of these three you either receive a workforce that follows the rules but drives with a Formula 1 car on gravel road, you have a workforce that is not aligned well and the roads get chaotic or you receive a workforce that drive in a direction that is not favorable for corporate strategic goals.*

## 5.6	Conclusions

Transformation processes are the pinnacle of strategic leadership because not only traditional rules and principles apply. Thus, common leadership models are insufficient to explain the impact strategic leaders may have and how they can turn around the ship. Nevertheless, to determine how to empower strategic leaders not only by choosing the right leader to do the job but also to choose the right leader to do the job right we propose a leadership model that encompasses strategic leadership capabilities as well as core responsibilities regarding three domains: corporate environment, value creation and people. Excelling at all three of these domains is a must to provide for sustainable business success. These responsibilities apply for senior executive leaders. However, strategic leadership is a role that—to different extents—applies to all leadership levels. The ultimate goal of strategic leadership is to provide for innovation and new market opportunities. History has it, that tradition itself is no guarantee for sustainable business success. Sustainable success in our view is a journey along the company life cycle that crucially consists of how a company deals with decline and which strategic decisions it makes.

Bibliography

Bass, B. M. (1990). From transactional to transformational leadership: Learning to share the vision. *Organizational Dynamics, 18*(4), 19–31.

Bennett, N., & Lemoine, G. J. (2014). *What VUCA really means for you. Crisis management.* Harvard Business Review.

Dinh, J. E., Lord, R. G., Gardner, W. L., Meuser, J. D., Liden, R. C., & Hu, J. (2014, Febuary). Leadership theory and research in the new millennium: Current theoretical trends and changing perspectives. *The Leadership Quarterly, 25*(1), 36–62.

Goldman, E. F. (2012). Leadership practices that encourage strategic thinking. *Journal of Strategy and Management, 5*(1), 25–40.

Goldman, E. F., Scott, A. R., & Follman, J. M. (2015). Organizational practices to develop strategic thinking. *Journal of Strategy and Management., 8*(2), 155–175.

Hambrick, D., & Pettigrew, A. (2001). Upper echelons: Donald Hambrick on executives and strategy. *Academy of Management Executive, 15*(3), 36–44.

Harrison, C. C. (2018). *Leadership theory and research: A critical approach to new and existing paradigms.* Switzerland: Palgrave Macmillan.

Hiatt, J. M. (2006). *ADKAR: A model for change in business, government and our community.* Loveland: Prosaic Inc.

Hitt, M. A., & Duane, R. (2002). The essence of strategic leadership: Managing human and social capital. *The Journal of Leadership and Organizational Studies, 9*(1), 3–14.

Kim, W. C., & Mauborgne, R. (2015). *Blue ocean strategy: How to create uncontested market space and make the competition irrelevant.* Boston: Harvard Business Review.

Kotter, J. P. (2010). *Leading change.* Boston: Harvard Business Review Press.

Lahteenmaki, S., Toivonen, J., & Mattila, M. (2001). Critical aspects of organizational learning research and proposals for its measurement. *British Journal of Management, 12,* 113–129.

Lewin, K. (1947). Frontiers in group dynamics: Concept, method and reality in social science; social equilibria and social change. *Human Relations, 1,* 5–41.

Magsaysay, J. F. & Hechanova, G. (2017, June). Building an implicit change leadership theory. *Leadership & Organization Development Journal. 38*(6), 834–848.

Mintzberg, H. (2001). The Yin & Yang of Managing. *Organizational Dynamics, 29*(4), 306–312.

Parris, D., & Peachey, J. A. (2013). Systematic literature review of servant leadership theory in organizational contexts. *Journal of Business Ethics, 113*(3), 377–393.

Porter, M. E. (1996). What is strategy? *Harvard Business Review, 74*(6), 61–78.

Ricci, G. L. (2011). *Performance and competitiveness in small and medium-sized enterprises: Study of the hotel sector of the central region of the State of Sao Paulo (Master's Thesis).* Sao Carlos, Brazil: Sao Paulo University.

Ricci, G. L. (2017). *Estudio de las redes de cooperación en el sector turismo en España y Brasil.* Propuesta de un modelo de gestión aplicado a los clusters de turismo. (Doctoral Thesis). Autonomous University of Madrid. Madrid, Spain.

Schoemaker, P. J. H., Krupp, S., & Howland, S. (2013). *Strategic leadership*: *The essential skills.* Harvard Business Review.

Simon, D. G., Hitt, M. A., Ireland, R. D., & Brett, B. A. (2010). Resource Orchestration to create competitive advantage. *Journal of Management, 37*(5), 1390–2063.

Stigter, M., & Cooper, C. L. (2016). *Solving the strategy Delusion: Mobilizing people and realizing distinctive strategies.* London: Palgrave.

Stock, M. R., Zacharias, N. A., & Schnellbaecher, A. (2016). How do strategy and leadership styles jointly affect co-development and its innovation outcomes? *The Journal of Product Innovation Management, 34*(2), 201–222.

Kai Gausmann is Manager for Leadership at Capgemini Invent. Before he worked at CPC Consulting ranked 1st as Hidden Champion by the German Capital magazine in the field of Change Management. Before CPC he worked in organizational development in the sport industry. He graduated as Sport Scientist at Göttingen University and has a specialization in Business Psychology.

Gysele Lima Ricci is a Post-doc Researcher at Bundeswehr Universität München. She is a Ph.D in Economics and Master in Production Engineering. She developed projects in different countries such Spain, Brazil and Germany and published more than 40 articles in recognized journals.

Chapter 6
How Do Leaders Embrace Stakeholder Engagement for Sustainability-Oriented Innovation?

Babak Ghassim and Lene Foss

Abstract In this chapter, we demonstrate how open innovation can be applied to sustainability contexts. In this regard, we specifically address the leadership challenges encountered in accessing the wide variety of knowledge from multiple external stakeholders. First, we demonstrate that implementing open innovation for sustainability requires specific organizational capabilities that are different from what a firm might already possess regarding its general innovations. Second, we argue that successfully leading open innovation for sustainability entails broadening the scope of external stakeholder engagement to collaborate with non-conventional stakeholders such as local communities. Some sustainability-oriented innovation initiatives may also benefit from transactional type of stakeholder engagement where frequent interactions assist companies to establish stable search platforms and secure timely access to external knowledge.

6.1 Introduction

During recent years, environmental and social performance has become increasingly central to firms' success in various industries, particularly in the resource extractive ones such as the minerals industry, whose business activities are closely intertwined with their impact on the social and natural environments (George et al. 2015). While the 'reactive' approach towards environmental and social sustainability positions them solely as costly practices driven by stakeholder and institutional pressures (Zollo et al. 2013), increasing awareness of the opportunities at the crossroads of these practices and shareholder value is giving rise to a more 'proactive' approach among company leaders (Hall and Wagner 2012). Such an integrated pursuit of economic, environmental and social sustainability has brought into focus the concept of 'Corporate Sustainability', which requires leaders to fulfill three critical tasks

B. Ghassim (✉) · L. Foss
The School of Business and Economics, UiT-The Arctic University of Norway, Tromsø, Norway
e-mail: babak.ghassim@uit.no

L. Foss
e-mail: lene.foss@uit.no

© Springer Nature Switzerland AG 2020
N. Pfeffermann (ed.), *New Leadership in Strategy and Communication*,
https://doi.org/10.1007/978-3-030-19681-3_6

in order to ensure success in the long term: economic prosperity, environmental protection and social equity (Wilson 2003). The underlying logic of an integrated corporate sustainability perspective is therefore its emphasis on undertaking practices that yield better socio-environmental performance and higher economic benefits.

This chapter builds on the recognition that 'innovation' is a key organizational factor in enabling firms to pursue integrated corporate sustainability practice (Hall and Wagner 2012; Nidumolu et al. 2009). Accordingly, the term 'Sustainability-Oriented Innovation' (SOI) (Adams et al. 2016) has emerged and is defined as: "Making intentional changes to an organization's philosophy and values, as well as to its products, processes or practices to serve the specific purpose of creating and realizing social and environmental value in addition to economic returns." (Adams et al. 2016, p. 181). We recognize that both academicians and practitioners pay considerable attention to the topic of 'what' makes some firms successful in simultaneously improving economic, environmental and social performance.

Consequently, this chapter offer insights to company leaders in the context of 'stakeholder engagement for SOI' (hereinafter, open SOI) by shedding light on how open innovation and developing internal capabilities can assist the leaders to fulfill their corporate sustainability objectives. Acknowledging the difficulty that firms encounter in undertaking SOI, research has shown that engaging external stakeholders is a prerequisite for the continuous creation and deployment of innovative solutions for tackling sustainability concerns (Hall and Vredenburg 2003; Segarra-Ona et al. 2017). Considering the variety of innovation types and the broad impact of SOI, company leaders are required to incorporate a diverse set of knowledge in their innovation processes, including knowledge about technologies, regulative standards, societal expectations and market demands (Clarke and Roome 1999; Ketata et al. 2015). Consequently, not only are the primary stakeholders, such as those within the value chain, relevant, but also the secondary stakeholders (e.g. not-for-profit organizations), who are deemed insignificant for general innovations can enable firms to overcome the complexity and uncertainty of SOI (Hall and Martin 2005).

Broadening the scope of external stakeholders in SOI does also challenge company leaders to develop particular capabilities, on top of those required for general innovations, in order to manage the knowledge inflow and effective learning. These capabilities range from stakeholder networking and competence mapping before the start of an innovation project, to relational capability and knowledge management during a project (Behnam et al. 2018; Kazadi et al. 2016). Openness towards external ideas and technologies does not by any means dispel the need for internal capabilities required for utilizing the external knowledge further down in the innovation processes.

For the purpose of this chapter, we will elaborate on the underlying processes of capability accumulation and open innovation in the context of open SOI. This will be achieved by explaining the main skills and organizational routines that are required when firm leaders aim to engage broader groups of stakeholders in their innovation processes, as well as specifying in what ways open innovation can improve the access to external stakeholders' knowledge. We illustrate our points with quotes from interviews in a recent study (Ghassim and Foss 2018). The remainder of the chapter is structured as follows: the next section provides an overview of the concepts underlying the phenomenon open SOI. Section 6.3 describes the internal capabilities and mechanisms of open innovation in the context of SOI. Finally, Sect. 6.4 presents the implications for company leaders on how to handle the challenges of developing internal capabilities and opening up the innovation processes in the sustainability context.

6.2   Conceptualizing Open Sustainability-Oriented Innovation

To explain the phenomenon of open SOI, the three generic concepts of corporate sustainability, external stakeholders and innovation are relevant. Consequently, the perspectives that lay the basis for this phenomenon are stakeholder engagement for sustainability, SOI and open innovation (Fig. 6.1).

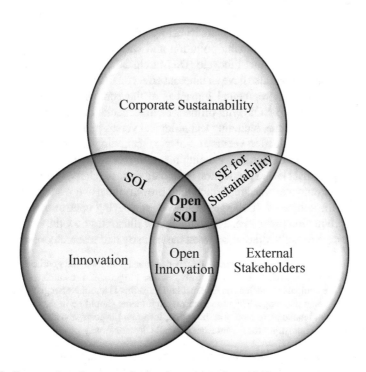

Fig. 6.1 Cross-section of concepts shaping the concept of open SOI

6.2.1 Sustainability-Oriented Innovation

Innovation is widely accepted as an important determinant of firms' economic success (Crossan and Apaydin 2010). Similarly, new technologies, products and organizational practices play a key role in addressing social and environmental issues (Hart 1995; Holmes and Smart 2009). This is reflected in the fact that the quest for corporate sustainability is increasingly resulting in innovation activities across different firms and industrial sectors (Nidumolu et al. 2009). Concerning the social aspect of SOI, innovation could appear on a continuum of purposes, from conflict resolution to the creation of social values (Murphy et al. 2012). In this regard, scholars in the field of corporate social responsibility (CSR) has emphasized that integrating social needs into organizational practices can enable leaders to find solutions with combined economic and social benefits, thus moving beyond purely philanthropic purposes (Jamali et al. 2011). By applying this perspective in a study of mineral firms in the UK, Bini et al. (2018) suggest that showing social commitment (to gain social license to operate) is an important driver for firms that set out to innovate their communication processes with societal stakeholders.

In the same vein, the significant pressure on firms to minimize their negative environmental footprint has led to increasing investment in technologies and products with the potential for minimizing pollution and waste throughout the production processes and overall product lifecycle (De Marchi 2012; Sharma and Vredenburg 1998). These innovations also have an inherent effect on the social aspect of corporate sustainability, since environmental impact is at the core of societal expectations (Suopajärvi et al. 2016). SOIs with primary environmental objectives cover a range of different classifications, including technological versus managerial (Peng and Liu 2016) and cleaner production versus end-of-pipe technologies (Muscio et al. 2017). Concerning the latter dichotomy, existing research (see for e.g., Bonte and Dienes 2013) highlights that cleaner production technologies have relatively greater potential to reduce environmental hazards, since they prevent intensive resource use and/or pollution at the source of discharge, instead of employing control measures at the end of the production processes. The quote below illustrates how the environmental issue is the core in collaboration between the industry and research organizations:

> We have good examples from research projects that manage to establish good cooperation with the industrial participants. This is mostly because the proposed research topic was of interest to the companies, such as environmental issues that is a real Norwegian challenge - or even a global challenge. Therefore, the research theme should be in such a way that companies could make more money at the end, or lowering the costs, and also improve the reputation of the industry. - Informant from Mineral Cluster Norway

Whereas SOI has received increasing theoretical and practical attention during recent years, it is still of relevance to ask what the specifics of SOI are and why they require changes in leadership approaches. The literature has so far focused on three facets of SOI, which can also act as barriers to leaders' involvement with these innovations: (1) balancing the multiplicity of sustainability dimensions and pathways; (2) the double externality problem; and (3) added complexity and uncertainty.

First, corporate sustainability, and hence SOI, require firms to adopt an integrated approach in which economic, environmental and social objectives are pursued simultaneously. In this respect, firms must develop innovation capabilities at different levels of process, product and social practices, in such a way that any improvement in one sustainability aspect does not, in any event, cause a negative effect on any other aspect of sustainability (Hart 1995). For example, if a mineral firm attempts to introduce asphalt aggregates with better possibilities for recycling and reuse, while continuing to produce high levels of air emissions, it would then face a challenge to make potential customers believe in its environmental responsibility, and thus fail to benefit financially from its product innovation. Accordingly, it is important to note that when talking about SOI, environmental and social improvements are not an 'accidental side effect' of general innovation practices, but should be at the core of a firm's business activities.

Second, the 'double externality problem' that is commonly used in the literature on environmental innovation (as a subset of SOI) can also apply to the broader context, such as SOI. In his influential paper, Rennings (2000) posits that such innovations produce positive spillovers in both the development and implementation phases, hence discouraging firms to invest in them. More specifically, in addition to the issue of knowledge spillovers (to competitors) during the development stage, which is common to all innovations, SOI produces an additional externality, as it generates social and environmental benefits (primarily for society) that are hard to be reaped in financial terms. Therefore, the role of regulative frameworks to punish harmful environmental and social impacts is crucial in incentivizing firms that may lose their competitive advantage in the market due to the higher costs resulting from SOI practices (del Rio et al. 2015). Nonetheless, company leaders operating in industries such as minerals have to address social and environmental issues, no matter what the strength of such regulative frameworks is, as low performance in these aspects can directly affect their survival in the long term.

Finally, and importantly, the added complexity and uncertainty associated with undertaking SOI differentiates it from general innovations (Hall and Vredenburg 2003). Complexity arises as a result of the socio-technical diversity inherent in sustainability contexts (Clarke and Roome 1999), where incorporating environmental and social considerations require knowledge about technologies, regulative standards and societal expectations (Adams et al. 2016). Uncertainty, on the other hand, points to the financial risks of SOI. Social and environmental improvements might be achieved at the expense of increasing the cost of processes and products, which could result in the market and system failures of these innovations (Foxon and Pearson 2008). SOI, with its potential impact on wider groups of stakeholders, may create conflict situations due to the opposing interests between the focal firm and its stakeholders, for instance local communities. Therefore, it is likely that the knowledge required for SOI is relatively more distributed among different actors in the innovation system, hence requiring the engagement of a diverse range of external stakeholders in innovation processes. This theme will be discussed in detail in the following sections.

6.2.2 Stakeholder Engagement for Sustainability: Beyond Managing Stakeholders

The emergence of stakeholder theory has given rise to studies that enquire into firms' relationships with external stakeholders and the consequences of such relationships. As stakeholder theory requires leaders to respond to the needs and expectations of a wide variety of stakeholders (Freeman 1984, 2010), scholars have paid considerable attention to investigating in what ways, if any, external stakeholders affect different aspects of corporate sustainability performance.

Within this body of work, two research streams are evident. In the first, research has focused on firm-level and institutional determinants of practices directed towards reducing/eliminating the negative influences of specific groups of external stakeholders on overall firm performance (see for e.g., Kassinis and Vafeas 2006). These studies frame the association between corporate sustainability and stakeholders based on Frooman's (1999) description of the resource interdependence between a firm and its stakeholders, in which the firm strives to manage those stakeholders (via undertaking sustainability practices) who can directly or indirectly influence its access to critical resources (e.g. financial, human, raw materials).

For example, Sharma and Henriques (2005) found that in the Canadian forest product industry, firms are most likely to adopt innovative environmental management practices when the managerial perception of threats coming from environmentalists and/or customers is high. Indeed, the substantial investments required to implement advanced environmental management (and its negative impact on short-term gain) impede firms from acting (environmentally) sustainably, unless, for instance, it is possible that customers will cancel their purchase orders.

The second research stream goes beyond such a pure focus on controlling stakeholders' negative influences, and instead tends to use the term 'stakeholder engagement' to indicate a more optimistic outlook of stakeholders' role in corporate sustainability (Aragón-Correa et al. 2008; Hillman and Keim 2001). In this case, stakeholder engagement is defined as "practices that the organization undertakes to involve stakeholders in a positive manner in organizational activities" (Greenwood 2007, pp. 317–318). Empirical studies in this area have employed the RBV, or occasionally its extension, the natural resource-based view (NRBV) (Hart 1995), to maintain that involving external stakeholders in efforts to alleviate environmental and social issues gives birth to valuable, rare and inimitable assets, which in turn assist firms in achieving higher financial performance.

In their study of automotive SMEs in Spain, Aragón-Correa et al. (2008) highlight that mutual understanding arising from collaborative relationships with external stakeholders enables firms to show more proactive approaches in environmental management and to achieve better financial performance relative to their competitors. Eccles et al. (2014) adopted a more inclusive view, by adding social issues to the sustainability aspects examined in the previous study. In this regard, they used a matched sample of U.S. companies and found support for their hypotheses, suggesting that firms with higher sustainability performance do engage external stakeholders more frequently in daily operations, and that the high level of trust between them is a source of persistent competitive advantage by avoiding costly conflicts.

While both of the research areas discussed above have contributed substantially to understanding of stakeholders' role in corporate sustainability, the latter is in harmony with this chapter, which assumes a positive contribution of stakeholders in innovation processes, with the aim of creating mutual benefits for firms and their external stakeholders. For a firm and external stakeholders that have an economic stake in its performance (such as suppliers), this benefit arises in the form of cost savings or increased income, whereas other stakeholders take advantage of social and environmental improvements, in the form of either decreased negative impacts on the natural or social environment, or increased values in these respects.

6.2.3 Open Innovation: Leveraging on External Stakeholders' Knowledge

Since the introduction of 'open innovation' by Chesbrough (2003) over fifteen years ago, the concept has received great momentum from scholars across different scientific disciplines, even outside business and management fields. As the concept has been under development throughout the years, definitions abound. Nonetheless, Chesbrough and Bogers (2014, p. 17) synthesize the original and the most recent descriptions, defining open innovation as "a distributed innovation process based on purposively managed knowledge flows across organizational boundaries, using pecuniary and non-pecuniary mechanisms in line with the organization's business model". Thus, it should be noted that openness implies inflows and outflows of knowledge, as well as a variety of practices for knowledge flow that might not necessarily involve monetary exchange.

At the core of open innovation is the understanding that knowledge is widely distributed among various stakeholders in the business environment, and that company leaders can, and should, use these external stakeholders as well as their internal knowledge base (Laursen and Salter 2006). This new paradigm of innovation management has challenged the traditional 'closed' and 'vertical' modes of innovation, by suggesting that the increased mobility of skilled workers, and less control of unwanted spillovers to other firms, are shrinking firms' margins from investment on internal resources, such as R&D (Chesbrough 2003). However, the crude distinction between firms which are or are not open has received criticism from scholars, who argue that the extremely closed mode of innovation does not occur in reality (Trott and Hartmann 2009). Instead, it is now widely accepted that different degrees of openness exist, and that firms can be placed on a continuum from closed to open innovators (Dahlander and Gann 2010). The quote from a CEO of a mineral exploration company illustrates this kind of openness:

Compared to the level of activities in Norway, I think there is a huge number of meeting places for exchanging ideas, with both knowledge institutions and industrial players. Moreover, the industry association has several committees that are focused on specific challenges of this industry. Committees provide the opportunity to meet experts and various persons in a particular subject, so they are very good arenas for both established projects and for new ideas. For example, based on the ideas we received in a respective committee, we initiated mineral exploration in a very unconventional area where we have collaborators from NTNU and a university in Denmark.

Firms can generally employ three core processes of open innovation (Gassmann and Enkel 2004): enriching their internal knowledge base through exploration and acquisition of knowledge from external sources (outside-in); using external pathways to exploit abandoned ideas and unutilized internal knowledge (inside-out); and joint knowledge development and commercialization by collaborating with complementary innovation partners (coupled). Gassmann and Enkel (2004) further elaborate that while all these processes represent an open innovation strategy, they are not equally important for all firms and in all business contexts. For example, the outside-in process seems to be highly important for firms in low- and medium-tech industries that expect knowledge spillovers from their machinery suppliers and/or customers. By contrast, the inside-out process better suits large and/or research-driven firms, whose aim is to commercialize innovations before competitors. Similarly, Chesbrough and Crowther (2006) evidence that firms in mature industries focus on the outside-in dimension of open innovation in order to complement their internally developed knowledge. In light of these contributions, open innovation in this chapter centers on the outside-in and coupled processes, as these include (wholly or partly) the flow of knowledge 'into' a firm.

The outside-in dimension is often categorized into acquiring and sourcing practices according to whether they are pecuniary or not (Dahlander and Gann 2010). Acquiring involves practices such as outsourcing R&D services and technology acquisition, through which a firm purchases knowledge (also in the form of embedded knowledge in technologies) and expertise from the market, such as from suppliers, universities and commercial research institutes. Sourcing, on the other hand, refers to monitoring the outside business environment and absorbing the available knowledge without exchange of money.

The coupled process requires firms to engage in a simultaneous 'give and take' of ideas and knowledge with external stakeholders, either via formal mechanisms such as strategic alliances or socially constructed relationships, such as informal networks (West et al. 2014). Compared to the outside-in process, the collaborative arrangements used in the coupled process can provide access to complex and tacit knowledge that is not usually available through search mechanisms or market transactions (Spithoven et al. 2013). However, due to the increasing cost of being involved in such collaborative relationships, which can weaken the positive effect of open innovation on performance outcomes (Faems et al. 2010), a combination of outside-in and coupled processes seems to be an appropriate strategy for firms to optimize their external innovation sources.

6.3   Prerequisites for Successfully Leading Through Open SOI

6.3.1   Internal Capabilities Required for Open SOI

As discussed in Sect. 6.1, although open innovation moves the locus of innovation outside organizational boundaries, it does not by any means dispel the need for internal capabilities required to utilize the externally acquired knowledge. As such, the literature on open SOI has so far investigated a wide variety of internal capabilities and their role in enabling firms to achieve different types of SOI. In the following section, we build on the theory of absorptive capacity and its three dimensions of recognition, assimilation and exploitation capabilities in order to synthesize the findings from the literature. Table 6.1 maps these findings, based on the three aforementioned capabilities and their micro-foundations.

According to Lane et al. (2006), recognition capability enables a firm to identify and understand external knowledge resources. In the second step, assimilation provides the ability to integrate external and internal knowledge, which could result in only a slight change, or in an entire transformation, of a firm's existing knowledge base. Finally, firms should be able to exploit the new knowledge by applying it to their daily operations in order to develop innovations. Overall, recognition, assimilation and exploitation capabilities allow a firm to convert external knowledge into innovative outputs.

The existing findings reveal various resources, routines and processes that underlie the recognition capability for SOI, including R&D (De Marchi 2012), competence mapping (Kazadi et al. 2016), employee training (Cainelli et al. 2015) and managerial social/environmental awareness (Ingenbleek and Dentoni 2016). Among these, the majority of articles consider internal R&D processes as the most prominent component of firms' prior knowledge required for identifying and understanding external knowledge. The technological complexity of SOI, particularly the more radical innovations such as cleaner production technologies, make R&D a more important

Table 6.1   An overview of the internal capabilities required for open SOI

Capability dimension	Recognition	Assimilation	Exploitation
Underlying skills and organizational routines	Internal R&D	Knowledge management	Stakeholder relationship management
	Competence mapping capability	Flexible structure and open culture	
	Employee training	Cross-functional coordination	
	Managerial social and environmental awareness	Boundary-spanning	

resource for these innovations than general innovations (Galliano and Nadel 2015). Others, such as Ghisetti et al. (2015), took a step further and found a moderating role for R&D in the relationship between external knowledge acquisition and innovative outputs, hence claiming that higher degrees of technological knowledge emerging from R&D can reinforce the positive effect that openness has on SOI.

A relatively smaller part of the literature that deals with recognition capability has extended the limited R&D-based view to absorptive capacity and found support for the necessity of other types of organizational routines for improving firms' knowledge base. In this regard, employee training allows company leaders to compensate for the lack of formal R&D knowledge by updating their personnel on changes in environmental, social and market areas, alongside more general technological knowledge (Bos-Brouwers 2010). Besides educating employees, the way managers interpret environmental and social issues can have a significant influence on their engagement with external stakeholders. Thus, managers' response to these issues in the form of directing firms' activities towards innovation in products and processes is predicted by their awareness and understanding of social responsibilities and environmental protection (Ingenbleek and Dentoni 2016). Increasing environmental and social awareness among company leaders can also help their respective firms to establish stakeholder relationships that are based on mutual understanding and common language (Eccles et al. 2014), as crucial components of competitive advantage in corporate sustainability contexts.

Next to recognition capability, the literature also highlights the existence of various microfoundations for assimilation capability. Although it is widely agreed in the literature that intra-organizational relationships support the integration of external and internal knowledge, researchers suggest different processes and routines to augment such relationships, which can be differentiated in terms of their formality. The first group includes formal organizational processes such as knowledge management (Ayuso et al. 2011) and cross-functional coordination (Ghisetti et al. 2015), whereas the second considers informal processes such as boundary spanning and nurturing open culture (Holmes and Smart 2009).

For example, Dangelico et al. (2017) highlight that facilitating collaborations between specialized environmental units and functional departments (e.g. R&D and marketing), as well as within the functions will increase the probability of designing products that address environmental and economic sustainability. Instead, in the case of firm-NGO collaborations, boundary-spanners act as conduits of knowledge in an informal way, as they explore external opportunities and 'travel around' different functional departments to exchange ideas and solutions (Holmes and Smart 2009). It is important to note the fact that formal and informal mechanisms of integration do not work in all firms and in all situations, hence factors such as levels of hierarchy and trust should be taken into account in choosing the most appropriate process for assimilation capability. However, jointly pursuing formal and informal processes, for example knowledge management and nurturing open culture, seems to be an appropriate strategy. The quote below illustrates how a company active in construction minerals makes use of informal (open culture) and formal (hierarchical authority) to augment the assimilation capability:

> It [dissemination of knowledge between organizational functions] is mostly a natural process inside our company… My impression is that it is the top and middle management level that both do and push for knowledge exchange and new things.

The final capability, exploitation, has received minimal attention from researchers in the field of open SOI. Existing studies base their unit of analysis at the project level and argue that in an ongoing SOI project, a firm should be able to retain its relationships with external stakeholders in such a way that secures the exchange of knowledge until the desired project outcome is achieved (Kazadi et al. 2016). As more tensions could arise (particularly between firms and secondary stakeholders such as NGOs) in the later stages of innovation projects, when firms aim to apply the integrated knowledge into the development of tangible outputs, trust and commitment to shared goals play a key role in the ultimate success of innovations.

6.3.2   Open Innovation Processes in the Context of Open SOI

In this section, the findings from the literature are discussed according to the two prevalent processes of open innovation in the context of SOI, which are outside-in and coupled mechanisms, and the different types of stakeholders included in each one.

The research conducted by Arnold (2017) highlights that SOI can particularly benefit from four types of open innovation tools: innovation workshops, sustainability-related web communities, ideas contests and dialogue. She defines these tools as enablers of collaboration between a firm and its external stakeholders, particularly customers, NGOs and society at large. In her view, special attention should be paid to the level of interaction in these open innovation tools, which can consequently influence external knowledge transfer and learning abilities from this knowledge. For example, workshops and web communities allow company leaders to establish intensive interaction with external stakeholders and to have access to their tacit knowledge about environmental and social issues. Accordingly, Hansen et al. (2011) focus exclusively on ideas contests as an open innovation tool with a medium level of interaction, in order to examine its suitability for generating SOI. In this regard, they develop a matrix crossing market and environmental impacts of innovations, in which the most advanced SOIs are placed in the upper right-hand cell. However, their findings do not show a great contribution from such contests for SOI, especially concerning the environmental impacts of innovations.

Concerning outside-in open innovation, the widespread belief in the literature on open SOI is that both external knowledge sourcing and the acquisition of knowledge embedded in technologies/R&D services are beneficial for the propensity of firms to achieve SOI outputs (Cainelli et al. 2015; Ketata et al. 2015). Nonetheless, while firms should be able to source/acquire knowledge from a diverse range of external stakeholders, some studies have not found supporting evidence for the positive contribution of certain specific stakeholders, such as suppliers (Segarra-Ona et al. 2017), customers (De Marchi 2012) and research organizations (Bonte and Dienes 2013). Such contradictory results can be explained by the various ways SOI is operationalized, as well as the variance in terms of empirical settings.

Based on these contradictory results, scholars have recently started to build a contingent link between external knowledge sourcing and SOI. In one of these studies, Mothe and Nguyen-Thi (2017) assert that although sporadic sourcing activities may result in SOI outputs, firms that persistently utilize external knowledge in their innovation are in a relatively better position to devise innovative outputs. Indeed, by conducting external knowledge sourcing over time, firms also develop a set of processes and routines (a capability) to diversify their channels of access to external knowledge. Other studies, such as Ghisetti et al. (2015), turn our attention to the deteriorating effect of excessive external knowledge sourcing on SOI, since too many external activities limit a firm's resources required for the subsequent stages of knowledge assimilation and exploitation. Thus, instead of a straightforward relationship between outside-in open innovation and SOI, managers should be aware of the limits for and conditional effects of their firms' reliance on external knowledge.

In contrast to outside-in open innovation, which is usually shown to comprise multiple types of stakeholders, the coupled process has been mostly conceptualized as restricted collaboration with specific stakeholder groups, mainly value chain partners and universities (Bonte and Dienes 2013) and NGOs/local communities (Holmes and Smart 2009). The point of departure of these studies is that knowledge in the context of corporate sustainability is not only distributed (hence requires open innovation in general), but is also complex and embedded in socially complex relationships and thereby can be effectively exchanged via two-way interactions between a focal firm and its stakeholders.

6.4 Implications for Policymakers and Company Leaders

This chapter has highlighted the complexity of open SOI by demonstrating its reliance on various mechanisms for external stakeholder engagement and internal capability building. As such, the findings can inform policies at national and local levels in designing appropriate structures for innovation in industries that are subject to sustainable development. Moreover, the study has important implications for firms regarding how to tackle the aforementioned complexity by embracing the value of stakeholder engagement.

6.4.1 Implications for Policymakers

In general, Norwegian politicians perceive the minerals industry and its further development as a double-edged sword. One the one hand, and besides their widespread use in everyday products, minerals are required for the development of a wide range of renewable technologies and green infrastructures, which are highly relevant to the rising political support for sustainable development (Heldal et al. 2016). Conversely, the environmental and social issues arising from mineral exploration and production reduce political interest in the industry because the legal and informal power of indigenous people, youth organizations, environmentalists and labor unions can damage the reputation of governing political parties. This has led to occasions when such opposing entities have been responsible for stopping or postponing exploration and production operations, even after the government has granted the required licenses. Therefore, the Norwegian governments' desire throughout the years to develop the minerals industry has mostly remained a verbal promise, but not put into practice.

The findings of this chapter provide an important message for policymakers if they want to overcome this situation: they should facilitate firm-stakeholder relationships in order to create the momentum for SOI. Owing to its potential in integrating economic, environmental and social sustainability, SOI can assist the minerals industry to pursue environmental and social imperatives, without compromising its profitability. In this regard, the overarching policy implication from the chapter is the need to design and implement supporting schemes that not only address the external mechanisms (e.g. proximity dimensions), but also the firms' internal capabilities (e.g. employee training), in order to close the knowledge gap between mineral firms and their stakeholders.

As far as the external mechanisms are concerned, specific attention should be paid to ensuring that there is sufficient recognition of various stakeholder groups who provide technical, market, social and legal knowledge. From an innovation supply perspective, the minerals industry is heavily dependent on the acquisition of technologies and technical services from suppliers, universities and research centers. Therefore, providing stable financing possibilities to create industrial clusters and university-industry linkages is of utmost importance for securing the flow of technical knowledge to the industry. Taking a demand-side perspective, the government should support existing intermediary organizations (e.g. the Association of Norwegian Mineral Industry) to strengthen their links with national and international agencies such as Innovation Norway and the European Innovation Partnership (EIP) on Raw Materials, which will accelerate the industry's link with potential markets for raw materials. Moreover, to create effective communication between mineral firms and environmental/societal stakeholders, we suggest that policymakers direct their efforts towards establishing transparent mechanisms for stakeholder engagement (van der Have and Rubalcaba 2016), which entails using established frameworks for evaluating environmental/societal performance. A big advantage of these frameworks (e.g. Towards Sustainable Mining in Canada or Finland) is that they offer key indicators for measuring the impacts of the minerals industry, and thereby create mutual commitment to shared sustainability objectives in which none of the parties will be able to override the agreed terms.

We also found that there is a need for policymakers to make a clear distinction between formal and informal institutional environments in promoting SOI. Increasing the coordination between formal structures such as environmental and innovation policies is necessary for investment in and the diffusion of sustainability-oriented processes and products, as a lack of such coordination could result in the market and system failures of these innovations (Ghisetti et al. 2015). On the other hand, policies that aim to promote social innovations should address cultural norms and values by, for example, nurturing trust-based relationships between mineral firms and local communities. In this regard, local governments can act as neutral entities to facilitate the trust building process and close the normative gap between the minerals industry and societal stakeholders.

Considering firms' internal capabilities, the findings point to a critical need for policies that aim to augment employee training programs, as well as the breadth and depth of higher education in disciplines related to the minerals industry. Indeed, what differentiates policy requirements in the context of SOI from general innovations is that governmental support for the former should include more than the R&D subsidies and financial incentives offered through generic policy schemes. An exemplary scheme in this respect is SkatteFUNN, the tax incentive scheme in Norway that is designed to stimulate R&D activities throughout all industries. As training programs in areas related to broader sustainability approaches such as environmental management systems require substantial human and financial resources, implementing an incentive system similar to SkatteFUNN could encourage firms to devote their resources to development areas in which immediate financial benefit is not evident.

6.4.2 Implications for Company Leaders

The call made in the chapter for policies that address both internal and external firm aspects of open SOI resonates directly with the need for firm-level strategies and practices that consider these two aspects. On the one hand, it is no longer an alternative for managers to isolate their firms from external stakeholders' knowledge. However, the findings presented throughout the chapter, and the quotes from company leaders in the Norwegian minerals industry also indicate that shifting focus to external stakeholders does not imply ignoring the internal capabilities required for utilizing the external knowledge.

Besides the importance of practices such as employee training that enable mineral firms to understand external knowledge and to assess its relevance, managers' attention should be drawn to the importance of setting specific objectives when dealing with social and environmental issues. The insights from the interviews specified that engaging a wide range of stakeholders, particularly those without any interest in the long-run financial condition of firms, will most likely expand the scope of social and environmental expectations. This will then pose a significant challenge for mineral firms to find a balance between their own and these stakeholders' interests, which might consequently lead to ineffective knowledge exchange and failure to take any innovative action. Instead, designing clear objectives and communicating them to external stakeholders not only facilitates mutual understanding, but also enables managers to better locate the required external knowledge as the objectives become narrower.

Concerning internal capabilities, another important implication for firms is that they should strengthen their organizational routines for knowledge assimilation, which simply implies dissemination and integration of externally acquired knowledge internally. In this regard, efforts are particularly needed to accelerate knowledge sharing across different organizational functions by means of assigning formal knowledge coordinators. While the use of informal practices of knowledge assimilation such as peer-to-peer interaction is more prevalent in the minerals industry, creating a balance between formal and informal structures is well suited to managers who want to optimize their organizational proximity to external stakeholders. This is because such a combination can assist firms to control their external knowledge transactions through hierarchical frameworks, while at the same time keep a certain level of flexibility to ease access to novel ideas and solutions.

The final remark about practical implications revolves around the external (to the firm) aspect of SOI mechanisms, specifically highlighting the necessity to consider both the relational and transactional types of stakeholder engagement in acquiring external knowledge. What we have seen so far in this respect is an unbalanced focus on reinforcing networks, industrial clusters and R&D alliances, which all aim to nurture collaborations between firms and external stakeholders. Although not reducing the significance of these relational mechanisms, this study, in agreement with Mothe and Nguyen-Thi (2017), strongly advises managers to establish stable search platforms to secure timely access to external knowledge. On some occasions, collaboration may lay the basis for such a platform when firms draw on their previous relationship with a specific stakeholder to continuously look for relevant knowledge in ongoing SOI processes. Other examples include creating/maintaining links with universities via employees who graduated from the same institute, or recruiting new employees from competitors or supplier companies.

Acknowledgements   We thank Wim Vanhaverbeke for encouraging us to write this chapter and for his valuable comments. Lene Foss thanks Malin Kvalheim for her assistance.

Bibliography

Adams, R., Jeanrenaud, S., Bessant, J., Denyer, D., & Overy, P. (2016). Sustainability-oriented Innovation: A systematic review. *International Journal of Management Reviews, 18*(2), 180–205.

Aragón-Correa, J. A., Hurtado-Torres, N., Sharma, S., & García-Morales, V. J. (2008). Environmental strategy and performance in small firms: A resource-based perspective. *Journal of Environmental Management, 86*(1), 88–103.

Arnold, M. (2017). Fostering sustainability by linking co-creation and relationship management concepts. *Journal of Cleaner Production, 140,* 179–188.

Ayuso, S., Ángel Rodríguez, M., García-Castro, R., & Ángel Ariño, M. (2011). Does stakeholder engagement promote sustainable innovation orientation? *Industrial Management & Data Systems, 111*(9), 1399–1417.

Behnam, S., Cagliano, R., & Grijalvo, M. (2018). How should firms reconcile their open innovation capabilities for incorporating external actors in innovations aimed at sustainable development? *Journal of Cleaner Production, 170,* 950–965.

Bini, L., Bellucci, M., & Giunta, F. (2018). Integrating sustainability in business model disclosure: Evidence from the UK mining industry. *Journal of Cleaner Production, 171,* 1161–1170.

Bonte, W., & Dienes, C. (2013). Environmental innovations and strategies for the development of new production technologies: Empirical evidence from Europe. *Business Strategy and the Environment, 22*(8), 501–516.

Bos-Brouwers, H. E. J. (2010). Corporate sustainability and innovation in SMEs: Evidence of themes and activities in practice. *Business Strategy and the Environment, 19*(7), 417–435.

Cainelli, G., De Marchi, V., & Grandinetti, R. (2015). Does the development of environmental innovation require different resources? Evidence from Spanish manufacturing firms. *Journal of Cleaner Production, 94,* 211–220.

Chesbrough, H. (2003). *Open innovation: The new imperative for creating and profiting from technology.* Boston, Mass: Harvard Business School Press.

Chesbrough, H., & Bogers, M. (2014). Explicating open innovation: Clarifying an emerging paradigm for understanding innovation. In H. Chesbrough, W. Vanhaverbeke, & J. West (Eds.), *New frontiers in open innovation.* New York, United States: Oxford University Press.

Chesbrough, H., & Crowther, A. K. (2006). Beyond high tech: Early adopters of open innovation in other industries. *R&D Management, 36*(3), 229–236.

Clarke, S., & Roome, N. (1999). Sustainable business: Learning—Action networks as organizational assets. *Business Strategy and the Environment, 8*(5), 296–310.

Crossan, M. M., & Apaydin, M. (2010). A multi-dimensional framework of organizational innovation: A systematic review of the literature. *Journal of Management Studies, 47*(6), 1154–1191.

Dahlander, L., & Gann, D. M. (2010). How open is innovation? *Research Policy, 39*(6), 699–709.

Dangelico, R. M., Pujari, D., & Pontrandolfo, P. (2017). Green product innovation in manufacturing firms: A sustainability-oriented dynamic capability perspective. *Business Strategy and the Environment, 26*(4), 490–506.

De Marchi, V. (2012). Environmental innovation and R&D cooperation: Empirical evidence from Spanish manufacturing firms. *Research Policy, 41*(3), 614–623.

del Rio, P., Penasco, C., & Romero-Jordan, D. (2015). Distinctive features of environmental innovators: An econometric analysis. *Business Strategy and the Environment, 24*(6), 361–385.

Eccles, R. G., Ioannou, I., & Serafeim, G. (2014). The impact of corporate sustainability on organizational processes and performance. *Management Science, 60*(11), 2835–2857.

Faems, D., De Visser, M., Andries, P., & Van Looy, B. (2010). Technology alliance portfolios and financial performance: Value-enhancing and cost-increasing effects of open innovation. *Journal of Product Innovation Management, 27*(6), 785–796.

Foxon, T., & Pearson, P. (2008). Overcoming barriers to innovation and diffusion of cleaner technologies: Some features of a sustainable innovation policy regime. *Journal of Cleaner Production, 16*(1), S148–S161.

Freeman, R. E. (1984). *Strategic management: A stakeholder perspective* (p. 13). Boston: Pitman.

Freeman, R. E. (2010). *Strategic management: A stakeholder approach.* Cambridge: Cambridge University Press.

Frooman, J. (1999). Stakeholder influence strategies. *The Academy of Management Review, 24*(2), 191–205.

Galliano, D., & Nadel, S. (2015). Firms' eco-innovation intensity and sectoral system of innovation: The case of French industry. *Industry and Innovation, 22*(6), 467–495.

Gassmann, O., & Enkel, E. (2004). *Towards a theory of open innovation, three core process archetypes.* Paper presented at the Proceedings of the R&D Management Conference, Lisbon, Portugal.

George, G., Schillebeeckx, S. J. D., & Liak, T. L. (2015). The management of natural resources: An overview and research agenda. *Academy of Management Journal, 58*(6), 1595–1613.

Ghassim, B., & Foss, L. (2018). Understanding the micro-foundations of internal capabilities for open innovation in the minerals industry: A holistic sustainability perspective. *Resources Policy.*

Ghisetti, C., Marzucchi, A., & Montresor, S. (2015). The open eco-innovation mode. An empirical investigation of eleven European countries. *Research Policy, 44*(5), 1080–1093.

Greenwood, M. (2007). Stakeholder engagement: Beyond the myth of corporate responsibility. *Journal of Business Ethics, 74*(4), 315–327.

Hall, J. K., & Martin, M. J. C. (2005). Disruptive technologies, stakeholders and the innovation value-added chain: A framework for evaluating radical technology development. *R&D Management, 35*(3), 273–284.

Hall, J. K., & Vredenburg, H. (2003). The challenges of innovating for sustainable development. *MIT Sloan Management Review, 45*(1), 61–68.

Hall, J. K., & Wagner, M. (2012). Integrating sustainability into firms' processes: Performance effects and the moderating role of business models and innovation. *Business Strategy and the Environment, 21*(3), 183–196.

Hansen, E. G., Bullinger, A. C., & Reichwald, R. (2011). Sustainability innovation contests: Evaluating contributions with an eco impact-innovativeness typology. *International Journal of Innovation and Sustainable Development, 5*(2–3), 221–245.

Hart, S. L. (1995). A natural-resource-based view of the firm. *Academy of Management Review, 20*, 986–1014.

Heldal, T., Schiellerup, H., & Aasly, K. A. (2016). *Minerals for the green economy.*

Hillman, A. J., & Keim, G. D. (2001). Shareholder value, stakeholder management, and social issues: What's the bottom line? *Strategic Management Journal, 22*(2), 125–139.

Holmes, S., & Smart, P. (2009). Exploring open innovation practice in firm-nonprofit engagements: A corporate social responsibility perspective. *R&D Management, 39*(4), 394–409.

Ingenbleek, P. T. M., & Dentoni, D. (2016). Learning from stakeholder pressure and embeddedness: The roles of absorptive capacity in the corporate social responsibility of Dutch Agribusinesses. *Sustainability, 8*(10).

Jamali, D., Yianni, M., & Abdallah, H. (2011). Strategic partnerships, social capital and innovation: Accounting for social alliance innovation. *Business Ethics: A European Review, 20*(4), 375–391.

Kassinis, G., & Vafeas, N. (2006). Stakeholder pressures and environmental performance. *Academy of Management Journal, 49*(1), 145–159.

Kazadi, K., Lievens, A., & Mahr, D. (2016). Stakeholder co-creation during the innovation process: Identifying capabilities for knowledge creation among multiple stakeholders. *Journal of Business Research, 69*(2), 525–540.

Ketata, I., Sofka, W., & Grimpe, C. (2015). The role of internal capabilities and firms' environment for sustainable innovation: Evidence for Germany. *R&D Management, 45*(1), 60–75.

Lane, P., Koka, B., & Pathak, S. (2006). The reification of absorptive capacity: A critical review and rejuvenation of the construct. *Academy of Management Review, 31*, 833–863.

Laursen, K., & Salter, A. (2006). Open for innovation: The role of openness in explaining innovation performance among U.K. manufacturing firms. *Strategic Management Journal, 27*(2), 131–150.

Mothe, C., & Nguyen-Thi, U. T. (2017). Persistent openness and environmental innovation: An empirical analysis of French manufacturing firms. *Journal of Cleaner Production, 162*, S59–S69.

Murphy, M., Perrot, F., & Rivera-Santos, M. (2012). New perspectives on learning and innovation in cross-sector collaborations. *Journal of Business Research, 65*(12), 1700–1709.

Muscio, A., Nardone, G., & Stasi, A. (2017). How does the search for knowledge drive firms' eco-innovation? Evidence from the wine industry. *Industry and Innovation, 24*(3), 298–320.

Nidumolu, R., Prahalad, C. K., & Rangaswami, M. R. (2009). Why sustainability is now the key driver of innovation. *Harvard Business Review, 87*(9), 56–64.

Peng, X., & Liu, Y. (2016). Behind eco-innovation: Managerial environmental awareness and external resource acquisition. *Journal of Cleaner Production, 139*, 347–360.

Rennings, K. (2000). Redefining innovation—Eco-innovation research and the contribution from ecological economics. *Ecological Economics, 32*(2), 319–332.

Segarra-Ona, M., Peiro-Signes, A., Albors-Garrigos, J., & De Miguel-Molina, B. (2017). Testing the social innovation construct: An empirical approach to align socially oriented objectives, stakeholder engagement, and environmental sustainability. *Corporate Social Responsibility and Environmental Management, 24*(1), 15–27.

Sharma, S., & Henriques, I. (2005). Stakeholder influences on sustainability practices in the Canadian forest products industry. *Strategic Management Journal, 26*(2), 159–180.

Sharma, S., & Vredenburg, H. (1998). Proactive corporate environmental strategy and the development of competitively valuable organizational capabilities. *Strategic Management Journal, 19*(8), 729–753.

Spithoven, A., Vanhaverbeke, W., & Roijakkers, N. (2013). Open innovation practices in SMEs and large enterprises. *Small Business Economics, 41*(3), 537–562.

Suopajärvi, L., Poelzer, G. A., Ejdemo, T., Klyuchnikova, E., Korchak, E., & Nygaard, V. (2016). Social sustainability in northern mining communities: A study of the European North and Northwest Russia. *Resources Policy, 47,* 61–68.

Trott, P., & Hartmann, D. (2009). Why 'open innovation' is old wine in new bottles. *International Journal of Innovation Management, 13*(4), 715–736.

van der Have, R. P., & Rubalcaba, L. (2016). Social innovation research: An emerging area of innovation studies? *Research Policy, 45*(9), 1923–1935.

West, J., Salter, A., Vanhaverbeke, W., & Chesbrough, H. (2014). Open innovation: The next decade. *Research Policy, 43*(5), 805–811.

Wilson, M. (2003). Corporate sustainability: What is it and where does it come from? *Ivey Business Journal (Online)*, 1.

Zollo, M., Cennamo, C., & Neumann, K. (2013). Beyond what and why: Understanding organizational evolution towards sustainable enterprise models. *Organization & Environment, 26*(3), 241–259.

Dr. Babak Ghassim is currently working as a research advisor at UiT The Arctic University of Norway. He took his Ph.D. in February 2019 at the School of Business and Economics, UiT, where he studied sustainability-oriented innovation in the Norwegian minerals industry from an open innovation perspective. His research addresses the firm-level capabilities and inter-organizational factors that assist firms in benefiting from their stakeholders' knowledge towards pursuing various innovation pathways aimed at a broader sustainability approach. He has several years of experience working with companies in resource-extractive industries and has been a dedicated member of the Norwegian Research School in Innovation (NORSI). Babak has published his work in journals such as Journal of Cleaner Production and Resources Policy.

Dr. Lene Foss is Professor of Entrepreneurship and Innovation at UiT The Arctic University of Norway, School of Business and Economics. She teaches and publishes within the fields of university entrepreneurship, entrepreneurial education, entrepreneurial networks, gender in entrepreneurship and enterprise development. Her research has 932 google citations. Foss has been associate editor in Journal of Small Business Management and serves now on its international board. She is editorial consultant in International Journal of Gender and Entrepreneurship, and board member at Norwegian research school in innovation (NORSI). Foss has been visiting research fellow at Cambridge Judge Business School, Cambridge and Säid Business School, Oxford, both UK as well as University of North Carolina at Chapel Hill, US.

Chapter 7
A Manager's Introduction to AI Ethics

Mark Esposito, Joshua Entsminger and Lisa Xiong

Abstract Researchers and managers need to better understand how different ideas of ethics shape how artificial intelligence can be effectively applied and leveraged as a source of competitive advantage. More fundamentally, managers need to understand how the competitive use of artificial intelligence cannot be differentiated from the specific ethical claims reflected in the design, operation, management, and maintenance of the technology. We pose how AI can be given dimensions to provide better insight into ethical requirements. We pose, however, that there is no ethical panacea, rather that the effective global management of ethical and unethical AI requires a larger common meta-ethical approach.

7.1 Introduction

Since 2016, MIT has been hosting the Moral Machine, the world's first and most comprehensive game exploring the kinds of decisions people would prefer an autonomous car, a car endowed with a version of artificial intelligence (AI), to make in life and death scenarios.[1] The most basic summation of the game can be seen in the following

---

[1] Grossman (2018).

M. Esposito (✉)
Hult International Business School, Cambridge, USA
e-mail: mark@mark-esposito.com

Thunderbird Global School of Management at Arizona State University, Phoenix, USA

J. Entsminger
IE University, Segovia, Spain
e-mail: joshentsminger@fas.harvard.edu

Ecole des Ponts Business School, Paris, France

L. Xiong
EMLyon Business School, Lyon, France
e-mail: Xionglisanc@gmail.com

Judge Business School at University of Cambridge, Cambridge, UK

© Springer Nature Switzerland AG 2020
N. Pfeffermann (ed.), *New Leadership in Strategy and Communication*,
https://doi.org/10.1007/978-3-030-19681-3_7

scenario: suppose you are in an autonomous car and a pedestrian steps out into the street such that there are two choices; first, the car can hit the pedestrian killing them; second, the car can swerve into a wall or road block, killing you the passenger.

Naturally, one of the first debates on autonomous cars from this ethical dilemma was, "what should the car do?" However, quickly, as AI researchers began to add contributions to the debate, the question became, "who should decide what the car should do?"[2] Should the responsibility of specifying behavior for an autonomous agent be the responsibility of the driver? Or, should it be the responsibility of the car designer and manufacturer? Should it be the responsibility of the jurisdiction in which the car is used, subject to national rules; or should such rules be more locally decided?

It is within these latter points that the nature of ethical AI and AI ethics is exposed. When an individual makes a choice, they are often considered to be responsible for that choice, where different theories of ethics have different implications for how the individual and society views the ethical consequences of such choices. With AI, however, any such choice made by the AI agent or system, autonomous or not, is at the behest of whatever it has been specified to do, of how it has been trained to understand what it has been specified to do, and how it is allowed to carry out any decision.

What the case of the autonomous car serves to show is that the issue of choice, and its liability, with AI is complex; but the use cases for AI far exceed the boundaries of autonomous cars. AI is becoming an essential component of business operations in the digital age, and with this essential position follows a new found need for clarity in the design and operation of AI and AI driven activities.

Managers need to better understand decisions on the use and design of business and corporate ethics constitute essential strategic decisions in the age of AI. This chapter is intended to serve as a brief introduction and guide to help managers navigate this strategic problem.

Business ethics concerns the assessment of responsibility and attributes of decisions made relating to issues outside of the firm; corporate ethics concerns the above within the firm. However, different definitions of ethics will hold direct consequences. We define ethics as the principles by which a given set of decisions related to specific behaviors and consequences are evaluated.[3] Whereas moral claims concerns the ability to define the decision making process relative to intentions and outcomes as good, bad, or irrelevant.[4]

As such, is it not within our scope to provide ethical insight on the particular rule making system to employ, as the nature of the arguments exceeds the viability of inclusion in this chapter. Rather, what we intend to prescribe is that there is no such thing as a competitive advantage absent of some claims on what kinds of behaviors to perform and not perform, and therefore are culpable to different ethical arguments;

---

[2]Esposito et al. (Forthcoming).

[3]See Dewey et al. (2012).

[4]See for inspiration, as the specific concept of relevant in relation to moral propositions derives from Matheis's work (2016).

wherein, in the AI driven age, the divide between sustainable competitive advantages and clarity of ethical claims is shrinking.

There is no such thing as ethically neutral approach to AI; rather, any design and operation of AI, particularly in relation to decision making, makes a claim on what kinds of data should inform output, and how it can be employed.

7.2   Varieties of Intelligence/Uncertainty/Ambiguity

Different kinds of AI will have different functional implications for the design and assessment of tasks, and therefor for the design and assessment of ethical strategies. As such, managers need to be able to locate the specific AI application they are attending to within the larger universe of AI and its development. More fundamentally, managers need to better understand what constitute an AI at all.

7.2.1   What Is AI?

For the purposes of this paper, we consider AI to be any system designed to extract operational rules from data to inform and improve future operations. It is the context and method by which these rules are derived that lay the foundations for defining how a given AI solution relates to ethical implications. For AI, fundamentally, simply describes a class of solutions to specific kinds of learning problems, where the solutions tend to be inspired by and designed after the operations of the human brain. First versions of image recognition problems were designed to mimic the operations of the visual cortex, giving rise to neural networks as a solution for a system to derive rules on discriminating an object from other objects and the background so as to consistently recognize it, but without being told how to recognize the object.[5]

By reflection, AI helps us to understand what it is about our daily lives that we take for granted. When training an employee, a manager does not need to teach them how to understand words, how to understand images, social context, how to see, how to walk, talk, identify problems, discriminate wrong solutions from right solutions—all these things are given.

However, AI is less a technology than the overall category of approaches to learning problems. As different approaches evolve, the specifics of how managers consider the a given solution, its requirements, its relationship to advantage, and its parallel ethical demands, must therefore change in kind.

---

[5]However, within the field, there is a divide between those who believe the current set of solution and problems genuinely constitute AI or whether such neural networks, or any other similar solution (machine learning, deep learning, etc.) is simply imitation.

7.2.2   Dimensions of AI

After being able to discriminate AI from simple solutions comes the need to identify the aspects and attributes of a given AI system. While AI as a field will develop, and the competency and functionally of any given AI solution will develop in turn, the general dimensions of solutions and their potential scope of implications can be initially defined by using the questions below. To that end, managers need to conventionally consider the following:

First, is the solution narrow or general? All current AI solutions, and most for the foreseeable future, can be considered narrow solutions—whatever learning gained by the application of an AI system to a specific task, such as recognizing a cat from a dog, cannot be applied to other tasks, such as recognizing a formula 1 car from a sports car.[6]

This is a problem called transfer learning, the ability to effectively transfer lessons and meta-lessons from the experience with attending to a given task to the ability of a system to solve other similar, or dissimilar, kinds of tasks.[7] Whereas, a general AI solution can transfer learning from one task to the ability to solve other tasks, meaning the scope of its decision making expands in proportion to its ability to learn and act.

Second, is the solution embedded or applied? Overall, the nature of implications and control over an AI solution is problematized by the integration or embedding of an AI solution in a product, outside of the scope of easy access and maintenance by a firm, or the application of an AI solution to a problem by a team maintaining control over the means of use. A solution embedded into a product, even a rice cooker for instance, has a different scope of implications than a solution applied in a controlled environment.

Third, is the solution mediated or unmediated? The next step concerns how the solution is itself connected up to decision making capabilities. In such cases, whether its input is used to inform decisions by a human decision maker, or is can make autonomous decisions. We can describe these as human-in-the-loop problems.

However, the concern is not simply whether or not a person is the end point of AI outputs, with the final decisions making power in their hands, but the composition of information impacting their decisions. For instance, is the information unbalanced in favor of AI input? This then demands a more nuanced analysis of the precise distribution of AI inputs across a firm, and the nature of decision making intermediation with those inputs, and the aggregate balance of input from AI and human agents to understand the scope of baseline and ethical risks.

Fourth, is the solution independent or restricted? Following directly from the problem of mediation, managers need to understand the means by which an AI can exercise decisions, and how. Even if an AI is autonomous, this does not a priori

---

[6]"Deep Learning beyond Cats and Dogs: Recent Advances in …". Accessed 2019. http://www.cs. ucf.edu/~bagci/publications/catsdogs.pdf.

[7]Managers should learn to discriminate between systems that learn how to perform a specific task, and system that can learn "how to learn" how to perform that or other tasks.

imply that the ethical decisions are independent. For instance, is the firm designing an autonomous car that learns how to make ethical decisions for itself by virtue of the owners preferences, or has a set number of rules it will apply autonomously.

An independent solution can exercise decisions without a human intermediary, but goes a step beyond autonomous in how it is designed to learn. While machine learning can be differentiated by supervised and unsupervised, this question concerns less which of the two methods is used than whether they are a preset and confined number of functional things it can do—whether to slow or speed up the car, and how it can make that decision.

Each of these questions serves to add understanding of the specific system at hand; but each also implies a different layering of the distribution of potential ethical implications, in the use of AI, and in the treatment of the AI itself. Currently, AI is not intelligent, nor is it alive—for all its complexity. As such, there are no moral and ethical responsibilities regarding the treatment of the solution itself—showing an AI violent and damaging images does not hurt the solution as it would hurt a human. Unlike the case for Facebook's content moderators demanded to supervise and delete violent images, leading to reported cases of PTSD among some moderators.[8]

However, we can suppose that AI is in fact sentient, and even in such cases, the nature of harm between sentient AI agents and humans would not be symmetrical—as there is different context, associations, and behaviors to be impacted through different kinds of exposure to harm, such as violent images of humans against humans or animals.

7.2.3   The End of Ambiguity

What the above implies is that all decisions in the use and design of AI systems and solutions are intentional and specific, regardless of technical difficulties. Consider attempting to embed the golden rule—do unto others as you would have them do unto you–into an AI system. While the principle of the rule is often clear, the nature of each specific interpretation is what is at stake.

For instance, in the trolley problem, where an operator must choose between doing nothing and incidentally killing 5 people, or changing direction, thus being responsible for killing one person.[9] We can assume that the lesson can be cemented if no additional information is attached. However, if additional information regarding who the people are is attached, then the specifics now change, making a moral accounting problem.

However, how to specifically account so as to, for instance, maximize welfare within the context of the problem, implies a great deal of external knowledge and value judgements. Then, assuming this is resolved, the ability to answer this question may not transfer to the ability to effectively address other ethical scenarios.

---

[8]Garcia (2018).
[9]Goldhill and Goldhill (2018).

Where ethical decisions emerge and are responded to in real-time, an AI system cannot abide such an ambiguous approach to decision making, as the ability to effectively interpret a problem and identify a solution is something integrated into the solution itself. Whether intentionally or unintentionally, how an AI system is designed makes a direct claim on decisions and ideas in relation to ethical and other behavioral problems. There is no such thing as ambiguous AI.

Yet this issue of ambiguity is exceptionally problematized by the fact that most AI solutions are black boxes—this means the specifics of how a system came to a given set of rules by which it makes some assessments and not others, some predictions and not others, is unintelligible and nearly impossible to precisely define.

The implication can be stated as: if someone is to hold a decision maker to account, and the decision maker is an AI agent, the agent could not feasibly explain itself. Any problem, or set of problems, for which explainable is a core and defining issue, beyond simply liability even, then AI should not be applied without human mediation, and even then, additional measures will need to be assessed. Proposals for such assessment have already emerged, such as counter factual analysis of how decision chains could have been assembled by an AI.[10]

Furthermore, this means that how ethics is assessed, argued, represented, and made coherent is a functional and strategic element of the business model and value chain for firms leveraging and applying AI. As such, while ethical definitions and models can be promoted, this will ultimately be a decision subject to the firm, or other kinds of models of responsibility.

7.3 Leveling Responsibility

After the general dimensions and attributes of the AI system help to clarify the implications, each of the above decisions will at some point reflect a design decision, and the scope of responsibility therefore needs to be further clarified.

When dealing with AI solutions, managers, and the firm at large, need to minimally attend to four levels of problems: operational, design, institutional, and aggregation. These levels concern how responsibility is distributed and delegated across the AI production and use chain; more so, they serve to define how awareness of AI usage and production is distributed across an organization and the potential inconsistencies among competing attitudes towards AI and ethics that can emerge.

---

[10]Katwala (2018).

7.3.1   *Operational Problems*

This level concerns the organization of how AI is applied. Managers should consider at least two dimensions of operational ethics—the specific problem to which AI is applied, and how that problem was selected.

The design problem, going back to the black box problem, creates a series of clear and present ethical issues if AI is unmediated in its connection to making decisions. First and foremost, there is no guarantee that the AI will make consistent ethical decisions even if directed to make ethical decisions; secondly, there is potentially no guarantee the AI will make ethical decisions at all.

The first aspect, concerning the problem to which AI is applied, can be seen in the infamous cases of Amazon lending use of its facial recognition software to ICE in the USA, and google leveraging tensorflow for US drone programs.[11,12] A core element of a firms identifies is how it defines the scope of its responsibility to the use of its products.

However, the issue of facial recognition reveals the problem of issue selection, and divides between data selection and problem identification can lead to competitive disadvantages. How Apple, for instance, designed the problem of facial recognition was as access—turning the face into a password. Beyond cybersecurity issues, such as how if your facial data is stolen that is not something you can change easily, making all future facial recognition based permission systems for you subject to external risk.

7.3.2   *Design Problems*

While design is conventionally first, design in AI, as the actual construction, is second to the ability to effectively understand and navigate the operational environment of a problems identification and handling.

This level concerns the decisions and frameworks by which AI production is handled and organized. Ultimately, the decisions at this stage will impact all other stages, and any other strategic decisions at other levels will feed back into the design stage cyclically in relation to how the update and maintenance of the system is managed.

The simplest point concerns the design and assessment of the problem such that a given data set is used. What can emerge is that the data set used can lead to biased responses. A facial recognition system trained only using faces of one ethnicity might lead to problems for other ethnicities, leading to further challenges.

Furthermore, when attempting to embed general ethical principles in the design stage, managers, and researchers for that matter, need to understand that some kinds of inferences and predictions regarding specific behaviors and intentions should be

---

[11] The intercept (2019).

[12] Day (2018).

understood on a spectrum of transferability, creating a burden of additional research and clarification. This means there is a ceiling to the efficacy of general rules for efficiently leveraging past experiences and data to precisely navigate complex social situations.

7.3.3 Institutional Problems

This level concerns how the AI system is applied to and informed by institutional environments. Managers should consider not only the problem to which AI is applied within an organization, but what that AI application, and how its managed, might serve to reinforce. This problem can be seen particularly in cases of AI in predictive policing, where the use of AI to mediate how resources are allocated in a city by predictions of violent crime, or the usage to predict whether or not a submitted call or form is fraudulent, takes over a specific set of tasks with the hopes of reducing human bias.[13] However, by removing the specific cognitive tasks involved, it can impact employee awareness and attention of how the problem was formulated, leading to a new set of biases in what kinds of cases are prioritized, what kinds of tasks and activities are prioritized and valued, and other such problems.

The institutional level concerns most heavily how problems are organized and defined, and how some kinds of answers and habits become preferred in kind.

Additionally, the institutional level demands an awareness of what we can call problems of reverse fit—when some kinds of activities and products become disproportionately preferred within an organization due to their ease and feasibility within existing managerial and supervisory methods, not by virtue of their overall efficacy and aspects.[14] Meaning, the attempt to fit solutions to existing frameworks rather than alternative frameworks which may be more optimal or robust in context of such new solutions. This can lead to path dependency in an organization on inefficient models of design and behavior, which can then impact how problems are prioritized and defined, leading to additional problems and biases in operational and design levels.[15]

Beyond these elements, the institutional level also concerns how AI is applied to impact the overall institutional environmental and culture of an organization. For instance, the use of AI in employee review by behavioral assessment through eye tracking to assess relative concentration might be insufficiently trained with individual patterns (design problem), leading to an operational problem (false insights), leading to an institutional challenge (employees prioritizing tasks which reinforce assessment), which can lead to further organizational culture challenges.

---

[13] "Does Predictive Policing Work?—Instituto Igarapé." Igarape.org.br. https://igarape.org.br/does-predictive-policing-work/.

[14] For inspiration, see Scott (2008). The issue is identified in relation to government organization of forestation in strict rows by virtue of state defined reasons.

[15] Esposito et al. (Forthcoming).

As such, an effective managerial approach will demand taking a wider view of the overall organization, its informational flows, and behavioral flows, in order to mitigate future systemic challenges.

7.3.4 Aggregation Problems

This level concerns the assessment of the systemic consequences of a given set of decision on the above levels when supplied at scale, as well as its relation to competing or complementary solutions by other firms.

Consider again the case of the autonomous car. Let's assume that two cars made in two different countries are given different operational standards. Now let's assume each car is about to be involved in an accident. If the goal of car 1 is to maximize welfare in car 1, while the goal in the car 2 is to maximize the chance of driver survival in car 2, how should the cars behave.[16]

AI systems with different, and potentially competing, goals will come into contact with one another, and how they engage with one another will derive from firm decisions at the preceding levels. This problem, however, is a point of awareness that needs to be considered within the scope of managerial responsibility, as the nature of the interaction between solutions can prominently play a role in how problems are effectively identified, and institutional behavior attended to, so as to avoid and manage the otherwise unintended consequences of applied AI.

7.4 Ethical Advantages

What the above, in terms of levels and dimensions, implies is two-fold: first, that there is no ambiguous use or design of artificial intelligence; second, that the effectiveness of an AI solution is contingent on the effectiveness of the design and the clarity of its application.

As such, when attempting to leverage AI to boost or modify a firms competitive advantage and value propositions, how that firm defines its specific relationship to its guiding ethical problems will functionally define their relationship.

As such, it is imperative for managers to understand AI, and the specifics of its intended relationship to competitive advantage and product or service offerings. To be more explicit, if there is truly no such thing as ambiguous AI, then the functional decisions managers make regarding the design, operation, and institutional strategies imply a continuous chain of liability.

However, the ethical challenges of AI bring in a new set of dilemmas for managers. While a core decision is often whether to build or buy the AI solution itself, the next problem is whether the rules by which AI outputs are connected to decisions should be

---

[16]Ibid.

given to firms alone or controlled by external agents. In such cases, ethical decisions are fully in the hands of an external agent, meaning the scope of what they take into account and how they take it into account can decide live and death scenarios, or more entrenched social issues, such as bias in autonomous car pickup selection.

To return to the primary example of autonomous cars, this concerns most directly whether or not the ethical behavior of the core should be encoded by the firm, or whether or not such behavior should be given as a direct series of choices on preferences to the driver. Such as, whether the driver should be able to choose whether or not the car should prefer the life of a pedestrian to the driver, and in what cases. This "what cases" element is often how individuals make choices, that is precisely the ambiguity that is no longer functionally tolerated or feasible in the design and operation of AI systems. As different ethical requirements and demands from individuals implies not just a generalization of "do good" or "try not to do bad" but all the specifics of such claims in relation to the specifics of every given moment on the road.

In the AI age, there are some kinds of problems for which the solutions should not be left to firms alone—more directly, there are some kinds of cases where ethics should not be a point of differentiation among firms.[17]

7.5 Conclusion

If managers are going to apply an AI solution, the training and know-how by which manager can effectively navigate the implications of AI usage is essential to maintaining a competitive solution. Indeed, any firm leveraging AI as a core component may need something beyond improved manager understanding, in the form of a Chief Ethics Officer, whose remit is to understand the source and consequences of the application of different principles and decisions in the design and usage of AI.

This responsibility could not fall to the innovation, AI officers, or researchers alone, as the context for use cases exceeds the specifics of the design problem, demanding more time and attention paid to the use cases and the externalities of use. All of such information needs to inform the design and problem selection, but must be additionally connected to other levels of use analysis and corporate governance. There needs to be a clear assessment of whether or not a human is in the loop, needs to be, and where.

Perhaps the most fundamental change will come from the introduction of autonomy into AI solutions—this means the ability for AI solutions to learn and make decisions independently, without a human-in-the-loop, often when connected to an external robotic body as well.

We recommend that managers and employees at each of the levels defined have a clear reporting structure to identify points of coherence and divergence with ethical

---

[17]Esposito et al. (Forthcoming)

principles, so as to exercise effective foresight in identifying future issues in design and implementation.

However, as we have noted, there is no singular system of ethics; furthermore, within any given system of ethics there remain disagreements on how to manage specific disputes, issues, and incidents. As such, there will be no singular solution to the question of how to manage ethical AI. This in turn demands that there is an additional global burden to identify less the conditions of any ethical AI than the conditions under which any given AI system, and AI augment system, relate with one another—in other words, there is a need for a set of meta-ethical principles to help guide how to resolve and predict incidents across a variety of ethical systems.

To that end, along with the demands of each researcher and company to understand AI, there is a need for governments to do so as well. We believe this needs, initially, a global compact on AI ethics, where governments take it upon themselves to identify the specific incentives relative to their goals to drive coherence of approaches to the design and use of AI.

Suffice to say, the problem of AI ethics will be a defining feature of the future of competition. Indeed, one of the defining problems will be whether or not ethics should be considered an advantage at all.

In an AI driven age, there are some problems which cannot be left to the researchers and managers alone; rather, they must involve the input of key stakeholders, and, in some cases, the restriction of decision making for the AI system to the rules of the individual or area.

References

Day, M. (2018). Amazon officials pitched their facial recognition software to ICE. The Seattle Times, October 23, 2018. Accessed 2019. https://www.seattletimes.com/business/amazon/amazon-officials-pitched-their-facial-recognition-software-to-ice/.

Dewey, J., Deen, P., & Hickman, L. A. (2012). *Unmodern philosophy and modern philosophy*. Carbondale: Southern Illinois University Press.

Esposito, M., Tse, T., Entsminger, J., & Jean, A. (Forthcoming). Why algorithmic accountability is a problem for everyone. Project Syndicate (2019).

Garcia, S. E. (2018). Ex-content moderator Sues Facebook, saying violent images caused her PTSD. The New York Times, September 25, 2018. Accessed March 30, 2019. https://www.nytimes.com/2018/09/25/technology/facebook-moderator-job-ptsd-lawsuit.html.

Goldhill, O., & Goldhill, O. (2018). Philosophers are building ethical algorithms to help control self-driving cars. Quartz, February 11, 2018. Accessed March 30, 2019. https://qz.com/1204395/self-driving-cars-trolley-problem-philosophers-are-building-ethical-algorithms-to-solve-the-problem/.

Grossman, D. (2018). 'Trolley problem' survey reveals society's unwritten rules about who self-driving cars can kill. Popular Mechanics, October 25, 2018. Accessed March 30, 2019. https://www.popularmechanics.com/technology/infrastructure/a24222017/trolley-problem-self-driving-cars/.

Katwala, A. (2018). How to make algorithms fair when you don't know what they're doing. WIRED, December 11, 2018. Accessed March 30, 2019. https://www.wired.co.uk/article/ai-bias-black-box-sandra-wachter.

Matheis, C. (2016). Refuge and refusal: Credibility assessment, status determination and making it feasible for refugees to say "no". *Migration Policy and Practice*, 17–35. https://doi.org/10.1057/9781137503817_2.

The Intercept. (2019). Google hedges on promise to end controversial involvement in military drone contract. The Intercept, March 01, 2019. Accessed 2019. https://theintercept.com/2019/03/01/google-project-maven-contract/.

Scott, J. C. (2008). Seeing like a state how certain schemes to improve the human condition have failed. New Haven, CT: Yale University Press.

Mark Esposito, Ph.D. is Professor of Economics at Hult International Business School and at Thunderbird School of Global Management at Arizona State University. He has equally a role at Harvard University's Division of Continuing Education. He holds Fellowships with the Judge Business School's Center for Circular Economy as well as Mohamed Bin Rashid School of Government in Dubai. He has co-founded Nexus FrontierTech, a global AI firm. He advises governments and cities in the interface between policy, business and technology. In 2016 he was inducted in the radar of Thinkers50 as one of the world's rising thinkers as well as Global Expert for the World Economic Forum.

Joshua Entsminger is a fellow at Nexus Frontier Tech on artificial intelligence and policy, a fellow at the Public Tech Lab in IE's School of Global and Public Affairs, and senior fellow at Ecole Des Ponts Business School. He serves as a teaching assistant on Systems Thinking at Harvard Extension School. He additionally served as a research contributor to the Future of Production Initiative at the World Economic Forum and to the Social Innovation Initiative at IE Business School.

Lisa Xiong is an Associate Professor in Strategy and Organization at EM Lyon Business School and an Associate at the Judge Business School's Circular Economy Center at the University of Cambridge. She is a candidate to the Executive Doctorate of Business Administration at Ecole des Ponts Paris Tech, one of France's most prestigious Grande Ecoles.

Part II
New Leadership in Communication

Chapter 8
Vision Setting: How Leadership Communication Empowers Workers and Teams

Andrew Breen

Abstract The modern corporation—inspired by the success of digital technology companies—aspires toward a flatter structure empowering less senior members with influence and decision making over their direction. Vision has always been important in successful organizations. It unites and gives purpose to the workforce. More traditional, hierarchical, tops-down organizations had only the top management layers to align for strategic decision making. From that, they dictated the tasks, processes and goals for the rest of the organization to follow keeping the organization aligned. Decision making is now pushed out to the edges. Modern firms value intellectual property built by knowledge workers with powerful technology tools. Research and development is critical to remaining competitive. Creativity is stifled by rigid direction and processes. Thus, workers are empowered to build the future of the company. However, uncoordinated work can lead to scattershot product development. A vision—and leaderships ability to effectively communicate it—are critical in today's fast moving economy. Several techniques for communicating vision in a modern company are discussed.

8.1 Introduction

The digital technology revolution—from the 1990s through today—was built not on the backs of large organizations of laborers like their industrial predecessors, but on the iterative output of small creative product development teams seeking solutions to problems while tolerating high failure rates. These modern organizations have developed new philosophies and approaches to organizational design. They are flatter in nature—fewer layers, larger span of control—facilitating bottoms up decision making around common goals with lightly managed knowledge workers. In its most extreme, decentralized, loosely connected networks of people that do not

A. Breen (✉)
523 E. 14th St. Apt 5A, New York, NY 10009, USA
e-mail: abreen@stern.nyu.edu

© Springer Nature Switzerland AG 2020   95
N. Pfeffermann (ed.), *New Leadership in Strategy and Communication*,
https://doi.org/10.1007/978-3-030-19681-3_8

have a formal organization employing them (e.g. open source software, Kickstarter projects, Uber drivers) required new ways of setting and reinforcing the purpose and nature of work. How could that happen?

Digital technology brings powerful tools enabling small teams of advanced workers leverage to have a disproportionate impact as compared to workers historically. Prior to the 1980s, scaling an operation required large groups of controlled workers. Tools were primitive and generally only to make the worker more efficient and reduce errors. However with digital technology, coupled with mass and instantaneous distribution via the Internet, worker output achieved an unprecedented pace, impact and scale. Highly educated, creative workers were massively empowered. Their early leaders knew it because they rose from these ranks with a disdain for traditional tops down management and "pay your dues" organizational ascendance. In these modern digital firms, hierarchies with inaccessible leaders who only communicated commands through memos were not accepted. Intimate and informal communication drove the agenda. Flat organizations with wider span of control—the average number of direct reports a manager has—did not inhibit the creative, problem solving process due to micro-management. The power equation was flipped with those at the lower levels harboring power in their knowledge of tool utilization and instantaneous data to measure their impact. A key power center of the 20th century leader, data, was now democratized and available to all. These businesses suddenly worried more about finding skilled human resources than capital (Gaybrick 2018). Today, all businesses are digital businesses. It is difficult to identify any industry which does not require some digital presence if not wholesale adoption just to compete. Modern, digital-centric organizations are young, unbounded by tradition and loosely organized. To be successful, they need new methods to organize workers to accomplish their mission: a culture of innovation and clear communication of their vision.

8.2 The Criticality of Vision Setting in a Modern Organization

As we've discussed, modern organizations who employ highly skilled workers with powerful technology tools need to empower them to unlock their creativity to the benefit of the organization. The required culture manifests as less dictation of tasks and more setting of expected outcomes and guardrails to operate within. However, without a vision, even well articulated goals and guardrails can lead to uncoordinated work and fruitless pursuits. Vision setting is the north star.

Research and development (R&D) within an organization is distinctly not operational task work. This deep work requires regular episodes of non-linear creative thinking and problem solving. Let's look at an example of where it worked and then went awry: Facebook. Facebook scaled quickly once it achieved product-market fit with 100 million active users by 2008, just four years after its founding, to become the world's largest social network. Its founder and CEO, Mark Zuckerberg internally promoted the saying "move fast and break things". This charge was meant to inspire his product development teams to try new things and not worry about potential negative consequences that traditional corporate risk management would have forbidden. That, along with its mission to "make the world more open and connected" did not form a proper vision but had enough of a purpose and method to set the goals—yet without any indication of guardrails.

Growth continued unabated in this way until Facebook began facing increased scrutiny around people manipulating and harming others through the platform. Unfortunately, the "world" that was to be more open and connected included those who had less positive intentions who abused the platform which was moving too fast to bother to add the necessary checks and balances. As Facebook matured into a global public company with a large workforce, advertiser and user base, that mantra became an operational liability. This forced Zuckerberg to drop "move fast and break things" in 2014 replacing it with the more sanguine "move fast with stable infrastructure" (Statt 2014). Subsequently, with the revelation of election tampering and an influence campaign from data leakage by Cambridge Analytica in 2016 (Granville 2018) as well as reports on the negative mental health effects caused by social media (Pantic 2014), Zuckerberg refined the mission to "bring the world closer together" (Heath 2017). This refinement gives further clarity toward building positive interactions and stemming the negative ones that got Facebook in this position. Notably, Facebook still lacks a publicly articulated vision which defines the point of arrival for what they are going to do to support this mission. Could that portend further challenges on the platform?

Without a clear vision, even the most well-meaning team members are left to use their own compass to decide if their work meets with the new standard. With a network of 10s of thousands of employees, Facebook is challenged in pursuing their mission without a well articulated vision to guide the strategies and tactics (goals and tasks) of their products (and people). As was reported in late 2018 with the leak of its internal moderator guidelines and practices, it struggles to implement a system to properly handle questionable content across geographies (Fisher 2018).

Through this example we can see how a mission is important to establish purpose for why an organization exists, but a vision is necessary to guide its activities. A great vision clearly sets the agenda for both operational as well as R&D teams and is reinforced not only through executive edicts but also in the tools used to manage the business.

8.3 Creating a Vision

Before discussing how to create a vision, we need to define what a vision is. Let's start with defining what a vision is not. A vision is not an organization's mission statement. The Business Dictionary defines a mission statement as:

> A written declaration of an organization's core purpose and focus that normally remains
> unchanged over time. Properly crafted mission statements (1) serve as filters to separate
> what is important from what is not, (2) clearly state which markets will be served and how,
> and (3) communicate a sense of intended direction to the entire organization.

Unpacking that statement, we see a few key differences between mission and vision. First, a mission statement transcends time. It should be defined early in the organization's existence and be immutable unless the organization materially changes its purpose. Second, it serves as the purpose of the organization. Without a purpose—and making a profit is not a purpose, it's an outcome—why would an organization exist?

The definition goes on to state:

> A mission is different from a vision in that the former is the cause and the latter is the effect;
> a mission is something to be accomplished whereas a vision is something to be pursued for
> that accomplishment.

This gives clarity around a mission defining the why or purpose or cause for the organization. The vision serves the mission in defining the effect or outcomes sought. In having an outcome, a vision can be achieved. In that sense, while a mission should be static, a vision can and should be dynamic changing as pursuits are achieved or deemed infeasible. In addition, a vision should easily translate into and be reconcilable against goals whereas a mission is a litmus test of sorts, testing whether an activity should or should not be pursued.

The test for a vision statement can be defined as:

1. Defines what is being pursued to achieve the mission but not the mission itself
2. Clearly articulates the outcomes (both customer and business) sought during the period that is far enough out that a clear path cannot be drawn today
3. Establishes a high-level method by which it will be pursued
4. Defines a specific customer with an identified need.

For Facebook, with a mission of "bring the world closer together", a vision statement might look like:

> In twelve months, Facebook will become the trusted place where any person can come and
> engage in meaningful dialog and learnings about others where they control and have confi-
> dence in how they and their data is represented as well as knowing that the best content will
> rise through a collaborative filter of people and machine ensuring hate and misinformation
> is suppressed.

This vision statement serves the mission while establishing the customer and organizational outcomes—whereby individual goals can be established—as well as a method by which it will be pursued. It is not prescriptive as that will be handled in the array of strategies that can be formed in service of it.

Historically and under a 20th century management style, executive leadership, often with the help of management consultants, would go about defining the mission and vision for an organization. When done well, that process and resulting artifact served its purpose in a tops-down management culture where the principal goal was to ensure alignment with middle managers driving the work of laborers. Lower level team members were not expected nor encouraged to exercise judgement nor work in creative pursuits—simply produce the product or service of the company. Even middle management was not incentivized nor expected to act outside the stated operational goals of the company. Thus, a tops-down approach served its purpose as only executive management held the information and responsibility to define the organization's mission, vision and goals.

Would this be true for our new, tech-centric, creative and flat organizations? Likely not. As discussed, creativity is thwarted by micro-management dictating task work. Tasks are undertaken to achieve a goal. In a modern, digital organization, the teams are asked to lay out their team strategies and tactics (individual goals and tasks) as they have the best information and knowledge to do it.

Since strategy is defined to serve a vision, if team members, especially those involved in creative problem solving, are not involved in the creation of the mission and vision, will they feel empowered by the goals and tasks outlined for them (even if they are allowed to define those tasks)? Not likely. Management should only set the process for establishing mission, vision, goals and tasks and go about facilitating it further. Taking that further, modern organizations also allow their network, customers and partners, to contribute to the vision of the company. One example of this is Tesla. They use direct communication and a two-way dialog to prioritize features via Twitter. The result is an unrivaled brand loyalty to such a point that the company, facing a delivery crunch with the Model 3, asked their customers to come help new customers get their new vehicles. Thousands showed up one weekend in California for this purpose (Roberts 2018).

As modern digital companies do, this process should be iterated upon but at cadences which balance the consistency needed versus the flexibility desired in a fast-paced dynamic environment which has more and more unknowns each day.

Mission: purpose or **why** the organization exists, its cause	Generally immutable unless the organization notably changes
Vision: **what** the organization does to achieve the mission, the point-of-arrival	Reviewed and updated annually or as the vision is achieved or deemed infeasible
Strategy: **how** the organization is going achieve the vision	Set and reviewed quarterly and revised based on market learnings
Tactics: **activities** which execute the strategy	Daily to weekly review as the team learns through testing the offering

A vision is a miniature narrative that the team will return to guide micro-decision making. The difference between an idea and a vision is that a vision has others bought in and believing it. An idea gains followers through a clear articulation of the problem and its importance; a connection to the problem through experience or observation; and conviction in that pursuit of its solution is worth the time, effort and investment (personally, professionally and financially).

To that end, what better way to get others in your organization—especially the creative problem solvers who will be tasked with experimenting and executing toward it—than having them co-create the vision. By giving them a substantial voice and participation in the process, management makes them an owner of it.

An exercise to co-create a vision:

1. Since you already have a mission, clearly articulate this in a written document to all members of the organization while announcing the vision co-creation process and a date for an all-hands meeting to discuss it.
2. Identify a committee of non-participants to review responses and form it into a narrative for review by the leadership team. Make sure there are representatives from multiple organizational functions and levels. Identify a chair to ensure the process progresses on schedule but clearly establish that all will have an equal voice in the deliberations. Decisions will be by consensus.
3. In the all hands, clearly define what a vision is versus a mission providing examples from other organizations (good and bad); answer clarifying questions. Identify the review committee and let them set the parameters and scope of their work.
4. Ask that any and all non-committee team members who want to participate to fill out a their thoughts on a questionnaire over the following weeks as they have a chance to reflect on it.
5. Publish the questionnaire with the following questions:

    a. Who, what, etc. from above.

6. Incentivize the team members not only by their ability to dictate this but that the most influential contributors will get a chance to stand up in front of the next all-hands to lay out the vision (social recognition is important to creatives).
7. Present the resulting narrative in written form to the wider team member base and offer a fully transparent commentary period on an internal discussion forum.
8. Refine the vision based on the commentary and circulate to the team members before the follow-up all-hands.
9. Have the follow-up all-hands where leadership introduces the new vision, the committee briefs on the process and participants who most substantially contributed to the resulting vision present it.
10. Lay out the next steps to how goals are going to be set based off the vision from there.

This process may seem laborious and bureaucratic but with well-defined guidelines and scope, the process will engage and empower team members allowing new perspectives and data to "bubble up" to leadership. This vision then serves as the charge for the organization providing clarity of objective. Subsequent planning, budgeting and hiring can be reconciled against it. The team then plans out the tactics to get there understanding that it will be winding—especially in more risky, unknown pursuits. As Jeff Bezos of Amazon says: "We are stubborn on vision. We are flexible on details" (Masnick). Amazon considers all new product proposals in the form of its future press release being written in a memo. This longer form vision serves the same purpose and has become endemic in the Amazon culture.

8.4   Communicating the Vision

Great organizations have clarity of purpose with supporting vision and strategy to achieve it. While we've established the relationship between mission and vision, we must discuss the relationship between vision and strategy. Where vision is what outcome we're going to achieve to service the mission, the strategy is the "how" or plan for achieving the vision. Visions are still high level statements and do not prescribe how it's going to be achieved. Strategic pillars define what the plan ultimately revolves around. Often stated in 2–5 (most commonly 3) short activities, they should be clear and easily memorable. These pillars then become not only the way to evaluate plan activities but also an effective way to communicate the vision on a more recurring and intimate basis.

Using our Facebook example, their strategic pillars could be:

Eliminate all false information from the platform including identifying and removing the offending users	Allow for complete control of how a given user's data is used (or not)	Provide complete transparency as to the data we have relating to each user

Once a vision has been established, it's important to communicate and regularly reinforce it. This is especially true for new team members who join the organization. However, while pillars are important, they can lose their impact over time. Reinforcement, conventionally, would have been done through all-hands meetings, leadership memos and other management artifacts. In successful small companies, communication of the vision is clear and able to be regularly reinforced because the leaders having regular direct contact with those building and operating the business. This is critical in the case of a startup before the company has found product-market fit as information flows in both directions to ensure everyone understands the learnings and the strategy is adjusted rapidly.

This intimate communication impacts the work of all team members. The vision and strategy is connected to their daily work. This also reinforces the purpose of the organization and helps maintain engaged employees. Engaged team members are more loyal, hardworking and their organizations yield better results (EEngagement). One method popular in digital companies to connect to daily tasks to vision is Objectives and Key Results (OKRs). First modernized and used by Intel with Silicon Valley venture capitalist John Doerr, they were popularized by Google and are popular today in many digital companies (Wodtke 2016).

A key tenant of OKRs is that they are cascaded down from the leader(s) of an organization in the form objectives formed from the strategy—strategic pillars in the best cases. All subordinate OKRs are stated in their connection to their leaders'. OKRs were not the first to employ this cascading goal system. However, there are unique aspects of OKRs that make them distinct and more effective than most conventional goal systems which suffer from being set almost as a rite of annual passage in organizations and ignored until end of year performance reviews. To that end, they are often ambiguously defined, readily manipulated to ensure successful outcome and not reconciled against the organization's vision.

OKRs are set by the individuals, not their leaders (the leaders establish their OKRs which their subordinates use to establish and reconcile theirs). They must be specific and measurable. Typically, they are set and reviewed on a quarterly basis where the objective might not change but the key results do. Key results are activities to achieve an outcome toward the objective and typically should not take longer than a quarter to assess. At the end of the period, OKRs are assessed first by their owners and affirmed by the leader. The assessments are not on a more traditional binary scale but rated between 0 and 100. Since they are specific and objective, they can be partially achieved. Because of that, team members are encouraged to enter stretch goals. Most organizations expect that, across 3–7 quarterly OKRs, a team member might achieve a 60–70. If the measurement gets much higher, they might be purposely setting low goals. If they are too low they might set too many stretch goals or their might be a performance issue. Since most creative technical work is done in small groups, teams can share OKRs which is why they are popular in digital R&D teams.

Continuing our Facebook example, an OKR might look like:

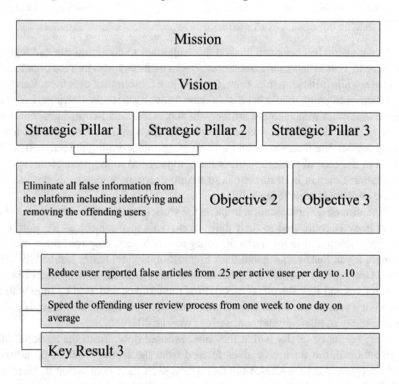

Another technique for communicating vision on a more intimate level connected to team work is through a "Markitecture". Markitectures were first utilized by enterprise technology sales people in the 1990s (Schmidt 2015) to communicate complex technical products to their customers. They were appealing because they were visual, organized and logically laid out. Since they were often selling to engineers, they looked like engineering architecture diagrams albeit with less purity and accuracy.

Used internally, these diagrams can convey how a team's work connects with other teams and the bigger picture company vision. In addition, new ideas can be reconciled against the established Markitecture for the organization. If something fits, it can lead to re-use of internal assets. If it does not, it at least raises the right strategic question as to whether the organization should do this work and do it within the parameters of the current organization versus another vehicle (e.g. spin-out or startup).

Facebook's markitecture might look like:

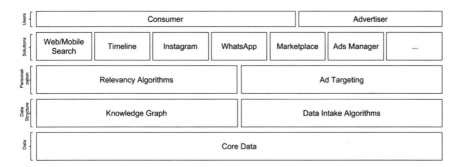

If high-functioning, the organization can use this to have the right level of discussion as to whether new ideas align with its vision. Vision communication can be challenging and needs reiteration. Simplicity and clarity are the key elements of visions that stand their many tests during their lifecycle.

8.5 Leading with Vision

Leaders galvanize their workforce through vision. The military maxim "taking of the hill" is more achievable when team members have complete clarity around mission objective and decision making ability around the tactics to achieve it given learnings and the operational situation on the ground.

However, traditional corporate structures tend to overvalue leadership. In an earlier era where information flowed up and decisions down this provided operational control and consistency. Managers spent years within the firm not only learning its products, markets and strategy but how to navigate the organization itself given the large complex hierarchies. As discussed, digital-centric companies are flat with decision making and empowerment pushed down. In this setting, managers have two primary roles: clearing any impediments in execution and setting and ensuring the plans reflect the vision and strategy guiding when learnings need to alter them. In addition, leaders and institutions are looked at more skeptically. "CEOs are trusted by seventeen percent of people now." says Richard Edelman, president and CEO of Edelman, a public relations firm. In Edelman's annual Trust Barometer report, the average employee is 3x more credible than a CEO as a source of information (Groysberg and Slind 2012). That's because, as discussed, the average employee actually has access to more detailed and immediate information on organizational and product performance. Thus, trust becomes the currency for leaders to establish along with respect for employees. Trust allows employees to contribute and speak out when vision or strategy clashes with their organizational view. There is no stronger example of this in a modern digital company than the employee protests that Google faced in the revelation of sexual harassment by a former executive. As uncomfortable as it was for leadership, because they had established a culture of openness and trust, they allowed employees to express their discomfort. Time will tell but this episode—and the changes Google made due to it—likely won them more employee engagement and loyalty (Wakabayashi et al. 2018). No employee should feel distrust or out of touch with the company vision nor their ability to manage their responsibilities toward achieving it.

Vision is not only critical at inception of a new product or business but also during its build-out and ongoing operations. Vision defines the outcomes sought in service of a mission. This is critical during research and development of a new initiative as the team needs room to discover new information which might have them pivot from the original idea. As long as the vision remains focused on outcomes—solving a problem for a specific user group—the team has the room to innovate. If they ultimately determine there is no problem or reasonable solution to that problem that is the time to review and change the vision.

For ongoing operations, vision provides an important stabilizing force for everything from marketing communications to customer support to product and service enhancements. In making micro-decisions in supporting their product, workers can reference vision and feel empowered to take the right action for the customer. There are two ways leaders use vision to establish culture in digital companies:

Culture staging

Jeff Bezos famously celebrates Amazon's "failures" (Kim 2016). While not truly celebrating what has not worked, Amazon values quickly learning what will not work so as not to waste time, effort or money on it. Bezos sets a cultural tone as a leader conveying that any employee can think creatively and take risks not only without retribution but to celebration. This empowers the workforce and gives it confidence. A self-fulfilling cycle.

Outcome setting

Elon Musk has been relentless in keeping his employees focused on a bigger picture vision even in light of a large number of internal (Model 3 production ramp) and external (short sellers, etc.) challenges in building Tesla. These issues can often distract a workforce. Musk has a unique ability to lay out what appears to be an impossible vision while keeping focused on the outcome and benefit they will achieve.

8.6 Conclusions

Vision is the outcome that the employees rally around. Vision is what the team reconciles new information and challenges against. Vision is what ultimately allows for a group of people—even a loosely connected and organized one—to execute a plan. Effective communication of it is reinforced through multiple layers and opportunities to convey it. Vision is constituted through its relationship to overall team mission and the strategy and tactics to execute it.

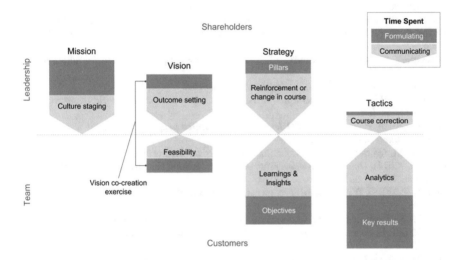

Bringing all the tools and techniques discussed herein together, we can see that vision communication must be regularly reinforced and affirmed with a bi-directional communication stream where the team is regularly seeking to support or refute its foundational hypotheses. Strategy and tactics provide the framework not only for execution but also for insights to arise and change the course. In their communication, both leaders and the team should never be so committed to disallow this upward propagation. Ideas are simply hypotheses to support or refute not a marker of any individual's contribution to the organization. Leaders must open up to this approach which runs counter to traditional management thinking but will support a culture of innovation.

In thinking about your own leadership communications, is there a connection of the tactics to a strategic pillar? Are the results of key performance indicators (KPIs) tied to a hypothesis that is being tested? Can the team easily recall the strategic pillars? Is there a process whereby insights from customers regularly—say quarterly—challenge the current strategy? Successful organizations are confident in their answers to these key questions.

Modern knowledge workers want to understand the purpose of their work. Talent without purpose will not be productive; they will either be paralyzed, disinterested or pursue their own interests. Providing constant, credible and transparent feedback framed through vision connects their work to a well-articulated mission for the organization.

Bibliography

EEngagement: *Towers Perrin: Employee Engagement/Global Workforce Study*. Towers Perrin (undated).

Fisher, M. (2018). Inside Facebook's secret rulebook for global political speech. *New York Times*.

Gaybrick, W. (2018). *Companies worry more about access to software developers than capital*. CNBC.

Granville, K. (2018). Facebook and Cambridge Analytica: What you need to know as fallout widens. *New York Times*.

Groysberg, B., & Slind, M. (2012). *Talk, Inc.* Harvard Press.

Heath, A. (2017). *Facebook has a new mission statement: 'To bring the world closer together'*. Business Insider.

Kim, E. (2016). *How Amazon CEO Jeff Bezos has inspired people to change the way they think about failure*. Business Insider.

Masnick, M. *Jeff Bezos on innovation: Stubborn on vision; flexible on details*. Boyden (undated).

Pantic, I. (2014). *Online social networking and mental health*. National Institutes of Health: US National Library of Medicine.

Roberts, P. (2018). Tesla's true-believer owners volunteer to help Musk complete deliveries. *Seattle Times*.

Schmidt, J. (2015). *It's easy to solve problems on a whiteboard*. Informatica.

Statt, N. (2014). *Zuckerberg: 'Move fast and break things' isn't how Facebook operates anymore*. CNET.

Wakabayashi, D., Griffith, E., Tsang, A., & Conger, K. (2018). Google walkout: Employees stage protest over handling of sexual harassment. *New York Times*.

Wodtke, C. (2016). *Introduction to OKRs*. O'Reilly.

Andrew Breen was most recently SVP of Digital for Argo Group (NYSE: ARGO), a multi-billion dollar commercial underwriter where he led the company's transition to a digital business and oversees Argo Ventures, their early stage investment fund. He leverages his 25 years in tech startups and large, regulated businesses to disrupt businesses through a product-centric tech approach. Andrew is formerly the VP of Product for American Express's Enterprise Digital division and has done eight tech startups across multiple industries including one that was sold to Palm where he started the wireless services division as part of the Treo smartphone initiative. Andrew is an adjunct professor at NYU's Stern School of Business and Courant Institute where he teaches a graduate course in technology product management and innovation. Andrew is a co-author of the 2017 book Strategy and Communication for Innovation.

Chapter 9
The Role of Communicators in Innovation Clusters

Bettina Sophie Blasini, Rani J. Dang, Tim Minshall and Letizia Mortara

Abstract Innovation clusters continue to be an important focus of economic development policies in many nations. Leading innovation clusters demonstrate that regional concentration strengthens the innovative capability and can lead to successful competitiveness on a global level, as demonstrated by regions such as Silicon Valley (US), Cambridge (UK) and Sophia Antipolis (France). However the successful creation of clusters still presents a challenge to policy makers as efforts to do so regularly fail. The development of innovation clusters has therefore received much academic and policy maker attention. While past research has examined a variety of factors as drivers for clustering effects, the role of communication within the cluster—and, specifically, the role of key individual communicators—in underpinning successful cluster development has received almost no academic attention. In this chapter, we will draw upon the relevant literature to develop a conceptual framework that will underpin research on this important topic by investigating the role of communicators in innovation clusters. Building on communication theories, the framework suggests that there are four influence-levels that shape and impact the role of communications in innovation clusters: the Individual Level, the Organizational Level, the Cluster Level and the Context. The interdisciplinary view on clustering effects contributes valuable insight to both communication studies and cluster theories. The framework developed within this chapter provides a structure to aid future research on the role of communicators within innovation clusters.

9.1 Introduction

Innovation clusters continue to be an important focus of economic development policies in many nations (Uyarra and Ramlogan 2012). Leading innovation clusters demonstrate that regional concentration strengthens the innovative capability and can lead to successful competitiveness on a global level, as demonstrated by regions such as Silicon Valley (US), Cambridge (UK) and Sophia Antipolis (France). However the successful creation of clusters still presents "[…] a unique challenge to policy

B. S. Blasini (✉) · R. J. Dang · T. Minshall · L. Mortara
University of Cambridge, Cambridge, UK

© Springer Nature Switzerland AG 2020

N. Pfeffermann (ed.), *New Leadership in Strategy and Communication*,
https://doi.org/10.1007/978-3-030-19681-3_9

makers" as efforts to do so regularly fail (Clark 2013: 6). The development of innovation clusters has therefore received much academic (e.g. Porter 1998; Karlsson 2008) and policymaker (e.g. European Commission 2008; OECD 2012) attention. While past research has examined specialized supply and demand, collaboration and competition, the infrastructure and mobility of goods as crucial drivers for clustering effects, the role of communication within the cluster—and, specifically, the role of key individual communicators—in underpinning successful cluster development has received almost no academic attention.

Perceptions gathered from interviews[1] in one of Europe's leading clusters indicates that this may be an important omission: "Public Communication is critical [...]. If it wasn't for the communication, the cluster wouldn't exist. It is very important that people within the cluster talk to each other and that the cluster communicates with the outside world" (PR Consultant, Cambridge, UK). For complex clusters, which are characterized by diverse internal and external stakeholders, communication was seen by the same interviewee as the "segment that holds it all together". Gathering and sharing information, communicators build interrelations between the actors and create a communication network. "We joined up and connected the companies in the Cambridge area and we also connected those companies with trade collaborators in other parts of the world. We are the catalyst for growth," explained an editor in chief of a newspaper in Cambridge. Connecting and informing relevant stakeholders, communicators describe their work as, in the words of one interviewee, "breaking down boundaries. That's the core of communication".

Such statements point to an important yet under-researched issue, which merits attention in order to enrich our understanding of how innovation clusters develop. In this chapter, we will draw upon the relevant literature to develop a conceptual framework that will underpin research on this important topic by investigating the role of communicators in innovation clusters. Building on communication theories, the framework suggests that there are four influence-levels that shape and impact the role of communications in innovation clusters: the Individual Level, the Organizational Level, the Cluster Level and the Context. The interdisciplinary view on clustering effects will contribute valuable insight to both communication studies and cluster theories. On a theoretical level this study provides a strong foundation for further research in the field of innovation communication and on a practical level it identifies aims and strategies of communicators in innovation clusters.

9.2 Economic and Sociological Cluster Theories

The geography of innovation shows a clustered economic landscape, characterized by a regional concentration of innovative capability. Established cluster theories discussed this phenomenon from different perspectives and identified factors, which

---

[1]Eight professional communicators working in the Cambridge (UK) cluster were interviewed as part of an on-going research project in April–May 2013.

support the local agglomeration. Marshall (1890, 1920) introduces the notion of external economies, an environment characterized by skilled labour, specialized goods, face-to-face contact and trust, which enables spill-overs and sparks innovative activity. Porter (1990, 1998, 2000) highlights the importance of external value and identifies supply and demand conditions, competition and collaboration and the context of the firm as most important for innovative capability. Krugman (1991, 1994) focuses on the dynamics of resource allocation across activities and location by identifying tangible and intangible goods that shape a complex economic environment. The research by Marshall, Porter and Krugman established a strong foundation to understand the dynamics of clustering-effects and influenced the academic perspective on how innovation takes places. While these papers have mostly been discussed in terms of their economic contribution, they also imply the importance of sociological aspects. Discussing externalities, Marshall, Porter and Krugman refer to information gathering, knowledge sharing and the resulting relations between the actors (as summarized in Table 9.1).

The sociological factors, as shown in Table 9.2, are kept implicit within in the economic theories and are not explained in depth. This leads to a blurred understanding of sociological externalities based on interrelations, common knowledge and information spill-overs. Focusing on the characterisation of interrelations within a network and the resulting information gathering and sharing processes, sociological theories complement the economic perspective on cluster dynamics. Network Theory, Knowledge-Based View and Gatekeeper Studies provide a valuable insight to the research field as shown in Table 9.2.

Both economic and sociological cluster theories highlight the importance of networking, knowledge gathering and information sharing. While economic theories imply such action as given processes, sociological theories acknowledge a deliberate and organized way of connecting and communicating. Gatekeeper studies point to the importance of specific actors managing information in order to build relationships inside and outside the cluster. Communication studies have a long tradition in analysing actors who pursue these aims on a professional basis, but have never been applied to the study of clusters. This discipline opens new perspectives of analysing

Table 9.1 The role of communication in economic cluster theories

Economic cluster theories	Role of communication
Marshall (1890, 1920)	Marshall introduces the idea of knowledge-spillover, which is based on the face-to-face contact of the people. According to Marshall, individuals move from firm to firm and exchange knowledge and ideas
Porter (1990, 1998, 2000)	In his Diamond-Model, Porter discusses knowledge about new innovations and early perceptions of new possibilities due to the close relationships between the actors
Krugman (1991, 1994)	Krugman's notion of centripetal forces can be understood as external economies, which are based on relationships and shared information

Table 9.2 The role of communication in sociological cluster theories

Sociological cluster theories	Author	Role of communication
Network theory	Granovetter (1973)	Close relationships that are characterized by strong ties are more likely to share knowledge than those who communicate infrequently or who are not emotionally attached. Weak ties, i.e. acquaintances, support the diffusion of knowledge, the creation of new opportunities and the emergence of new collaborations
	Burt (1992, 2001)	Structural holes define potential connections between units that are not connected and lead to non-redundant information, as the sources are more additive than overlapping
Knowledge based view	Polanyi (1958)	Tacit knowledge can be understood as "not codified" knowledge. As it defies easy articulation or codification, it is difficult to exchange over long distances
	Asheim and Gertler (2006)	Tacit knowledge depends on shared conventions and norms that have been fostered by a common institutional environment and relies on a mutual language and communication codes. Thus it is transferred through face-to-face interactions between individuals
Gatekeeper studies	Dang et al. (2011)	To access tacit knowledge and context-laden information, "listening posts" are created in order to build channels inside and outside the cluster
	Lezaric et al. (2008)	Gatekeepers are characterized by three functions: To search information from external sources To transcode and translate the meaning of information To transfer information and to disseminate accumulated and local knowledge

and understanding cluster processes. Thus the next section will introduce the different role of communicators as discussed in communication studies.

9.3   Communication Studies

Originally communication studies focused on journalism, which has been defined as public mass communication that targets a broad and heterogeneous audience in order to inform (Pürer 2003: 75). Over the last century, corporate communication, especially public relations, gained a lot of academic attention. Public relations can be also understood as public communication but addresses defined stakeholders to pursue specific messages. In their interplay, journalism and public relations target a broad audience and thus shape and influence the public opinion. The following two sections introduce the actors in these professions:

9.3.1   Journalists

The understanding of the journalists' role has been changing over time in academic research. Based on the News-Bias studies, early research regarded journalists as powerful "gatekeepers" who decide what is newsworthy according to their own principles. Further studies took social aspects into account: gatekeeping as part of a profession, influenced by social norms and values, political and social standards and criteria of production such as time pressure or constrained wordcount. Research on news factors and news values strengthened the perspective that journalists are not isolated actors, but part of a social process that is influenced by journalism routines.

Targeting the public, journalism has always been attributed a strong impact on society and has been discussed in terms of its framing, priming and story-telling potential. Recent research assumes that mass media structures knowledge and opinions of the recipients and define what they perceive as important (Rössler 1997). Thus journalists do not influence what recipients think, but what they think about.

9.3.2   Public Relations Consultants

Discussing the impact of journalists on the public opinion, public relations has become prominent in communication studies. Public relations is persuasive communication following a certain strategy to evoke publicity by both functioning as a source for journalism and targeting stakeholders directly. Addressing journalism public relations consultants have been understood as influential actors in the dynamics of public communication. Ever since Bearns (1985) stressed that public relations consultants determine journalism through timing and content, the relationship of influence has been a recurring theme in communication research. More recent studies show a rather balanced relation: the intereffication model by Bentele et al. (1997) analysed the daily collaborations between public relations consultants and journalists and showed a bilateral give and take interrelation (Bentele et al. 1997). Based on these results, they created the model of intereffication. "Intereffication" stems from the Latin terms "inter" and "efficare" and means "to enable each other". While journalists needs the basic information of public relations, public relations benefits by the broad and heterogeneous audience of journalism and its significance. Therefore analysing public communicators both journalists and public relations consultants should be taken into account.

9.4   Innovation Communication

Understanding the traditional roles of communicators in communication studies offers valuable insight to the role of communicators in innovation clusters. Going back to Porter, clusters can be understood as "geographic concentrations of interconnected companies, specialized suppliers and service providers, firms in related industries and associated institutions (e.g. universities, standard agencies, and trade association) in particular fields, that compete but also cooperate" (Porter 2000: 253). Consisting of heterogeneous members, clusters are characterized by many communicators taking part in public communication, pursuing different aims and strategies. To understand the role of communicators in innovation clusters, the young and developing research field innovation communication provides significant contribution.

Introducing the concept of "Innovation Journalism" in 2004, Nordfors sparked the academic debate about how communication might affect innovation innovations or innovation clusters (Nordfors 2004a, b; Nordfors and Ventresca 2006; Nordfors and Uskali; Nordfors 2009). His research highlights the leverage of communication in two ways: Journalists start a public discussion and thereby create a public agenda. Furthermore journalists may explain complex innovations and create meanings.

Building on Nordfors' research, Zerfass (2005) introduces a broad view on communication, which includes not only journalism but also public relations and interpersonal communication to meet the challenges of innovation clusters. According to Zerfass the complexity of innovation clusters calls for manifold public and bilateral relations driven by communication. Innovation journalists as defined by Nordfors play an important role in facilitating information flow, which allows collaborations and the identification of entrepreneurial opportunity. Innovation public relations consultants aim to systematically plan, implement, and evaluate communication strategies in order to create an understanding of and trust in innovations. Finally innovation-related leadership communicators seek to influence attitudes towards innovations by mediating meaning in asymmetrical, social relations (Zerfass 2005: 11).

Pfeffermann (2011) highlights the strategic perspective on communication by discussing innovation communication as a cross-functional dynamic capability of an innovative company or cluster. As defined by Teece et al. (1997) dynamic capabilities are the firm's capacity to integrate, build and reconfigure internal and external resources and competences to address and shape rapidly changing business environments (Teece et al. 1997: 516). Pfeffermann shows that communicators can achieve this aim by introducing ideas and concepts, generating and highlighting context-issues, presenting the organization's innovative capability, building up new stakeholder schemata or modifying existing ones (Pfeffermann 2011: 263). Though Pfeffermann's research is only valid for public relations consultants—journalists aim for neutral information—it gives new and valuable insight to the research field of innovation communication.

Nordfors, Zerfass and Pfeffermann provide important contributions from different perspectives to the research field innovation communication. By highlighting the potential of communicators in innovation clusters their studies complement cluster theories on different levels.

In summary, according to Nordfors, Zerfass and Pfeffermann communicators may:

- Name and explain innovations (Nordfors 2004a)
- Create a public news agenda (Nordfors 2004b)
- Formulate a shared vision of the cluster (Zerfass 2005)
- Connect stakeholders inside and outside the cluster (Zerfass 2005)
- Create collaborations and entrepreneurial opportunities (Zerfass 2005)
- Introduce new ideas and concepts (Pfeffermann 2011)
- Build new knowledge schemata and or modifies existing ones (Pfeffermann 2011)
- And thus creates and maintains a cluster's innovative capability (Pfeffermann 2011).

These highlighted issues point to the need for more research on the role of communicators in innovation clusters. The analysis of communicators requests an interdisciplinary approach to take both business and communication studies into account. In the next section the conceptual approach to analyse communicators will be introduced.

9.5   Conceptual Approach

Communication studies have a broad tradition in analysing communicators in various contexts. Its origins go back to sociological studies by Weber, Durkheim and Bourdieu who discussed action theory and identified the possibilities and limits of the individual's action. This can be seen as response to system theory, which dominated the field over a long period of time and drew attention away from the individual and towards field structures. Communication studies benefited from both research perspectives and developed complex perspectives on communicators, its personal potential of action and the influences by its environment. Based on the influential sociological studies, communication studies show different approaches to analyse the role of communicators by identifying the elements and composition of the "role".

The interest of communication studies can be explained by the potential influence of the communicator's role on the content of news. Thus some approaches focus on the media content, yet offer a valuable basis to understand and identify influencing factors on the communicator and of the communicator. Table 9.3 summarises the influential communicator studies by Shoemaker and Reese (1996), Weischenberg (1992) and Preston (2009).

9.5.1 Theoretical Foundation

The Hierarchy of Influences approach by Shoemaker and Reese (1996), the Zwiebelmodell by Weischenberg (1992), the Integrative Multi-Level Model by Esser (1998) and the Clusters of Influences approach by Preston (2009) contribute valuable insight to the influences on the role of communicators. Though they follow different research interests, they show similar patterns and recurring sets of structures as summarized in Table 9.4.

As shown in Table 9.4, influence approaches usually draw on four or five influence levels to cover the complex power structure in the communication field. Whether four or five levels are identified, depends on the structuration of influence levels: While Shoemaker and Reese (1996) and Preston (2009) differentiate between economic and cultural background, Weischenberg (1992) combines these influences in a Media System level. Furthermore, the level Media Routines appears in most of the approaches as a level on its own, only Esser mentions routines in terms of patterns and structures in the context of the Organizational Level. Besides these minor distinctions, the approaches show a common structure:

- Individual Influences
- Organizational Influences
- Communication System Influences
- Cultural, political, economical background Influences.

The hierarchical order of the levels also shows the same composition: The Individual Level is in the centre of influence structure, embedded in the Organizational Influence Level, covered by the System Influences, surrounded by the Macro Influence Level, such as cultural, political and economical conditions. While the early studies by Shoemaker and Reese (1996) and Weischenberg (1992) do not explicitly take interrelations between the levels into account, the more recent approaches by Esser (1998) and Preston (2009) highlight the reciprocal influences between the levels. According to Esser, the levels must be understood as "open" and not as "closed" systems (Esser 1998: 33).

To express in which context each of the levels works, graphic models provide important insight to the dynamic complexity of the communication field: By visualising, models simplify and structure the influences. Furthermore they create a profound foundation for an empirical analysis as they define the relevant factors and point out to interrelations. In the next section, a new model will be created in reference to

Table 9.3 Conceptual approach on the role of communicators

Conceptual approach	Influence levels and categories
Shoemaker & Reese (1996)	Individual Level: Personal aspects such as professional backgrounds and experiences, professional roles and ethics, personal attitudes, values and beliefs and the power within the organization Routines Level: Routinized and repeated practises, which can be viewed as both enabling and constraining Organizational Level: Organizational roles, organizational structure, organizational policies Extra-Media Level: Institutions in society, government, advertisers, public relations, influential news sources, interest groups, and other media organizations Ideological Level: Social interest and the construction of meaning
Weischenberg (1992)	Media-Actor: Demographic Data, social and political opinions, perception of the role, image of Recipient, professionalism and socialization Media-Message: Origin of information, reference groups, patterns of presentation and news, construction of reality effects and retroactive effects Media-Institution: Economical imperatives, political imperatives, organisational imperatives and technological imperatives Media-System: Societal conditions, historical and legal foundations, communication policy, professional and ethical standards
Esser (1997)	Individual Level: Subjective values, political attitudes, work motivation, self-perception, professionalism and demographic data Organizational Level: Job profiles and practices, organisational structure, distribution of competences, work processes, control and technology within the media organization Legal-Normative and Economic Level: Economic conditions of the media market, press law, self-control in media, ethic foundation, trade unions and associations, education of journalists Historic-Cultural Level: Freedom of the press media history, perception of the press, journalistic tradition, understanding of objectivity, political culture and socio-political conditions
Preston (2009)	Individual Level: Personal characteristics, background, values of the communicator, definitions and perception of their professional roles Media Routines: Taken-for granted institutional practices and norms, that frame and shape how individuals work and function within complex settings Organizational Influences: Organizational values, strategic goals, policies and power structures of the company Political-Economic Factors: Political and economic culture, distribution of power in society Cultural and Ideological Power: Norms, values and cultural background

Table 9.4 Structure of influence levels

Levels of influence	Shoemaker & Reese	Weischenberg	Esser	Preston
Individual Level	✔	✔	✔	✔
Media Routines Level	✔	✔		✔
Organizational Level	✔	✔	✔	✔
Political & Economic Level	✔	✔	✔	✔
Culture & Ideology Level	✔		✔	✔

the established models in order to build a theoretical framework for the analysis of communicators in innovation clusters.

9.5.2   Creation of a Conceptual Framework

The research on communicators in innovation clusters requires a new and specific framework. Unlike the frameworks discussed earlier, which focus on journalism only, this research includes different kinds of communicators to meet the challenges of innovation clusters. This points to the need for an interdisciplinary understanding of influence levels and categories. The framework proposed in Fig. 9.1 classifies four influence levels:

1. Individual Level
2. Organizational Level
3. Cluster Level
4. Context.

The centre of the model shown in Fig. 9.1 positions the research interest of the study, the role of communicators in innovation clusters. This role is influenced by the levels evolving around it: The individual influences, the organizational influences and the cluster influences. Graphically these influence levels are structured in a tetrahedron, which visualizes the reciprocal interaction between them. The openness of the levels is highlighted by the dotted lines that separate the levels. This visualization overcomes the problems of hierarchy as the influence levels take place on the same levels and therefore all interact. This interaction takes place in front of a societal background, which surrounds the influence-levels and the role of the communicator. It indicates that everything takes place in a certain context that has been established by various factors, such as historical, cultural, political and economical conditions, which have to be taken into account. In the following section each level will be explained in detail.

Individual Level

The Individual Level deals with the personal and unique attributes of the communicators. This level appeared in all models discussed earlier, as it covers important

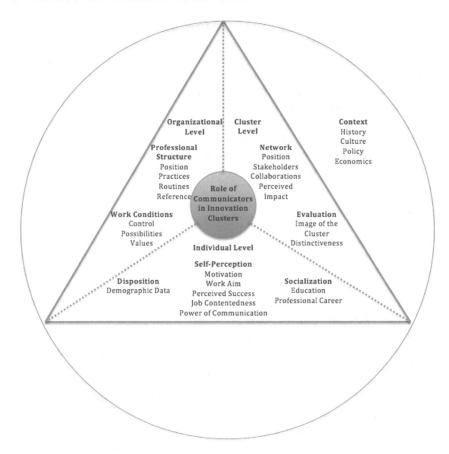

Fig. 9.1   The role of communicators in innovation clusters

information about the personal position of the interviewees and thus builds the foundation to understand their ideas throughout the whole study. As the level implies various attributes, the model suggests three categories: Disposition, Socialization and Self-Perception. Disposition covers the demographic data, such as gender and age. Socialization also refers to individual attributes, covering the communicator's education and professional career. This category provides information on the interviewee's background and thus creates a deeper understanding for the communicator's position nowadays. Besides individual facts as dispositions and socialization, the Individual Level also covers the category Self-Perception. This category structures complex aspects of individual perceptions regarding the interviewee's role. Work motivation will provide valuable insight to the attitudes and expectations of the communicator. Closely linked is the definition of the work's aim, which identifies beliefs and ideas of the profession in general. Subsequently it is very interesting to analyse the perceived success of these aims to check the broad aspirations with the reality.

This leads to the next aspect, the job contentedness, which deals specifically with the satisfaction in the profession and links back to realization of aims and motivations. Finally, the category Self-Perception points out to the importance of the perception of the power of communication. This aspect completes the idea about their profession, what they want to do, what they plan to do, if and how they achieve it and what greater power they might hold. As discussed in the literature review, the individual communicator has been seen as very influential in terms of creating a news agenda. Though the attributes and attitudes of the communicator is crucial, research pointed to the importance of the organizational setting of the communicator. Thus the next influence level will refer to the organizational categories and influence factors in detail.

Organization Level

The Organization Level is also a strong component in the established models as it puts the individual in a context. The organization is the closest environment of communicators and frames their role strongly. Thus the level is distinguished into two categories, the Professional Structure and the Work Conditions. The professional structure first analyses the position of the communicator within the organization. That implies the understanding of the hierarchy structures and power relations within the given company. The identification of the position leads to the professional practices. The description of the daily activities provides valuable insight to the work of communicators and in which ways they might influences their role. The professional practices are closely linked to professional routines, which can be understood as both enabling and constraining. Analysing practices and routines, the reference groups of the communicators play a crucial role in different ways: as origins of information sources, as guidance in terms of professional capabilities but also as peer group pressure. Thus, reference groups have an impact on the role of communicators and should be analysed in the organizational context. Building on the findings of the category Practices, the category Work Conditions investigates how the quality of the work is perceived by the communicators. This category implies internal control, which could be developed trough hierarchal structures, peer pressure or organizational tendencies. However the possibilities and perceived freedom will also be analysed. The organizational atmosphere is much influenced by constraints and tolerance, and leads to the professional values that characterize a company. Values are a broad field and could be based on a journalistic education, organizational codes or general ethic norms that influence the organizational context. In sum the Organizational Level acknowledges that the communicators work in an organizational context that shapes and conditions their professional role. The research interest requests to understand the role of communicators also in a broader context, which impacts both individual and organization: the cluster.

Cluster Level

The Cluster Level is highly specific to this research interest, as it has not been discussed in communication studies before. While the models summarized earlier in

Sect. 9.5 focus on communication structures, such as media-institutions or the journalism field, this study takes place in a very distinctive environment: innovation clusters. This requests distinctive categories, which characterize—and thus analyse—the specifics of clusters. Applying a journalistic model to a business research field, previous models only benefit in terms of level analysis. To identify relevant categories for the Cluster Level, the cluster theories provide valuable contributions. As discussed in the literature review, the benefit of clusters lies in the network structure, which makes information flow and knowledge spillovers possible. Thus the Network is the first categories of the Cluster Level. This category refers to the interrelations and connections in terms of communication. First of all, the position in a network is influential for the role of communicators. Whether they are located centrally or rather distanced might impact the way they communicate. The position is connected with the important role of stakeholders: it is important to analyse who the stakeholders of the relevant communicators are and how they work with them. Also, if there are different communication strategies for each stakeholder could be very interesting. Bearing in mind the insights from the literature review, which identifies internal and external stakeholders of clusters, communicators can be seen as central actors who gather and spread knowledge. At this point it is very interesting to analyse the perceived success of their work: the perceived impact. How the communicators perceive their communication strategies are influencing their stakeholders, provides interesting insight to their professional role and evaluates their work experience. This leads to the second category of the Cluster Level, the Evaluation of the cluster. In this category, the perceived image of the clusters will be analysed. The image communicators have about a cluster will influence their behaviour, their opinions—their role. While the image can be created by other people's views, the second aspect only measures the individual opinion about the cluster. Analysing the distinctiveness from the communicator's point of view completes the evaluation as this aspect summarizes the perceived most important characteristics about the cluster and thereby reveal the personal "image".

Defined as regional agglomeration, clusters are influenced by their specific environment. This environment must be understood as broad background which influences not only the cluster but also the organizations and actors of the cluster. Thus it is important to set the so far identified levels of influence in a certain context.

Context

The three influence levels Individual Level, Organizational Level and Cluster Level are embedded in a specific Context. As explained earlier, the context impacts the influence-levels and the role of communicators in innovation clusters. The Context is identified as history, culture, policy and economics. These conditions are highly specific to geographic areas and frame the cluster, organizations and actors. The history factor examines the historical evolution of the cluster and takes important milestones into account. History is closely linked to further factors, namely culture, policy and economics. While history focuses on the development of a cluster, the culture factor looks at cultural environment of the cluster, for example the relation to universities and research institutions. The policy factor investigates how political decisions and support influences the cluster and shapes its situation. Political initiatives might lead to financial support, which will be examined by the factor economics. This latter factor concentrates on the financial situation and external support. The Context is important for the cluster, the organization and the actors and conditions the role of communicators in innovation clusters.

Together, the *Individual Level*, the *Organizational Level*, the *Cluster Level* and the *Context* create a valuable framework to analyse and understand the role of communicators in innovation clusters. Taking various levels and perspectives into account, the framework offers a strong foundation for further studies, which will be discussed in the following section.

9.6 Conclusions

The preceding sections brought together a diverse body of academic literature on cluster theories and communication studies in order to find interrelations between the two research fields. The classic economic cluster theories by Marshall (1890, 1920), Porter (1990, 1998, 2000) and Krugman (1991, 1994) indicate the importance of communication, which informs but also connects actors inside and outside the cluster. While economic cluster theories refer to communication as given spillovers, sociological cluster theories acknowledge communication as an organized, deliberate process and thus complement the economic theories. The insights of Network Theory (Granovetter 1973; Burt 1992), Knowledge-Based View (Polany 1958; Asheim and Gertler 2006) and Gatekeeper Studies (Lezaric et al. 2008) highlight the importance of communicators who build connections to gather and share information. While there is no research on communicators in innovation clusters, the young and developing research field innovation communication discussed by Nordfors (2004a, b, 2009), Zerfass (2005) and Pfeffermann (2011) offers valuable insight and strengthens the demand for further research on the role of communicators in innovation clusters.

To achieve this aim, a conceptual framework has been created based on established conceptual approaches by Shoemaker and Reese (1996), Weischenberg (1992) and Preston (2009). Though these studies focus on different research interests, they share a common quest into the interrelation between a communicator and the context that influences in both constraining and enabling ways. As discussed in the literature review communicators are not isolated individuals but part of an organizational and social context. To analyse the role of communicators, the relevant variables are heterogeneous and complex. The discussed models identify different influence levels and allocate influence factors appropriate to the specific research question. Also this study's research interest requested an unique conceptual framework to take the specification of innovation clusters into account. The tetrahedron-shaped framework identifies four influence levels: the individual level, the organizational level, the cluster level and the context which cover influence factors that shape and condition the communicator. The framework offers a strong foundation to investigate the role of communicators in innovation clusters.

This study addresses an unexplored interdisciplinary research area at the interface of communication and business studies and thus makes a series of academic and practical contributions.

Innovation Communication

Innovation communication is a new research field, which developed over the last decade. So far, research concentrated on innovation communication in terms of either innovation journalism or innovation communication on a corporate level. This research focuses on communication at a cluster level and therefore views innovation communication from a complex perspective: a cluster's communication is published by many different actors in different positions, pursuing different aims with different strategies. While previous communicator studies focused on a specific profession in public communication, this study includes different professions, which create the public message of a cluster. This approach allows the comparison of different professional positions and will make patterns visible. The inclusion of heterogeneous professions also allows the analysis of interaction. As discussed in the literature review, the interrelations between journalists and public relations consultants are intense and influential. Including everybody who engages in public communication about the cluster provides a holistic and comprehensive perspective on innovation communication.

Cluster Theories

Addressing innovation communication at a cluster level, this research contributes also to cluster theories. While both economic and sociological cluster theories refer to communication, the role of communicators stays implicit. Though sociological theories highlight connections and the management of information, the range of actors remain unclear. Investigating the people who make the message sheds light on a research angle, which has not been explored. The conceptual framework allows an understanding of the position of the communicators—who they are, what they do, what aim they pursue and which strategies they apply. This insight addresses the research gap between the assumed importance of communication in clusters and the missing data on it. The conceptual framework offers a valuable foundation for empiric research on the role of communicators in innovation clusters. The formulated categories, influence factors and their interrelation lead to a systematic structure for a qualitative interview guideline with relevant communicators. The results will establish how communicators connect in clusters, how they gather knowledge and share information and thereby create certain roles of communicators. Understanding the meaning of communicators will contribute valuable insight to classic cluster theories in terms of clustering dynamics. In addition to established cluster factors such as specialized supply and demand, collaboration and competition and the infrastructure and mobility of goods, communicators and their information networks should also be seen as crucial drivers for a cluster's development. Based on this understanding, further future research will be able to focus on the impact of communicators on the cluster's success and investigate causal effects on a quantitative basis. It will be interesting to compare different clusters in different developmental stages and in different locations to make similarities and differences visible.

Practical Contributions

This understanding provides valuable practical implications. Innovation clusters are based on the premise of interrelations and common knowledge and demand the creation and maintaining of a network of information inside and to the outside of the cluster. This points to three major aims for communicators:

1. Create connections and dialogue with **internal stakeholders**
2. Create interrelations and communication with **external stakeholders** and the outside world
3. Create a **common language**.

Internal communication is crucial to connect the members within a cluster, to create a shared environment and a common vision. Only if members are connected and talk to each other will collaborations and entrepreneurial opportunities be possible. Knowledge spill-overs are rarely coincidence but based on a frequent dialogue between actors. As one journalist of one of the Cambridge High Tech Clusters explains "We actually joined up companies in the business community. We helped them collaborate with one another. Before you would have had companies operating in isolation. But you have got to have networks and connections. And we formed that role by publicizing companies and their role and what they were doing. We acted as catalyst for collaborations." It shows the important aim of communicators to connect and relate members of the cluster and thereby create a vibrant community. This practical insight refers back to cluster theory and provides valuable contribution to the studies of Marshall, Porter and Krugman who implied the importance of collaboration and shared knowledge by indicating the role of communication. Only if the cluster's members are aware of their environment, collaborations and entrepreneurial opportunities arise.

Furthermore **external communication** plays an important role as it connects the cluster to the outside world and allows collaborations with other clusters. External communication also has a representative role, aiming to attract potential members or financial supporters. A public relations consultant in Cambridge explained: "For Cambridge and the Cluster, to get this information out there is really key because you have people from around the world that are coming to Cambridge to look at it, and some are coming here to invest, that's because they are hearing about it, they are hearing the positive news. This is one of the key places in the world. If we didn't communicate that would they think to come to Cambridge? It's really important to get that message across". The strategic communication to the outside attracts attention from all over the world and thus positions the clusters in a global competition. This result can be drawn back to theoretical position of gatekeeper studies as discussed by Dang et al. (2011) and Lezaric et al. (2008), who highlight the importance of creating channels not only inside but also outside the cluster. By representing the cluster to the outside world, communicators attract potential members and financial supporters.

To communicate successfully to the inside and the outside of the cluster, communicators have to create a **common language** by explaining complex innovations in order to make them accessible to a broader audience. A public relations consultant in the Cambridge cluster describes this process as "translation": "This is translation. The raw material compared to what is then written, you have to translate what can be something quite complex and put it into language. That is part of the communication, getting the language right that you are not dumbing down the technology or the invention but in that you are still caring up consistent messaging so that people will pick up and start to understand." Only by means of public communication, internal and external stakeholders will be able to comprehend unknown innovations. Communicators have to make sure that the message is consistent and understandable so that it reaches the recipients successfully. This has been discussed by Nordfors (2004a, b) in terms of journalism. Anecdotal evidence shows that also public relations consultants are aware of the importance of "translation" and pursue this aim and thus contributes to the research on innovation communication.

The identification of communicators' aims in innovation clusters make strengths and weaknesses of clusters visible and point to specific improvements in terms of internal and external communication and the creation of a common language. Understanding the role of communicators in innovation clusters provides first insights to the interrelations of communication and a cluster's development. This understanding allows future research on the dynamics of communication and creates a new perspective on the evolution of innovation clusters.

Bibliography

Asheim, B. T., & Gertler, M. S. (2006). The geography of innovation. In J. Fagerberg, D. C Mowery, & R. R Nelson (Eds.), *The Oxford handbook of innovation* (pp. 291–317). Oxford: Oxford University Press.

Baerns, B. (1985). *Öffentlichkeitsarbeit oder Journalismus. Zum Einfluß im Mediensystem*. Köln: Verlag Wissenschaft und Politik.

Bentele, G., Liebert, T., & Seeling, S. (1997). Von der Determination zur Intereffikation. Ein Integriertes Modell zum Verhältnis von Public Relations und Journalismus. In G. Bentele & M. Haller (Eds.), *Aktuelle Entstehung von Öffentlichkeit. Akteure, Strukturen, Veränderungen* (pp. 225–250). Konstanz: UVK.

Burt, R. S. (1992). *Structural holes*. Cambridge, MA: Harvard University Press.

Burt, R. S. (2001). The social capital of structural holes. In M. F. Guillén, R. Collins, P. England, & M. Meyer (Eds.), *New direction in economic sociology* (pp. 201–247). New York: Russell Sage Foundation.

Clark, J. (2013). *Tech country report*. Retrieved from http://admin.bvca.co.uk/library/documents/Tech_Country.pdf. [13.05.2013].

Dang, R. J., Mortara, L., Thomson, R., & Minshall, T. (2011). Developing a technology intelligence strategy to access knowledge of innovation clusters. In M. Hülsmann & N. Pfeffermann (Eds.),

Strategies and communications for innovations. An Integrative Management View for Companies and Networks. Berlin: Springer-Verlag.

Esser, F. (1998). *Die Kräfte hinter den Schlagzeilen. Englischer und deutscher Journalismus im Vergleich*. München: Verlag Karl Alber.

European Commision. (2008). Towards world-class clusters in the European Union: Implementing the broad-based innovation strategy. *European Commission*. SEC (2008) 263.

Granovetter, M. S. (1973). The strength of weak ties. *The American Journal of Sociology, 78*(6), 1360–1380.

Karlsson, C. (2008). *Handbook of research on cluster theory*. Edward Elgar Publishing.

Krugman, P. (1991). Increasing returns and economic geography. *Journal of Political Economic Geography, 99*(3), 483–499.

Krugman, P. (1994). Complex landscapes in economic geography. *The American Economic Review, 84*(2), 412–416.

Lezaric, N., Longhi, C., & Thomas, C. (2008). Gatekeepers of knowledge versus platforms of knowledge: From potential to realized absorptive capacity. *Regional Studies, 42*(6), 837–852.

Marshall, A. (1890). *Principles of economics*. London: Macmillan.

Marshall, A. (1920). *Industry and trade*. London: Macmillan.

Nordfors, D. (2004a). The role of journalism in innovation systems. *Innovation Journalism, 1*(7/8), 1–18.

Nordfors, D. (2004b). The concept of innovation journalism. And a program for developing it. *Innovation Journalism, 1*(1), 1–14.

Nordfors, D., & Ventresca, M. (2006). Innovation journalism: Towards research on the interplay of journalism in innovation ecosystems. *Innovation Journalism, 3*(2), 1–18.

Nordfors, D., & Uskali, T. (forthcoming). *The role of journalism in creating the metaphor of Silicon Valley* (pp. 1–20).

Nordfors, D. (2009). Innovation journalism, attention work and the innovation economy. A review of the innovation journalism initiative 2003–2009. *Innovation Journalism, 6*(1), 1–46.

OECD (2012). *Cluster policy and smart specialisation. OECD science, technology and industry outlook 2012*. OECD Publishing.

Pfeffermann, N. (2011). Innovation communication as a cross-functional dynamic capability: Strategies for organizations and networks. In M. Hülsmann & N. Pfeffermann (Eds.), *Strategies and communications for innovations. An integrative management view for companies and networks* (pp. 257–292). Berlin: Springer-Verlag.

Polany, M. (1958). *Personal knowledge: Towards a post-critical philosophy*. London: Routledge & Kegan Paul.

Porter, M. (1990). The competitive advantage of nations. *Harvard Business Review*, 73–93.

Porter, M. (1998). Clusters and the economics of competition. *Harvard Business Review*, 77–90.

Porter, M. (2000). Location, clusters and company strategy. In G. Clark, M. Gertler, & M. Feldman (Eds.), *The Oxford handbook of economic geography* (pp. 253–274).

Pürer, H. (2003). *Publizistik- und Kommunikationswissenschaft. Ein Handbuch*. Konstanz: UTB.

Preston, P. (2009). *Making the news: Journalism and news cultures in Europe*. London & New York: Routledge.

Rössler, P. (1997). *Agenda Setting. Theoretische Annahmen und Empirische Evidenzen einer Medienwirkungshypothese*. Opladen: Westdeutscher Verlag.

Shoemaker, P. J., & Reese, S. D. (1996). *Mediating the message: Theories of influence on mass media content* (2nd ed.). White Plains, NY: Longman.

Teece, D. J., Pisano, G., & Shuen, A. (1997). Dynamic capabilities and strategic management. *Strategic Management Journal, 18*(7), 509–533.

Uyarra, E., & Ramlogan, R. (2012). *The effects of cluster policy on innovation. UK National endowment for science, technology and the arts* (NESTA). Working Paper 12/05.

Weischenberg, S. (1992). *Journalistik. Medienkommunikation. Theorie und Praxis*. Opladen: Medienetik, Medieninstitutionen.

Zerfass, A. (2005). Innovation readiness. *Innovation Journalism, 2*(8), 1–27.

Bettina Sophie Blasini is a Ph.D. student at the University of Cambridge under the supervision of Tim Minshall. She holds a Master of Arts in Mass Communication and Journalism from Ludwig-Maximilians University Munich and an M.Phil. in Innovation, Strategy and Organisation of Cambridge Judge Business School. Besides her studies, Bettina Sophie worked at the University of Munich and as a freelance journalist.

Rani Jeanne Dang is an Assistant Professor at the University of Nice Sophia-Antipolis, affiliated to CNRS, the National Centre for Scientific Research in France (GREDEG Research group) where she researches and teaches in the fields of innovation, entrepreneurship and strategy. She is also a Broman Research Fellow at the Institute for Innovation and Entrepreneurship, University of Gothenburg, Sweden. She holds a Ph.D. in Management from University of Nice/CNRS and was also a visiting researcher at the University of Cambridge, UK and a project officer at the Pôle de Compétitivité SCS (Secured Communicating Solutions), the cluster-led innovation policy in France. Rani's main research interest is the development of small firms through collaborations in regional ecosystems of innovation.

Tim Minshall is a Senior Lecturer at the University of Cambridge Centre for Technology Management. His researches, teaches, writes and consults on the topics of open innovation, technology enterprise, the financing of innovation, and university-industry knowledge exchange. He is a non-executive director of St. John's Innovation Centre Ltd, Cambridge and a Visiting Professor at Doshisha University Institute for Technology, Enterprise and Competitiveness in Japan. He has a BEng from Aston University, and a Ph.D. from Cambridge University Engineering Department. Prior to joining the University of Cambridge, he worked as an engineer, teacher, consultant, freelance writer and project manager in the UK, Japan and Australia.

Letizia Mortara is a Senior Research Associate at the Centre for Technology Management at the Institute of Manufacturing, University of Cambridge (UK). Her current interests and expertise include Open Innovation, Additive Manufacturing and Technology Intelligence. Letizia has a first degree in Industrial Chemistry from the University of Bologna (Italy). After working as a process/product manager for the IVM group specialised in coatings, stains and primers for wood, she moved to the UK where she gained her Ph.D. in processing and process scale-up of advanced ceramic materials at Cranfield University.

Chapter 10
Narratives and Optics: Communication Dynamics Political Leaders Face Today

William Howe and Joseph C. Santora

Abstract In this chapter, we focus on the emerging language used by the contemporary media in their considerations of political leadership and what that language says about the situations leaders face today as they seek to transmit their messages to the public or to specific constituencies. Analysis of media coverage of leadership indicates that these leaders communicate through a combination of "narratives" and "optics" and that a tension exists between the two—to communicate messages/visions through "narratives" that use the words/stories to which we are bound or through images/videos that seek to move beyond words/stories. We conclude that leaders are moving increasingly toward the immediacy of "optics" to communicate messages but must inevitably resort to the more protracted messaging of language-bound "narratives." Some thoughts are also offered about communication—by leaders but also in general—for the coming decades.

10.1 Introduction

In recent years, media coverage of leadership and political leadership in particular has focused on what leaders *say,* on the one hand, vis-à-vis what those leaders and/or their supporters *project visually*, on the other hand. Put in terms of the language that has emerged in the media and seeped into our culture, the focus is on leaders' *narratives,* their verbal statements or messages, vis-à-vis their *optics*, their images or visual performances. The two prongs of the focus may work in synergy with each other or, in some cases, may provide opposing perspectives. In addition, both may be intentional—consciously devised as strategic communication; shaped by possible biases of the specific media outlets (e.g., liberal or conservative newspapers,

W. Howe
3969 Mahaila Ave., 304, San Diego, CA 92122, USA
e-mail: wh@san.rr.com

J. C. Santora (✉)
4 Rue Louis Codet, 75007 Paris, France
e-mail: jcsantora1@gmail.com

© Springer Nature Switzerland AG 2020   127
N. Pfeffermann (ed.), *New Leadership in Strategy and Communication*,
https://doi.org/10.1007/978-3-030-19681-3_10

television, or popular social media); or simply perceived and interpreted by constituents or the public at large in a particular way.

One could argue, of course, that *optics* are elements of a broad, overall *narrative* and that a silent movie or a "photoplay" can provide a *narrative*. Likewise, it may be argued that a *narrative* is elicited by or evoked by *optics*. To be sure, the two communication types may overlap and provide mutually supportive messaging. Such was the case with the silent movies that were accompanied by subtitles more than a century ago (1891–1931), and such is the case when political leaders today deliver verbal messages and also cultivate a visual context for those messages. Nevertheless, in this chapter we argue that the two can generally be considered conceptually distinct and that they offer, because they are distinct and in tension with each other, a unique means of approaching communication in contemporary political leadership—as verbally expressed/interpreted messaging or as visually projected/interpreted messaging.

In addition, we propose that these two communication modes have become important means of understanding leadership in a broad sense today, regardless of the sector (i.e., for-profit or non-profit), the movement (i.e., social), or the leadership contexts in the United States or abroad. Furthermore, as society moves increasingly toward a more visual mode of communication (*optics*) and away from centuries of reliance on a verbal mode of communication (*narratives*), the predominance of *optics* ("a picture is worth a thousand words" carried to new levels by contemporary imaging) is becoming increasingly apparent. The social fabric today seems to yearn for the immediacy and materiality of images rather than the time-bound and immaterial nature of words, almost as though people in general, and leaders in particular, want to collapse time, leap beyond the problems of interpretation that have often characterized communication from the days of scriptural hermeneutics forward, and arrive at a visually immediate, sensorially graspable means of communicating. In brief, in the tension between the two forms of communication, *narratives* and *optics*, *optics* seems to be winning the battle about how leaders communicate and about how people desire to receive leaders' messages. Seeing is believing and, perhaps unfortunately, listening and/or reading are becoming far less effective in eliciting belief in the 21st century.

10.1.1   Research Approach

During the two-year period, January 2017–March 2019, we watched television news coverage presented by a major national/international network (i.e., MSNBC) carefully for approximately four hours per day on most days. This time period corresponds roughly with the first two years of Donald Trump's US presidency, a period during which presidential leadership in particular was certainly a principal focus of MSNBC and other cable networks (e.g., CNN, Fox News) as well as the print media, though other foci included US Congressional leadership (e.g., various hearings), US Supreme Court leadership (e.g., the Senate Judiciary Committee hearings on the confirmation of Supreme Court nominee Bret Kavanaugh), and the interactions of US leaders with European leaders such as German Chancellor Angela Merkel (2005–), French President Emmanuel Macron (2017–), and British Prime Minister Theresa May (2016–); Asian leaders, Japanese Prime Minister Shinzo Abe (2006–2007, 2012–) and North Korean Supreme Leader Kim Jong-un (2011–); and Russian President Vladimir Putin (2002–2008, 2012–).

We did not establish any predetermined themes to guide our viewing of the network coverage. We watched that coverage without any initial research intent in mind. Nevertheless, several themes pertinent to our lifelong interest in leadership emerged over time and particularly piqued our interest. Three such themes included:

- A new way of approaching the attribution theory of leadership with US President Trump as leader attributing leadership to himself and seeking credit for policies and/or actions he may or may not have caused (in contrast to Meindl and Ehrlich's 1987 original formulation of the attribution theory of leadership whereby followers attribute leadership to individuals).
- The purposeful, intentional, and relentless obfuscation of truth, facts, and empirical evidence by leaders to further advance their personal aims and desires and/or political agendas.
- Leaders' Image Management (IM) in a media age of sound bites and tweets.

Each theme seemed worthy of our attention and research in its own right. However, we decided to focus on what became increasingly apparent and intriguing to us and what seemed to embody significant communication questions: (1) To what degree do political leaders, as depicted by a cable news network (e.g., CNN, Fox, or MSNBC), communicate through intended or unintended "narratives," through intended or unintended "optics," or both communication forms?; and (2) What role do the news media play in displaying or interpreting the "narratives" and the "optics," and are those media reporting objectively, through a politically biased lens, or, perhaps more alarmingly, from the point of view of a leader who is using or manipulating them? Those two questions guided much of our television viewing approach.

The issue of "narratives" vis-à-vis "optics" became increasingly interesting and compelling to us during this two-year period. Nevertheless, we chose to forego any systematic collection of data—that is, documenting each mention of "narratives" or "optics" and their contexts. Our purpose took shape as an unfolding consideration of what the two terms—both used extensively by the media—mean, what they suggest about the ways in which leaders choose to communicate today, how the media are attracted to and perhaps even complicit with what leaders "intend" to say, and what the developing dynamic between "narratives" and "optics" may suggest about the way human communication in general may be evolving in today's political environment. In brief, we purposefully did not adopt any formal qualitative research protocols. Rather, our interest lay in exploring some of the possible implications of verbal ("narratives") communication versus visual ("optics") communication, and how political leaders may consciously choose either or both of these communication forms or be interpreted through them via the television news media.

In essence, this exploratory research, then, sought to probe the way(s) many people experience communication by political leaders or by those political media analysts and pundits who interpret the way(s) political leaders communicate in a special way. At the same time, our research sought to consider where human communication may be headed, with a specific focus on political leadership. When Guttenberg (1400–1468) invented the printing press some 600 years ago in the 15th century, he inaugurated the era of "narratives." Is it highly possible that today we are headed for a major revolution in communication—Communication 2.0, 3.0, or even 4.0—a significant revolution that transports us way beyond text and the Guttenberg "narratives"?

10.1.2 Narrative

By definition a "narrative" is "a spoken or written account of connected events: a story … a way of presenting or understanding a situation or series of events that reflects and promotes a particular point of view or set of values" (Merriam-Webster.com). Thus, a narrative has a close association with fiction or story-telling, and it is often value-laden or heavy with the advancement of an agenda or perspective. It does not claim to lay out empirical truth or to represent "reality." Indeed, some current literary scholars, for whom the notion of narrative has long been familiar, might argue that all communication in language is a narrative that offers only an interpretation (or a mis-interpretation) of reality, which in itself is a concept that has no ground or legitimacy (see Derrida 2016).

Traditional notions of narrative, which stem from literature and literary analysis, include elements such as plot, character, setting, and point of view. Literary genres (e.g., the novel, short story, fantasy, drama, autobiography, biography, and narrative poetry) are often noted as belonging to this tradition. More recently, narrative has come to include journalistic accounts, blogs, and sometimes even texts or tweets or a series of texts or tweets. At times, too, narrative today is associated with images, films, television shows, and videos, though for the purposes of this chapter, we are leaving such visual media to what we will develop as optics—visual presentations of an event or a series of events that include a leader or leaders—as opposed to narrative, defined as a "spoken or written account." To be sure, optics and narratives may have some common ground and are even sometimes used interchangeably today by semioticians who focus on meaning making through "signs" and on *both* verbal and non-verbal signs as communicating meaning (Barthes 2013). We believe it is useful, however, to keep them conceptually distinct—that is, to consider *optics as what is seen and narratives as what is spoken or written with language.* Given that differentiation, it is worth noting that cultural context may be crucial to any consideration of narratives. Some cultures rely heavily on oral narratives, other on written narratives, and, in some cases, on oral narratives that have become written narratives over time (see Homer, *Iliad* and the *Odyssey,* 8th BC/2011).

In the past several decades, there has been a serious discussion about the importance to human life of narrative or storytelling. "Evidence strongly suggests," Flanagan (1992) argues, "that humans in all cultures come to cast their own identity in some sort of narrative form. We are inveterate storytellers" (p. 198). Within the broad areas of "philosophy of mind" or psychological approaches to narrative, a person's entire identity is frequently conceived as a narrative. Lakoff and Johnson (2003) have even suggested that humans make use of "conceptual metaphors" to simplify data and present compelling narratives for broad, meaningful phenomena (e.g., Trump's "wall" or "swamp" as metaphors that convey his narrative succinctly and powerfully).

Through its association with fiction and stories, narrative, as used today, assumes that what is expressed is not true—not necessarily false, but certainly far from empirically verifiable and usually promoting an idiosyncratic point of view. "Any creation of a narrative," Pasupathi says, "is a bit of a lie" (quoted by Beck 2015). Clark (2012) even suggests that narrative has evolved "from the world of literature to that of politics," and he associates it with the kind of "misinformation" that political parties and their leaders use to advance their own interests: "The long journey of narrative [from literature to politics] … arrived so conspicuously in the barrio of spin doctors, speech writers, and other political handlers."

More recently, scholars and writers have specifically associated narrative with what leaders—and political leaders in particular—do to present their positions. Mayer (2014), for example, argues that leaders use stories to bring people together, create common understanding, and promote collective action. Similarly, Tolchard (2017) claims that "narratives or larger stories about the way the world is, are essential to political candidates," while Ewing (2016) sums up the infectious use of narrative by the media to describe what political leaders and their followers/voters do today:

> Study the current [2016] election for a week or two and you'll notice one word turn up again and again in the commentary: *narrative*. Politicians control the narrative, they reinforce the narrative, they seize the narrative, they reshape the narrative, they build the narrative, and that's before the voters get their say, at which point they might defy the narrative, overturn the narrative, confirm the narrative, or perhaps just get heartily sick of the narrative and stay home ….'The narrative' really does matter. Candidates need to find a story about themselves and the country that feels credible, that they have permission to tell, and that voters want to come true. The one who can do that, and who uses the media effectively to put that story across, will win ….Using narrative well is not persuading people of the story you want to tell. It's about finding the story they already believe – or are close to believing ….

The narrative, then, can be something the leader creates, something the media offer as an interpretation of what the leader says or writes, something followers (or perhaps opponents) interpret in their own way and that may or not reflect their personal narratives, or all these simultaneously.

The narrative, it seems then, resides within the mindsets as well as in the language that reflects the mindsets of the leader, the media, and the followers, and it may or may not be consistent across those three. Since language is always interpreted, what the leader intends as a narrative may or may not reflect how the media perceive and interpret the narrative and then present it to the followers or the general population, who may or may not resonate with the narrative as it is expressed to them by the leader and/or the media. In short, what we have is storytelling that involves the original storyteller (the leader) and layered interpretation of the story by the media and the followers/population. It is even quite possible, of course, that the media may create the narrative by presenting its interpretation more forcefully or more repeatedly than the leader's original story, assuming a story, composed as it is of language and tropes, can ever have a distinctive origin; likewise, it is possible that the followers/population may create the narrative by disseminating it, in one or more interpretations, via social media.

Thus, narrative can get bogged down in layers of interpretation, misinterpretation, and consequent misunderstanding. Because it is comprised of language and because language is expressed as a time-bound medium, narrative can be messy and confusing. Moreover, everyone knows that it is a *story*—a "lie," a fabricated account—from the start, after which it may become many different stories or as many stories as there are listeners or readers (Fish 1982). Yet, narratives can be powerful and highly influential when used effectively and repeated again and again, almost like a refrain in a poem or song. Stockley (2011) notes some political narratives that took hold and had a huge impact upon entire nations and the world: "Ronald Reagan versus the evil empire. Margaret Thatcher versus the Argentinian generals and, later, the miners. Tony Blair versus Gordon Brown .... The West versus Al Queda." Or in today's political environment, Trump versus the "fake news"; Trump versus the "investigators"; or the use of the words "wall" and "swamp" as powerful Trumpian metaphors that call up entire narratives. An interesting side note: the *Journal of Narrative Politics* was launched nearly five years ago in 2014 to explore narrative "as a mode of knowing."

Stockley (2011) suggests that there are four features that characterize political narratives:

> First, the story and the events must affect people and their world views. They must evoke an emotional reaction .... Second, political storytellers should explain the world to their listeners and enable them to understand their place within it .... Third, a true political storyteller will give people hope – or at least, reassurance about themselves and their future .... Fourth, politicians need to remember who owns the narrative.

Those four characteristics sound much like what leaders should do to create an effective vision (see Bennis and Nanus 1985). They represent perhaps the most positive view on narrative, whereas, in fact, much narrative in political life today generally means a story that is a fabrication and an attempt to persuade people to accept a biased point of view. In that sense (and contra-Stockley), a narrative seeks to bring people to an understanding of their place in the world, and that world tends to coincide with the world of the leader expressing the narrative. Such a narrative may give people hope, but it is all too likely to provide a false sense of hope (e.g., in a wall as a means of solving the complexities of the US immigration problem, or in tax breaks for the middle class when these tax breaks may actually benefit the most affluent people).

In general, narrative, as used in the realm of political leadership, has become synonymous with "story", "fiction", "falsehood", "fabrication," a self-aggrandizing account that may have little relationship to "empirical truth," "facts," or "reality." Nevertheless, "narrative" is a term as well as a concept used today as much as any other word to describe what political leaders say, as if those leaders are weaving rhetorical worlds of their own with the intent of persuading followers/population to accept the rhetoric literally and as more truthful, more factual, and more real than anything the opposition may be saying.

People, it seems, are often eager to believe and to find some meaning in what political leaders say, something that resonates with their values or with their sense that current policies and practices are unfair, corrupt, undemocratic, or inegalitarian. As Ellerton (2016) puts it:

> What we value most in politicians is not that they tell the truth, but that they agree with us, or at least that the worldview they espouse resonates with our own …. We care much more that our narratives provide us with meaning than that they are true …. The problem is that often the truth does not speak for itself – it has to be interpreted through a narrative. This means facts alone are not enough.

Such a conclusion has profound implications for the way political leaders communicate today. It suggests that those leaders should *follow* the people and express what the people may want to hear or read, rather than what may be "true" or "right" or perhaps in the best interest of the people or society. Moral or ethical considerations may take a back seat to the interests of the leader, the political party, or the donors who are funding the leader or the party. For example, President Trump has even asserted that his narrative may be more truthful than what his opposition presents through evidentiary investigation and facts: "Don't believe the crap you see from these people, the fake news … What you're seeing [optics] and what you're reading [narrative] is not what's happening" (broadcast on most major cable networks, July 24, 2018). This leadership communication approach seeks to construct a reality that is dramatically at odds with what other people may see, hear, read, or believe, suggesting that politics is up for grabs today, may be unaccountable to empirical data, and may question empirical reality over socially-constructed reality. At times, science, evidence, data, and facts may become secondary to the story that the leader creates.

10.1.3   The Use of "Narrative"

The use of "narrative" by the media underscores its meaning as a fabricated, though often powerful and compelling story (Selected examples below are listed in chronological order from January 2019 to February 2019, with parenthetical comments).

- "He (Trump) was telling a narrative that he could stomach" (Tim O'Brien, on MSNBC, January 26, 2019) [Narrative is a palatable account that may reluctantly acquiesce to opposing people or groups]
- "[It is] a false narrative" (President Trump, tweet of January 31, 2019, on the news media) [Note the redundancy here, as if a narrative, which is itself a story or fabrication, is doubly false. This may lend legitimacy to the idea of narrative as a contrast to "false narrative"]

- "He (Roger Stone) seems to be able to get his narrative out there" (Shelby Holliday, on MSNBC, February 2, 2019) [Narrative is a personal interpretation or story]
- "If they [facts] are not convenient to his political narrative, he simply ignores them" (Ben Rhodes, on MSNBC, February 5, 2019) [Narrative may be inconsistent with facts and may intentionally ignore facts]
- "… Mueller builds a narrative" (Phil Rucker, on MSNBC, February 5, 2019) [Narrative is a constructed or interpreted account of evidence]
- "A narrative that advances whatever one wants" (Mya Wiley, on MSNBC, February 8, 2019) [Narrative can be a personal, self-aggrandizing expression]
- "[He] could be painting a narrative" (Berit Berger, on MSNBC, February 15, 2019) [Narrative here becomes an expressive art form, painting]
- "[It is] a narrative that this administration has tapped into" (Michael Steele, MSNBC, February 15, 2019) [Narrative may exist within the minds of the followers/population and be used by political leaders]
- "[He] has created a narrative about bad people coming over the border" (Ali Velshi, on MSNBC, February 17, 2019) [Narrative is one interpretation of events or situations]
- "I'm not interested in one narrative against another. I'm interested in the truth" (Tulsi Gabbard, on MSNBC, February 20, 2019) [Narrative is something different from truth]
- "… a narrative… a version of facts that didn't exist" (Andrew McCabe, on MSNBC, February 20, 2019) [Narrative is interpretation of facts, not an expression of facts themselves]
- "That narrative is easy to fall into" (Zerlina Maxwell, MSNBC, February 24, 2019) [Narrative is set out as an effort to manipulate, a trap]
- "It's a way to control the media narrative" (MSNBC, February 24, 2019) [Both leaders and the media may advance narratives, supporting or opposing each other]
- "I touted the Trump narrative for over a decade" (Michael Cohen, Congressional hearing, February 27, 2019) [Narrative is one specific story among possible stories].

At times, leaders who construct a narrative may seek to "change the narrative" because of challenges, shifting circumstances, or "pushback" from a significant opposition group. In that case, the narrative may be amended, altered substantially, or perhaps even denied. Once the followers/population have found a leader's narrative meaningful and consistent with their personal narratives, they may give the leader a degree of flexibility that allows the leader to shift the narrative, with that new narrative becoming legitimate in its own right. Sometimes the narrative shift is referred to as "moving the goalposts," a unique football-related metaphor that some people may accept as a change, while others may oppose it, depending on their "team" affiliation.

Changing the narrative may occur at will for political purposes, as if a given narrative becomes too constraining or too restrictive for effective messaging. Though some people may try to hold the leader accountable to an embedded narrative and ask for consistency, a charismatic leader may be able to change the narrative to suit personal or political purposes. This is equivalent to a novel, similar to Cortázar's

Hopscotch (1987), where the narrative can be read in different ways and different orders, or to Beckett's *Unnamable* (2009), which continuously begins again and revises itself.

10.1.4   Language Related to Narrative

Though "narrative" may be a dominant way of discussing what political leaders (and the media) use to express meaningful messages, there are related expressions that could be described as part of the same messaging system. Some legal language, for example, has insinuated itself into political discourse and become part of this system. Messages are often "litigated," "re-litigated," or "adjudicated." "Messaging" itself is sometimes used synonymously with "narrative" and, like "narrative," is conceived as in opposition to facts and "truth": "I'll give you the facts and then the messaging" (Garret Haake, MSNBC, January 16, 2019).

At times, narrative is reduced to its basic components: "words." As Trump has argued in terms of the "chyrons"—the narrative of words that scroll across the bottom of TV monitors, perhaps in support of or in opposition to the "optics" we see on the screen—words, in fact, do matter: "It's those words, those sometimes beautiful, sometimes nasty words that matter" (Russo 2019). Furthermore, chyrons (narratives), in conjunction with the visual images above them (optics), are together a concise representation of what we propose in this chapter—both the narrative and the optics are important today in conveying a message. A chyron demonstrates that the two may be present simultaneously, though chyrons may or may not relate to the visual images we see in the news, and chyrons may be a message from a leader or from the network itself. Nevertheless, chyrons may serve as evidence to support the ultimate conclusion to be drawn from this chapter: While the narrative may matter, the optics (the immediacy of visual images) seems to capture most of our attention and to diminish the narrative.

10.1.5   Optics

In contrast to narrative, grounded in language and the amount of time it takes to hear or read language, optics may provide a far more immediate, visceral, and engaging means of communicating today, especially for political leaders who are often highly sensitive to image management. We live in an age in which text seems to be dwindling in length and impact, giving way to shorter and shorter forms and even to abbreviated formats. Long speeches and documents are rapidly being replaced by far shorter "texts," tweets, and even emoticons that convey messages in succinct visual fashion. Even email messaging, still highly prevalent in 2019, is losing significant ground to Twitter (with limitations on the number of characters per message), Facebook (with wide use of images and videos), Instagram (with photos and videos), and other social media that reflect a culture which values immediacy, visual as opposed to verbal messaging, and time-saving devices that obviate centuries of reliance on books and book repositories (libraries).

"Optics," both the word and the visual messaging it represents, is used increasingly by the media, by commentators, and by political leaders themselves. Optics can be a kind of messaging that political leaders, well aware of the power and immediacy of visual stimuli, use intentionally to influence followers/general public, or it can be a form of unintentional messaging constructed through images media have captured and/or edited and then present to the public. Furthermore, optics can be perceived as positive or negative, supportive of a leader and the message or undermining that message. It may be contrived by the leader and followers or captured spontaneously by the media. Further, it may include a wide variety of circumstances (e.g., public appearances, informal gatherings, interviews, meetings, hearings, or a round of golf), a variety of venues (e.g., NATO headquarters, the White House, a stadium, or a golf course), and a variety of people (e.g., members of the media, supportive followers, hecklers, or family members).

The word "optics" evolved over centuries with the development of lenses and various theories of light and vision. It derives from the Greek term for "appearance, look." In the 20th century it became part of scientific discourse: "wave optics" and "quantum optics." Zimmer (2010) suggests that the use of the term in politics first emerged in Canada, where the French *optique* can mean "perspective or point of view" as well as the science of optics; thus, it assumed a decidedly political connotation. He argues that it gives "a scientific-sounding gloss to P.R. and image making." The Oxford English Dictionary (OED.com) offers the following definition for this contemporary use of the word: "the way in which a situation, event, or course of action is perceived by the public. Freq. in political contexts." Burkeman (2012) disparages the term and asserts, as we have for the word "narrative," that it stands in stark contrast to "reality," at least as used in the United States: "Optics is not just ghastly jargon coined by D.C. insiders. It also unwittingly describes politics' disconnect from people's reality. ... [It] is about impressions, appearance, the way someone might interpret what is seen. In that, *optics* isn't necessarily about facts."

To be sure, "optics," like "narrative," has taken on negative connotations in the area of political leadership, most specifically as "bad optics." It is in this sense that Western media have described the seemingly friendly visual encounters between Trump and Russian President Vladimir Putin or between Trump and Korean Supreme Leader Kim Jong-un. Despite its negative connotations at times, "optics" remains a powerfully embedded term in current media discourse on political leadership. It is used to describe situations political leaders may intentionally create and cultivate (e.g., Trump at rallies) or situations in which leaders are, quite ironically, thrust into circumstances that provide a contrast to what they may desire (e.g., Trump with Democratic House Speaker Pelosi (D-Calif.) looming behind him at the 2019 "State of the Union Address"). Needless to say, the media may shape the optics for leaders by framing them in either supportive or undermining ways.

10.1.6   The Use of "Optics"

We have selected five recent examples (again in chronological order) of the use of "optics" to demonstrate how visual depictions of leaders can be positive or negative, or cultivated or haphazard:

- "It's the optics as opposed to the real injuries to people" (Heidi Heitkamp, on MSNBC, January 23, 2019) [What one sees in the media is set in contrast to reality and facts]
- "The optics are going to be cuckoo for Tuesday night's State of the Union Address" ("Nation Prepares for State of the Union Tweetstorm," *The Boston Globe*, February 5, 2019) [Optics may precipitate a narrative of tweets, again with the visual presentation of Pelosi looking over Trump's shoulder leading to a tweeting narrative by Trump as an attempt to counter the optics]
- "The optics were terrific" (Rick Tyler, on MSNBC, February 9, 2019) [Reference to Elizabeth Warren's announcement of her candidacy for president of the United States in Lawrence, MA, the site of one of the most important strikes in American history, with flags waving and diverse supporters behind her]
- "He understood how the optics are supposed to work for his benefit" (on MSNBC, February 24, 2019) [On Trump's conscious use of optics to support his political agenda]
- "How bad are the optics for the President?" (Alex Witt, on MSNBC, February 24, 2019) [On Trump and his association with dictatorial "leaders"].

In early 2019, optics was perhaps most powerfully in evidence when US Governor Ralph Northam (D-VA) was discovered to have a photo on his medical school yearbook page showing one person in "blackface" and another wearing a Ku Klux Klan (KKK) outfit. That alarming visual was the subject of media attention for many days and created considerable controversy in Virginia politics and throughout the United States. Similarly, a photo of Trump with porn star Stormy Daniels has added fuel to the many controversies about his leadership. On the more positive side, media coverage of US Senator Amy Klobuchar (D-MN) announcing her candidacy for president of the United States in a severe Minnesota snowstorm suggested to many people that she who hails from America's "heartland" is indeed a leader who can face and overcome many challenges.

10.1.7   Language Related to Optics

Since optics involves visual expression that tends to be immediate and may be broadcast to millions of viewers worldwide, it provides a highly powerful intended or unintended message. At its most powerful level, it can explode in "viral moments" that are disseminated across multiple networks and social media and may radically alter the standing of a current or aspiring political leader in a very short time. Such was the case, for example, of Trump telling an interviewer in a short video that he could "grab 'em [women] by the p****. You can do anything." Surprisingly, Trump survived that video by distracting attention from it with the debacle of former US presidential hopeful Hillary Clinton's emails, an example where narrative deflected optics.

But humans tend to rely on vision and visual stimuli. As Wilson (2012) has argued,

> … early prehuman primates … came to depend more and more on vision and less on smell than did most other mammals. They acquired large eyes with color vision, which were placed forward on the head to give binocular vision and a better sense of depth. (pp. 23–24)

Many expressions used by the media and commentators today when discussing political leadership demonstrate this tendency to emphasize the visual:

- *Look*: Many media commentators as well as leaders initiate their remarks or narratives with "look," a nearly unconscious expression which suggests that interlocutors are called upon to pay attention to the narrative that follows rather than simply hear it. Commentators or leaders use "listen" infrequently.
- *Paint*: Narratives are sometimes described as "painted," with the leader "painting" a visual picture through the messaging and thus creating a subtle shift from narrative into optics. One commentator put it this way recently:
  "We're seeing a narrative being painted …" (on MSNBC, January 26, 2019), with the emphasis on the visual seeming to subsume the verbal.

Such expressions convey an immediacy that is underscored by other expressions which seek to shorten time or even eliminate time altogether, as if time—the narrative mode—can be transformed into unadorned and instant messaging—optics. The following eight expressions can be useful in gaining a better understanding:

- *at the end of the day*: This expression provides a quick picture that sums up the entire day or an entire period of time.
- *plain and simple"* or *"pure and simple"*: This expression demonstrates a desire to elude narrative elaboration or the need for rhetorical flourishes.
- *weaponizing*: This expression offers a message that has the power, immediacy, and explosiveness of a bullet or a bomb.
- *in real time*: This expression indicates delivery that is immediate or concurrent with an event.
- *the fact of the matter*: This expression doubles an attempt to provide the real, empirical, materialistic, immediate perspective that seeks to move beyond an extended, non-factual narrative.
- *period!*: This expression urges us to understand that a narrative can be summed up and decided ultimately by the immediacy of a conclusive, final punctuation mark. It reflects the same teleological perspective as "at the end of the day."
- *the crux of the issue*: This expression suggests that complex discussions (narratives) can be reduced to a central, summary point.
- *calculus*: This expression suggests that a narrative expression can be reduced to a more immediate, mathematical expression.

It is imperative that leaders and the media convey messages quickly and succinctly today, moving in some ways beyond a language—narrative—that may be inadequate to what they want to convey. Though they may need to rely on language as we all do at this point in human evolution, they may seek to use messaging that transcends time-bound words and provides images—optics—which express entire narratives. Even if they must resort to a narrative, they can make use of expressions noted above—expressions that make use of words but attempt to bypass words.

The emphasis on immediacy in optics is echoed in society at large today with phone receptionists, who are famous for using expressions such as "Give me a quick second to check" or "Can I put you on a brief hold?" These expressions underlie the ever-increasing desire to communicate rapidly and without too many of the constraints of language, a desire that may be especially pronounced among political leaders and the media that cover political leaders.

10.1.8   Twenty Thoughts for Consideration

- Narratives and optics are often seen as fabricated and divorced from facts, reality, and "truth," though many followers may accept their messages without question.
- Narratives and optics are forms of communication subject to multiple interpretation (i.e., points of view, conflict, and disagreement) by various stakeholders: the media and commentators as well as followers and the general population.
- Narratives and optics may convey leaders' carefully constructed, purposeful messages, but they may also convey unintended, ironic, or ambiguous messages.
- Narratives and optics involve two-way communication that includes the leaders, on the one hand, and followers and the general population, on the other hand. Both groups are dynamic and interactive, with leaders influencing the followers/population and the followers/population influencing the leaders.
- Narratives, as expressed by leaders, often reflect what lies within the followers/population. As such, leadership may flow from the followers/population as much as it does the other way around. Polling is crucial in this regard.
- Narratives aspire to transcend words and become non-verbal optics in many ways.
- Narratives may involve layered interpretation that includes the leader, the media, and the followers/population, as well as multiple news outlets and multiple social media platforms.
- Narratives are far more important for the meaning they express than for whatever "truth" they may convey or claim to convey. Likewise, the way leaders construct "reality" in their messages may be far more important than any empirical "reality."
- Narratives can be shifted or changed over time, and optics can be adjusted to promote revised messaging.
- Optics, through repetition and patterning, may serve as a non-verbal form of narrative that is created over time.
- Optics have become increasingly important for people who rely upon visual input as a result of their evolutionary development.

- Optics may be far more compelling and effective in today's world of diminishing text and the increasing visual media and social media.
- Political leaders, recognizing the power of the media to present their messages and to edit and interpret them, may seek to cultivate special relationships with specific media outlets.
- Political leaders must use both compelling, succinct narratives and carefully created optics to communicate with followers and the general population.
- Political leaders may attack opposing narratives and opposing optics, thereby creating additional narratives and optics.
- Political leaders can manipulate the media and their followers through the use of strategically designed narratives and optics.
- Political leaders need to establish credibility and trust to make effective use of narratives and optics.
- The media play a significant role in interpreting and presenting narratives and optics, and in shaping them for specific followers and the general population.
- The followers/population may interpret narratives one way (or be encouraged to do so by the leader), whereas the media may interpret those same narratives in a significantly different way, creating a tension that may persist over time.
- Much of the language used in political discourse today seems to strive for the kind of immediacy available through optics.

10.1.9   Where Are We Headed? The Possible Future of Leadership and Communication

We have suggested that non-verbal communication may loom as increasingly important in the future and that optics—or perhaps a more positive way of framing visual communication than the by-now-baggage-heavy term "optics"—may further diminish the need for narratives and verbal communication.

In *The Social Conquest of Earth*, Wilson (2012) argues that humans have evolved to "become the experts at mind reading …. We express our intentions as appropriate to the moment and read those of others brilliantly…. From infancy we are predisposed to read the intention of others…. [We] acquired a 'theory of mind,' the recognition that [our] own mental states would be shared by others" (pp. 226–228). In *Consilience: The Unity of Knowledge*, Wilson (1998) hypothesizes a new kind of language that leaves verbal language far behind:

> The observer reads the script unfolding not as ink on paper but as electric patterns in live tissue. At least some of the thinker's subjective experience – his feeling – is transferred. The observer reflects, he laughs or weeps. And from his own mind patterns he is able to transmit the subjective responses back. The two brains are linked by perception of brain activity …. The communicants can perform feats that resemble extrasensory perception (ESP) …. The first thinker reads a novel; the second thinker follows the narrative. (p. 129)

In *The Origins of Creativity*, Wilson (2017) argues for the need to "escape the bubble in which the unaided human sensory world remains unnecessarily trapped" (p. 92).

The next great revolution in communication, by leaders or others, may resemble the extrasensory perception (ESP) to which Wilson refers, something that takes us beyond text (narratives) and even beyond our heavy reliance on visual input (optics). As he and others (see Ramachandran 2012; Eagleman 2012) have suggested, we are complex connecting animals who have developed, through the use of our complex connecting brains and their billions of constantly connecting neurons, intricate non-verbal means of communicating with each other. Through the use of "mirror neurons," for example, we are capable of "reading" the behaviors of others and understanding the actions and intentions of others. In processes that closely resemble "empathy," mirror neurons help us develop a "language" unlike anything related to traditional languages and unlike the basic optical data we process with our eyes.

Research on mirror neurons in neuroscience, neurophysiology, cognitive science, and cognitive psychology (see Ramachandran and Blakeslee 1997; Pineda 2009) is in its initial stages. But such research could potentially uncover exciting pathways about how humans could communicate with each other in ways far subtler than language (narratives) or visual images (optics). We seem to be moving away from the use of narratives over thousands of years (spoken first and then written later) to an increasing focus on visual forms of communication such as photos and videos, most of which provide textless messaging. In the future, it may be possible—though this is controversial—to develop through what philosophers (see Doherty 2008; Wellman 2014) have called "theory of mind" or the capacity to infer others' thoughts, beliefs, desires, and mental states, and to convey our own in turn.

Clearly, further research on the human brain and nonlinguistic interpersonal communication may lead in the coming decades to a communication revolution that will take us beyond our long reliance on speaking and writing, and beyond language as we know it. Given the alacrity of communication changes we have experienced in the past century—from wire-bound to wireless, print to electronic media, email to texts and tweets, language-bound to visual modes—we are certainly poised for many new changes to come. Those changes will undoubtedly affect leaders, the media, the followers (people), and the planet.

Humans might more appropriately be called "*homo connectans*" than "*homo sapiens*". We are a connecting species—our brains constantly firing with billions of neuronal connections like a physiological/neurological embodiment of our need to connect with each other and with everything we experience in our lives. Put another way, we have an intense desire, built in through thousands of years of evolution, to exist in *community*, and central to such an existence is our need to *communicate*. Over time, we will almost certainly refine our current modes of communication and develop new ones—still grounded in language or perhaps moving beyond it—that are more immediate and powerful, as well as more honest and transparent. The leaders of the future, currently limited by narratives (language) and optics (visual imagery) and by the fabrications those concepts seem to involve, could find themselves in a new era of communication where their very intentions, thoughts, and behaviors would be laid bare for all to know. In such an era, leadership might be completely accountable and responsible, or, if everything is open and available and shared, perhaps leadership—all too often still understood as what individuals express and do—would be unnecessary and truly the kind of interactive, collaborative endeavor that practitioners, scholars, and educators have called for in the past fifty years.

Bibliography

Barthes, R. (2013). *Mythologies*. New York: Hill & Wang.
Beck, J. (2015, August 10). Life's stories. *The Atlantic Daily* (available online).
Beckett, S. (2009). *Three novels: Molloy, Malone Dies, The Unnamable*. New York: Grove Press.
Bennis, W., & Nanus, B. (1985). *Leaders: The strategies for taking charge*. New York: Harper & Row.
Burkeman, O. (2012, August 28). Why the 2012 campaign "optics" really don't look good. *The Guardian* (available online).
Clark, R. P. (2012, February 20). How "narrative" moved from literature to politics. *Poynter* (available online).
Cortázar, J. (1987). *Hopscotch*. New York: Pantheon.
De Man, P. (1983). *Blindness and insight*. Minneapolis, MN: University of Minnesota Press.
Derrida, J. (2016). *Of grammatology*. Baltimore, MD: Johns Hopkins University Press.
Doherty, M. J. (2008). *Theory of mind*. Abingdon-on-Thames, UK: Routledge.
Eagleman, D. (2012). *Incognito: The secret lives of the brain*. New York: Vintage.
Ellerton, P. (2016, October 9). Post-truth politics and the US election: Why the narrative trumps the facts. *The Conversation* (available online).
Ewing, T. (2016, May). Narrative in politics. *The Library* (*available online*).
Fish, S. (1982). *Is there a text in this class?*. Cambridge, MA: Harvard University Press.
Flannagan, O. (1992). *Consciousness reconsidered*. Cambridge: MIT Press.
Homer. (8th BC/2011). *Iliad & Odyssey* (Trans. S. Butler). San Diego, CA: Canterbury Classics.
Lakoff, G., & Johnson, M. (2003). *Metaphors we live by*. Chicago, IL: University of Chicago Press.
Mayer, F. (2014). *Narrative politics: Stories and collective action*. London: Oxford University Press.
Meindl, J. R., & Ehrlich, S. B. (1987). The romance of leadership and the evaluation of organizational performance. *Academy of Management Journal, 30*(1), 91–109.
Pineda, J. A. (2009). *Mirror neuron systems: The role of mirroring processes in social cognition*. New York: Humana Press.
Ramachandran, V. S. (2012). *The tell-tale brain: A neuroscientist's quest for what makes us human*. New York: W. W. Norton.
Ramachandran, V. S., & Blakeslee, S. (1997). *Phantoms in the brain: Probing the Mysteries of the Human Mind*. New York: William Morrow.
Russo, A. (2019, January 21). Trump harbors huge obsession with chyrons, ex-staffer says. *HuffPost* (available online).
Stockley, N. (2011, May 27). Political narratives—A few basics. *Personal Blog* (available online).
Tolchard, H. (2017, December). The power in a political narrative. *Berkeley Political Review* (available online).
Wellman, H. M. (2014). *Making minds: How theory of mind develops*. London: Oxford University Press.
Wilson, E. O. (1998). *Consilience: The unity of knowledge*. Boston: Vintage Books.
Wilson, E. O. (2012). *The social conquest of earth*. New York: Liveright Publishing Corp.
Wilson, E. O. (2017). *The origins of creativity*. New York: Liveright Publishing Corp.
Zimmer, B. (2010, March 4). Optics. *The New York Times Magazine* (available online).

Dr. William Howe (Ph.D. Stanford) is a learner-centered educator and an independent researcher. He has taught at and was a dean and vice-president at several universities in the United States. He has served on various committees and commissions on academic quality and research. Dr. Howe publishes on leadership and leadership education.

Dr. Joseph C. Santora (EdD Fordham) is at Hult International Business School, Ashridge Executive Campus, Berkhamsted, UK. He has taught at various business schools throughout the world and has been a consultant for nonprofits and corporations. Areas of research interest include leadership, executive succession, nonprofits, family business, and coaching.

Chapter 11
Innovation, Leadership and Communication Intelligence

Ian C. Woodward and Samah Shaffakat

Abstract In this chapter, we consider the relationship between effective leadership, communication, innovation and creativity within organizations and teams. In a dynamic business world where innovation is a critical driver for competiveness and growth, we argue that closing the gap between ineffective and effective leadership and communication approaches matters. To assist, we provide two interrelated "tools" that can improve effective leadership communication practices at every stage of the innovation cycle—from ideation through to implementation. These lead to clear, open and compelling communication interactions that underpin innovation and engagement at inter and intra—organizational levels. Our focus is on increasing the chances of successful innovation outcomes by using effective leadership and communication approaches, combined with "communication intelligence" and "fair process".

11.1 Introduction

Why do highly innovative companies like Apple, Google and Gore expend considerable effort and resources communicating the value and utility of their products and services? A simple answer—because by communicating effectively, they capture the minds and hearts of their customers. These firms emphasize effective communication that is clear, open and compelling inside their businesses.

Effective communication means achieving the desired outcomes and objectives of communication exchanges, in a specific context or situation, that leads to shared understanding and satisfaction for the participants in those exchanges (Woodward et al. 2016). Effective communication underpins every phase of successful innovation: from tapping innovation resources and investment; through the ideation process; through implementing change; to marketing the innovative products and services produced; and engaging in interactive customer feedback. On the other hand, ineffective communication or miscommunication cuts idea generation short, confuses creative exploration, wastes resources and effort, contributes to implementation disasters,

I. C. Woodward (✉) · S. Shaffakat
INSEAD, Fontainebleau, France

© Springer Nature Switzerland AG 2020   145
N. Pfeffermann (ed.), *New Leadership in Strategy and Communication*,
https://doi.org/10.1007/978-3-030-19681-3_11

and demotivates or disengages people. A key challenge for leading people towards innovation outcomes is how to close the gap between ineffective and effective communication interactions for all those involved.

Therefore, in this chapter, we discuss the relationship between innovation, creativity, leadership and effective communication. We examine the research on leadership and communication approaches as well as the behaviors and language that facilitate, engage and mobilize innovation, creativity and collaboration within organizations and teams. We then present two interrelated "tools" that will substantially improve the potential for effective leadership communication in practice, in every phase of innovation. The fundamental "tool" is building "communication intelligence" amongst innovation stakeholders, to increase interaction that is clear, open and compelling. In addition, we recommend adopting the '*INVOLVE*'—"fair process" leadership communication practices across the various innovation phases to positively motivate and engage people.

11.2 Innovation, Creativity and Work

Knowledge and information are two key constituents of dynamic innovation and change (Pfeffermann 2011). The focus on innovation and creativity is critical in a modern business world where organizations are under continuous pressure to perform and deal with the paradigm shift of knowledge work in a digital age (Dalkir 2013; Drucker 2009). To maintain competitive edge, organizations must meet this shift (Mayfield and Mayfield 2004), find efficient ways to promote innovation at different levels (Mayfield and Mayfield 2008), and understand the nature, opportunities and threats of disruption (D'Aveni 1999).

Moreover, to create and maintain continuous innovation flow, employees need to be motivated to innovate, and have the skills and capacity to do so (Mayfield and Mayfield 2008). Employee creativity lays the groundwork for organizational innovation (Oldham and Cummings 1996). Creative employees bring forth solutions to problems, defend their ideas and provide an action plan for how to put these ideas into practice (Gumusluoglu and Ilsev 2009a, b). Innovation is core to many established management approaches such as total quality management (Osayawe Ehigie and Clement Akpan 2004), Kaizen (Imai 1986), and organizational learning (Senge 1990).

By its nature, innovation requires 'out of the box thinking', doing new things or doing old things in new ways. This includes introducing novel and better ways of carrying out work tasks (West et al. 2003). Zaltman et al. (1973) see innovation as any "idea, practice, or material artifact" taken up by an individual, group, or organization in order to bring about change. From this perspective, types of innovation will differ depending on the *level* of focus (Amabile et al. 1996; Mayfield and Mayfield 2008). At the organizational level, innovation concerns the domains of strategy, structure, organizational processes, and new market, product or service selection. Innovation at the group level can include designing and creating new products, processes and administrative routines. Individual level innovation includes idea generation, process effectiveness, and improving individual work (Stoker et al. 2001). At the leadership level, it can include creating, articulating and persuading about innovative ideas or initiatives (Elkins and Keller 2003; Jung et al. 2003).

Creativity, the basis of innovation, is sometimes assumed to be limited to certain areas or professions such as sciences or arts (Mumford et al. 1997). However, creativity is required in any jobs with tasks that pose complicated, unclear problems where effective performance is contingent on developing new and effective solutions (Ford 2000; Mumford and Gustafson 1988). This is the setting of modern business, where leaders and their teams need to deal with largely "adaptive" problems—problems that have no straightforward solution or quick fixes available. Solving adaptive problems requires innovation—and a transformation in beliefs, ideologies, values and ways of working (Heifetz 1994).

Creative work, in general, comprises processes to both produce ideas, and to implement them (Vincent et al. 2002). The idea generation process (ideation) includes: defining a problem, collecting information, conceptual structure generation and combing these concepts to form a new category (Mumford et al. 2003). For idea execution, the key processes include: idea evaluation, vetting and testing, and formulating and designing a plan (Lonergan et al. 2004). Creative work revolves around individuals who must actively look for, and manipulate knowledge and concepts (Byrne et al. 2009). This requires expertise, and years of experience (Qin and Simon 1990; Weisberg 1999). For a successful solution, problems will need expertise from several areas in various forms, which makes creative work collaborative (with communication exchanges), as well as individually focused (Cagliano et al. 2000).

Creative work constitutes several stages, consuming time and energy. For example, successful idea execution requires continuous effort supported by a good amount of intrinsic motivation (Collins and Amabile 1999). Creative work demands organizational resources along with the time and commitment of several people and groups. As such, politics and persuasion are likely to come into play to secure resources for successful project completion (Dudek and Hall 1991). As innovation moves into implementation, leaders must also focus attention on "active monitoring and tailoring the plan" to cope with the inevitable challenges faced in in the field (Byrne et al. 2009, p. 264).

Furthermore, creative work entails risk (Mumford et al. 2002), as the idea might not be generated at all or might not be sound enough, and as such the resulting product might not fit the market need (Cardinal and Hatfield 2000). Therefore, context plays a huge role and the leadership of creative efforts should consider not just the organizational strategy but socio-technical aspects as well (Byrne et al. 2009). Cultivating an innovative environment fosters risk-taking and provides an opportunity to employ more creative techniques in the workplace (Gumusluoglu and Ilsev 2009a, b), which requires trust. Perceptions of trustworthy and engaging management are enhanced by leadership communication openness or transparency (e.g. Butler 1991; McCauley and Kuhnert 1992). To succeed, innovation and creative work activities require effective leadership approaches and effective communication interactions at every point.

11.3   Innovation, Leadership Approaches and Communication

Much research on leadership and innovation examines the results of leader behaviors on outcomes such as effectiveness or efficiency rather than the innovation outcomes (De Jong and Den Hartog 2007). A special issue of *The Leadership Quarterly* (2004, Vol. 15, No. 1)—'leading for innovation' examined the creative efforts of leaders. Mumford and Licuanan (2004) summed up the research by emphasizing the various communication roles played by leaders, such as facilitating problem definition, as well as encouraging open discussion on different concepts or ideas, that also allows followers to understand the source and meaning of these. They highlighted that many traditional leadership approaches might not fully fit the innovative leadership required into the future.

Leadership support and guidance are critical in facilitating innovation at the early stage, as these enable successful team processes (Tannenbaum et al. 1996). The skills, attitudes and knowledge of a leader affect group climate and norms (Hackman 2002) and through monitoring, coaching and feedback, a leader creates a supportive environment, which helps the team to innovate (West et al. 2003) and perform successfully (McIntyre and Salas 1995).

Zaccaro et al. (2001) highlight a series of factors essential for team success, and they see leadership as the most critical. The degree to which the leader draws team objectives, and organizes and manages the team to make sure that these objectives are attained, adds significantly to team innovation (West et al. 2003). According to Yukl et al. (1990), leaders who clearly communicate instructions, such as deadlines, standards and priorities, were more successful in leading innovative teams.

In examining the research on leaders of creative efforts, Byrne et al. (2009) noted that leaders are likely to structure the work environment "by creating groupings of technical expertise"(p 259), and promoting effective communication between groups operating in a flat structure. They also, however, note the value of leadership coordination to assist the actors in the creative process. Earlier research noted the role of leaders structuring activities as well as fostering teams of diverse people who communicate effectively with one another (Mumford et al. 2007). The positive relation between innovation and effective, engaging leadership is confirmed by a number of studies in R&D settings (e.g., Keller 1992; Waldman and Atwater 1994).

Interestingly, Bel et al. (2015) examined the interplay between communication, leadership styles and the probability of successful innovation, and found a positive link between innovation and firm size, regular communication and result-oriented leadership. However, they also found that although organizations require "both strong leadership and sufficient communication to overcome inertia; frequent communication – particularly amongst strong managers and in larger firms – can cause leaders to pull the firm in different directions, resulting in disagreement and a failure to successfully innovate" (p. 1). This research suggests that as the organization size increases, it will be essential to achieve coordinated and collaborative communication with more emphasis on the effectiveness, rather than quantity, of communication.

Leaders also have a strong impact on employees' work behaviors, which also includes innovation behaviors (De Jong and Den Hartog 2007). The impact can be direct through identifying and addressing followers' intrinsic motivation essential to creativity (Tierney et al. 1999) or indirect by creating a safe climate for exploring different approaches (Amabile et al. 1996). Innovative behaviors in the workplace depend on interpersonal interaction (Anderson et al. 2004; Zhou and Shalley 2003). As Basadur (2004, p. 103) writes about the most effective leaders of tomorrow as the being the ones who "will help individuals (…) to coordinate and integrate their differing styles through a process of applied creativity that includes continuously discovering and defining new problems, solving those problems and implementing the new solutions." We argue that the leadership approaches deployed and underpinned by effective communication including appropriate language are crucial for innovation outcomes. So, what kinds approaches are likely to contribute to this?

11.3.1 Organic, Transformational and Charismatic Leadership

One contemporary leadership approach related to collective innovation is the so-called "organic" paradigm (Avery 2004). These are "leaderless" or "leaderful" organizations or teams, where leadership may not be "vested in" a single individual (p. 63); and leadership roles and tasks might shift amongst different people as teams self manage over time. It involves organizations where motivated people mutually work together and "sense-make" through collaborative communication. An analogy is a jazz quartet where rhythm, melody and harmony flow dynamically through the improvisations of the different players to create a whole creative musical performance, and where the quartet members are shifting and signaling seamlessly between leading and supporting roles.

In organic work settings, effective communication exchange amongst organizational members is extensive to make sense of "rapidly changing circumstances", and to share vision, knowledge and information (Avery 2004, pp. 63–64). Organic leadership is important, as this modern, ad hoc approach is process oriented, enabling people to quickly innovate and adjust in a fast changing business environment while mutually solving adaptive problems. The growth of organic organizations (including matrix or latticed structures) is accelerating, particularly in the technology or entrepreneurial arenas (e.g. see Gore case study by Bell, in Avery 2004). Collaborative and transparent behaviors with active listening are essential for clear, open and compelling communication in these organic situations.

By contrast, the two major leadership approaches that research over time demonstrates as particularly associated with innovation are: transformational and charismatic leadership. Both are built on the assumption of effective leadership communication and interaction.

By intellectually challenging followers, espousing innovation and communicating a strong vision with a clear sense of emotional purpose, transformational leaders nurture a climate where employees are motivated to search for innovative approaches (Ling et al. 2008). Transformational leaders promote creative ideas, based on the "championing role" they adopt (Howell and Higgins 1990). In these settings employees can exceed performance expectations and be stimulated to take on innovative work techniques. Charismatic leadership with people engagement also influences the organizational climate (Koene et al. 2002). Charismatic leaders demonstrate innovative behaviors that deviate from the regular norms, and in doing so, they permit 'out of the box' idea generation for those involved with technological innovations, such as R&D teams (Conger and Kanungo 1987).

There are many research examples of the impact of transformational leadership on creativity and innovation. For example, Sosik et al. (1998) in their research on 36 undergraduate students found that transformational leadership enhanced creativity in a 'group decision support system' context. Research by Howell and Avolio (1992) on 78 managers in a Canadian financial institution found transformational leadership behaviors was positively related to the business-unit performance and this relationship also needed clear leadership support for innovation. Research by Gumusluoglu and Ilsev (2009a, b), shows a positive correlation between transformational leadership, followers' creativity, and organizational innovation also influenced by psychological empowerment, intrinsic motivation, and the perception of support for innovation. Likewise, Chen et al. (2014) in their study of 151 CEOs and matching senior management team members from Chinese Manufacturing firms, show transformational leadership to positively influence product innovation performance (conceptualized as the degree to which a new product or service achieves its market share, sales, investment return and profit objectives).

Although transformational leaders at times can adopt a more directive communication style, they also actively seek followers' involvement by emphasizing the significance of collaboration in performing joint and collective tasks, offering a chance to share and learn, and delegating responsibility to their followers to perform any necessary work to ensure effective performance (Bass 1985). In doing so, they create an empowering environment where followers pursue innovation in their work tasks. Amabile et al. (1996) show that autonomy enables employees to be more creative, as it enables them to believe that they have greater personal authority over how to go about accomplishing their tasks. Empowerment further caters to the intrinsic motivation of followers, which (as already discussed) contributes to innovative behaviors (Jung and Sosik 2002).

11.3.2 Leadership, Motivating Language and Framing

Research studies show that employees' behaviors can be impacted by a leader's conscious use of speech [such as (Mayfield et al. 1995). 'Leader talk' enables leaders to seek, and gain, trust and acknowledgement of subordinates (Reina and Reina 1999).

Appropriate leader language skills in motivating and conveying vision are significant (Goleman et al. 2001). Transformational and charismatic leaders try to engage stakeholders around vision, ideas, innovation and change. Two effective communication attributes for these leadership approaches are the notions of motivating language and communication framing.

The motivating language framework involves classifying leadership communication language into three types: direction-giving language, empathetic language and meaning-making language. Direction-giving language occurs when leaders reduce uncertainty by elucidating roles, performance expectations, goals and responsibilities. Empathetic language happens when leaders go beyond the mere economic exchange between them and their followers, to care for their peoples' emotional well-being. Meaning-making language takes place when leaders convey and express the organizational norms, culture, behaviors, and values that are unique and relevant for each organization building affiliation, supporting change management and organizational socialization which Jablin (2001) refers to as "Entry and Assimilation".

Mayfield and Mayfield (2004) highlight motivating language as part of the innovation relationship between leader and followers, arguing that when leaders strategically communicate through motivating language, follower innovation increases. Direction-giving language, empathetic language and meaning-making language encourage innovation through a combination of, for example: catering to followers' intrinsic motivation and understanding of what the task entails; and delineating reward policies and organizational goals, as well as risk-taking methods (Mayfield and Mayfield 2002). Additionally, this lays the groundwork for leadership training interventions to enhance workers' innovativeness (Zorn and Ruccio 1998) and provides ways to maximize employee outcomes such as satisfaction, performance, retention and so on (Graen et al. 2004; Mayfield et al. 1995).

Fairhurst and Sarr (1996), who view leadership as a 'language game', contend that framing is the most important skill in this game. "Just as an artist works from a palette of colors to paint a picture, the leader who manages meaning works from a vocabulary of words and symbols to help construct a frame in the mind of the listener" (Fairhurst and Sarr 1996, p. 100). They further explain how framing helps leaders to motivate actions and secure backing for their vision (Fairhurst and Sarr 1996; Fairhurst 2011). These visionary leaders frame the purpose in a way that is relevant and meaningful (Conger 1991). As Snow (2004) states, 'Collective action frames, like picture frames, focus attention by punctuating or specifying what in our sensual field is relevant and what is irrelevant, what is "in frame" and what is "out of frame", in relation to the object of orientation' (p. 384). Particularly important framing methods for encouraging engagement with innovation are metaphors, stories, examples, and catchphrases.

Innovation literature also identifies the importance of framing. For example, Pfeffermann et al. (2008) note, "framing innovation/s for successful commercialization, innovation communication might be an important managerial function; understood as a firm's capital that tends to enhance competitive advantage" (p. 41). Such innovation communication (combining communication and innovation capital) would include framing innovations, "to facilitate the adoption process." (p. 41).

We argue that for innovation and creativity, an effective leader communicator will use a combination of language that motivates, and messages that are framed, to appeal to the innovation interests of participants. This supports effective communication that is clear, open and compelling. A leader's language can give others a sense of direction and logic; harness emotive appeal; and place the rational and emotional basis in the context of why people should be doing or committing to something. Using words and phrases that resonate on these different levels in a relevant way should increase the motivation of stakeholders to engage in the innovation process and outcomes. Equally, getting innovation participants to describe objectives or ideas in their own language and choice of words should increase "ownership". It is not transmitting the message effectively (albeit this is crucial for a leader), but interacting, that builds commitment and understanding. Engaging in dialogue with open and constructive questioning, can clarify the vision, objectives and priorities. Besides, framing the messages using stories and examples should "bring to life" both the objectives of an innovation undertaking, and its potential impact. A resonating catchphrase or motivating language can encapsulate innovation intentions and culture. Just consider Apple's own "Think Different" campaign; the description of Toyota process innovation in "Lean Thinking"; or Elon Musk's quote, "Failure is an option here. If things are not failing, you are not innovating enough."

Our review highlights the need for leaders to encourage innovation and creativity, engage stakeholders in the process, engender commitment to change, and enable an environment of creative work and knowledge sharing. In all, effective communication is a key ingredient for the leadership approaches related to innovation outcomes. Some scholars view communication as an essence of leadership rather than a mere technique (Barge 1994; Macik-Frey 2007). Salacuse (2006, p. 23), maintains, "Indeed, leadership could not exist without communication." In a similar vein, Barge (1994, p. 21) comments "leadership is enacted through communication." Literature is replete with studies that highlight the significance of communication interactions in effective leadership (e.g. Den Hartog and Verburg 1997; Fairhurst 2011; Tourish and Jackson 2008). When leading for innovation, these communication interactions occur in different ways (e.g. speaking, listening, reading, writing, behaving, interpersonal relationships) and through different formats (e.g. face to face, or technology and media). We believe that the leader's role modeling of effective communication matters, and so does the environment for communication interaction and innovation work that they foster in their organization or team.

So, the question arises, how is this achieved and sustained? The answer lies in building on appropriate leadership approaches, behaviors and motivating language or framing, by deploying effective leadership communication that is essentially clear, open and compelling at its core. This will help to close the gap between ineffective and effective communication. To assist with increasing the capacity to do this, we present and recommend two interrelated "tools" that will substantially improve effective leadership communication at every phase of innovation. The fundamental "tool" is building "communication intelligence" amongst innovation stakeholders, especially when role modeled by leaders.

11.4 Innovation Leadership and Communication Intelligence

Effective leadership communication that reverberates with people is both relevant and comprehensible. The communication exchanges should resonate (such as with motivating language, vivid examples or cooperative questioning). These exchanges are interactive, dynamic and contextual. These build trust through the quality of the relationship experiences, and these relationships are stronger when communication contact is inclusive and accessible.

We contend that leaders should utilize their "communication intelligence" to engage stakeholders in innovation and creativity, by demonstrating effective leadership communication that resonates, clarifies, and connects. This will lead to communication that is clear (comprehensible and meaning based), open (inclusive and interactive) and compelling (motivating and relevant).

"Communication Intelligence" (CI) is a model that fully integrates eight elements to achieve effective leadership communication (Woodward 2015). CI combines four mindsets (the things people need to think about for effective communication); and four clusters of communication techniques and qualities (the ways people need to undertake communication activities to be effective). People with high levels of "communication intelligence" use all the mindsets and use techniques from across all four clusters, particularly those that are natural for them. Yet, they increase their communication effectiveness by learning, then using, techniques from other clusters that are less natural for them.

There are four CI Communication Mindsets (the 'what', 'why', 'where' and 'who'): Awareness, Message, Presence, Communication Formats; and Four CI Clusters of Communication Qualities and Techniques (the 'how'): Rational, Structural, Expressive and Visual (see Fig. 11.1). For a leader's communication approach to be completely understood, these eight elements sit underneath a person's cultural background (because different national cultures have unique communication characteristics), and their individual personality trait of extraversion/introversion—as all eight elements are present in all cultures (contextually adjusted); and are found in both extraverts and introverts (Woodward 2015) (see Fig. 11.1).

CI applies across all kinds of communication situations—public, group, interpersonal, and intrapersonal (with self), and are present across the various communication mediums used by leaders (from email to presentations; from personal conversations to team discussions; from blogs to video posts). Combining these CI elements produces effective interaction between leaders and people, both within and outside their organization or setting, and generates a platform for achieving relevant meaning, connection and results. It also underscores the multiplicity and complexity of effective leadership communication interactions—especially for dealing with adaptive and creative processes.

The four CI "mindsets" for leaders, interrelated to innovation and creativity, are:

- **Awareness**: of self, others, context and purpose. This refers to a person's ability to be deeply aware of the communication needs and preference styles of those

Fig. 11.1 Communication
intelligence framework for
leaders. *Source* IC
Woodward (2015)

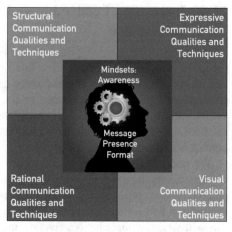

Eight Communication Intelligence Elements:
Four Mindsets and Four Technique Clusters

involved in communication as well as oneself; and also be aware of the situation, context or purpose of the communication activity (e.g. informing, inquiring, influencing, persuading, entertaining, motivating, inspiring, brainstorming). For example, in innovation leadership communication, the "awareness mindset" is reflected in: a desire to involve others, and to take account of the diversity effects (e.g. culture, gender, generational cohort) that would influence the communication interactions (transmission and interpretation); understanding the likely motivators, language and themes that will be relevant to others; and considering cultural context or physical environment when planning activities such as brainstorming or innovation evaluation discussions.

- **Message**: the core meaning and content of the communication, supported by structure and appropriate details, as well as message relevance and clarity. For example, in innovation leadership communication, the "message mindset" is reflected in: clear message framing and exchange amongst stakeholders, striving for comprehension, clarity and relevance; opening up the free flow of ideas and analysis messaging including an appropriate balance of listening, inquiry and advocacy (asking and telling) that leads to constructive dialogue and shared understanding; and innovation vision messages that resonate quickly.

- **Presence**: beyond the words—the nonverbal (e.g. body language and gestures), paraverbal (e.g. sound/tone of voice) and visual, symbolic or expressive features. For innovation leadership communication, the visible presence of leaders and followers during their interactions will influence the engagement and environment for creative and open thinking. This is reflected in, for example, open body postures; respectful and measured, yet expressive voice tone; and visual communication devices (charts and technology collaboration tools) that are stylistically owned by the participants in the innovation process. Moreover, some form of visualization and "personal energy" is an essential ingredient in creative brainstorming and ideation activity.

- **Format**: the choice and use of communication formats, media and repertoire that are "fit for purpose and situation" (e.g. behaviors, spoken, written, listening, thinking/reflection, novels, text, email, instant messaging, video and the like). For example, in innovation leadership communication, the "format mindset" is reflected in continuously adopting or adjusting communication activities, media and technologies that are available, accessible, appropriate and useful for each stage of the innovation process and assist "ease of collaboration".

The four CI clusters of communication techniques and qualities for leaders interrelated to innovation and creativity are:

- **Rational**: techniques and qualities that affect the logic, factuality, knowledge level, intellectual substance, idea clarity and simplicity of language for comprehension in communication. For innovation, "rational" qualities would include: being objective; using verifiable evidence and key facts; suppressing and recognizing bias in thoughts and words (especially when separating idea generation from analysis in the brainstorming and decision stages); clarifying complex ideas and concepts into simple words for understanding; and providing precise summaries of action items and priorities for innovation implementation.
- **Structural**: techniques and qualities that affect the language or sound clarity, consistency, order/flow, construction, thoroughness, levels of detail and accuracy in communication. For innovation, "structural" qualities would include: methods for agenda setting, organizing, disseminating and exchanging information; discussion preparation; sequencing participative debate; utilizing deliberate "unstructured" times for communication exchange to allow free-low dialogue and openness without power control; using rhetorical tools such as "catchphrase", repetition and triads (lists in three for summation); and ensuring innovation implementation plans are appropriately and accurately documented.
- **Expressive**: qualities that affect the expression, emotion, interactivity, personalization and authenticity of communication. In innovation leadership communication, "expressive" qualities would include: storytelling; using inspiring and motivating language; displaying appropriate expressive nonverbal and paraverbal communication (such as body gestures and voice tone) in support of ideas; active listening (where mind, verbal and non-verbal communication are focused); demonstrating personal commitment and enthusiasm; and exhibiting behaviors engendering a sense of trustworthiness, risk-taking, openness and collaboration.
- **Visual**: qualities that affect the appearance, visuality, conceptuality, creativity and symbolism of communication. In innovation leadership communication, "visual" qualities would include: active idea generation; producing graphical, design or visual representations of ideas and messages; demonstrating future facing messaging to allow people to imagine success after problem solving and ideation; and articulating the 'big picture'. [Adapted from Woodward (2015) and Woodward et al. (2016)].

Combining the CI elements creates clear, open and compelling communication. This is valuable for all leadership approaches to engage people with innovation and to encourage everyone to have "communication intelligent" interactions. One additional "tool" will help to increase involvement levels in innovation processes, as well as build innovation capability and confidence over time. This is '*INVOLVE*'—the "fair process" leadership communication practices, which are usable across different phases of innovation engagement.

11.5   INVOLVE—Fair Process Leadership and Communication

Our earlier discussion on leadership approaches raised the fundamental issue of successfully managing processes for innovation and creativity. "Fair process leadership" (FPL) is one framework that does this successfully at individual, team and organizational levels. FPL is an integrative framework that supports effective leadership, particularly in situations emphasizing process engagement and transparency with stakeholders, as well as objective evaluation (Van Der Heyden 2013). These characteristics are important for innovation and creativity cultures, as "fair process" promises a high level of commitment and trust, which are necessary ingredients for high performance (Kim and Mauborgne 1997; Van Der Heyden and Limberg 2007).

In simple terms, "fair process" exists where the participants in any decision-making process understand the process that will be followed, as well as the associated rules and modes of engagement and communication, and perceives these to be fair with respect to all participants. Fair process principles provide a "means", rather than an "end", towards more engaged decision-making and improved implementation (Van Der Heyden and Limberg 2007), which we contend are essential elements for successful innovation.

Van Der Heyden et al. (2005) developed a "*Fair Process Leadership*" model with *a process description,* consisting of 5 steps (the 5 "*E*"s) and a description of *fair play behaviors* that leaders need to demonstrate throughout these steps (the 5 "*C*"s). This model as further espoused by Van Der Heyden (2013) represents an interlinked cycle for decision-making, implementation and continuous review (see circular model within Fig. 11.2). It can be directly applicable for teams and organizations seeking innovation built on deep engagement and commitment including organic teams. In additional work, Woodward et al. (2016) posited three communication practices to enhance "fair process" as an actionable concept, these are the '*INVOLVE*' practices (also see Fig. 11.2). How might fair process principles translate into leading, working and communicating within innovation processes? We adapt and cite from the relevant research below.

Leaders and teams can adopt the '*INVOLVE*' communication practices to enact "fair process" to facilitate innovation, creativity and engagement. The core principle is a simple and compelling message: '*INVOLVE*'. This is the deep belief and con-

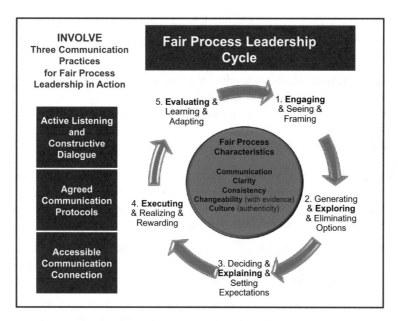

Fig. 11.2 INVOLVE—fair process effective leadership communication. *Adapted* from Van Der Heyden (2013) and Woodward et al. (2016)

viction to encourage effective communication that demonstrates fair process for all those involved. Three specific communication practices put this principle into action at each stage of the innovation cycle:

- **Active Listening and Constructive Dialogue** [productive behaviors for interaction, deliberation, option generation, analysis, decision-making, explanation and evaluation; as well as an appropriate balance of inquiry (asking) and advocacy (telling) with demonstrable active listening];
- **Agreed Communication Protocols** [mutually developed and transparent communication rules with commitment and follow-through that are culturally appropriate—these should not be bureaucratic, but guidelines to coordinate the development and exchange of knowledge as the creative ideas emerge and are taken forward—e.g. the expected rules or ways of behaving for doing unstructured or structured activities, or the expected norms for using technology collaboration systems]; and
- **Accessible Communication Connection** [useful, convenient and readily available communication activities, formats and media to facilitate participation and engagement with internal and external stakeholders—these are the communication format choices that make sense at any point in time, and are subject to change during the innovation cycle to foster collaboration]. (Adapted from Woodward et al. 2016).

The five complementary and mutually reinforcing behavioral characteristics of FPL identified by Van Der Heyden (2013), and to which the '*INVOLVE*' communication practices relate, are:

- **Communication**: the ability to give all actors a "voice" without fear or pressure of retaliation once that "voice" is exercised;
- **Clarity**: the transparency of behaviors, interactions and exchanges by the actors of the process;
- **Consistency**: the uniformity in the treatment of actors, issues, and steps, including over time;
- **Changeability**: the possibility of 'correction', changing actors' beliefs, and also possibly changing the chosen course, as a function of new evidence; and
- **Culture**: the commitment to 'do the fair thing' not just superficially, but deeply and authentically. [Adapted from Van Der Heyden (2013)].

With these behaviors in place, all participants can dynamically enact the five stages of the FPL model (Van Der Heyden 2013) for innovation and creativity—deploying the '*INVOLVE*' practices along the way:

- *Engage*: Establish an innovation process to involve relevant people; seek inputs to framing issues and generating ideas; seek constructive challenge to views; make contributions to the process design and priorities before the decisions are actually made, when influencing this is still a possibility.
- *Explore*: Generate and explore all options and their potential outcomes thoroughly and comprehensively. Allow an open and dynamic ideation process, by not closing options early and keeping idea generation separated from analysis. Then through constructive debate and analysis eliminate those options that are neither promising nor capable of successful implementation, and take forward the most prospective options.
- *Explain*: Make a clear innovation decision, where the leadership (or group in an organic setting) explains its rationale. Effective communication will take sufficient time and energy to develop understanding, especially for those impacted outside the decision-making group. All the innovation participants should be thoroughly briefed, fully committed, and hold clear and compelling messages for stakeholders outside the decision group. Roles, responsibilities and priorities for successful implementation and execution are articulated clearly; and the challenges, expected benefits, rewards, or appropriate sanctions for poor execution are enunciated.
- *Execute*: Ensure all relevant individuals implicated by the innovation decision are clear on what they are supposed to do and their focus for implementation. Adjust and adapt if outcomes are not according to plan, while informing and involving others to sustain coordination in execution; and maintain rewards (or sanctions) in line with expectations formed and announced previously.
- *Evaluate*: Seek critical feedback from relevant stakeholders on the decision, the plan and the process followed to get there; share lessons learnt based on evidence; utilize this knowledge for future innovation process work. [Adapted from Van Der Heyden et al. (2005), Van Der Heyden (2013) and Woodward et al. (2016)].

We argue that when individuals, teams and organizations demonstrate "fair process" and the '*INVOLVE*' practices for innovation processes there will be transparent, respectful, constructive and objective communication behaviors. This is characterized by communication described as: "open and authentic rather than hidden or opaque; inclusive rather than dictating; and clear rather than confused" (Woodward et al. 2016). Such communication encourages creative thought with open minds and comprehensible articulation, that is "communication intelligent", and clear, open and compelling.

11.6 Conclusion

Today's world is volatile, uncertain, complex, ambiguous and diverse (VUCAD). It is intensely competitive, with change as the "increasing constant." Innovation (from process engineering and new products, to technology creation and new business models) is an imperative for contemporary business. In global organizations, there are initiatives to increase learning, share knowledge, and to develop new capabilities for leaders to engage people and ideas. These are intended to deliver positive and dynamic business outcomes (Beechler and Woodward 2009). This is especially so in entrepreneurial and emerging organizations which are seeing rapid growth in the 21st century (Koryak et al. 2015); as well as global business opportunities for technology innovation, transfer and investment (Audretsch et al. 2014). Effective communication interactions are essential to empower these innovation initiatives and exchanges. This applies equally to positional leaders and their followers, and where leadership is dispersed and organic. Leaders are required to champion innovation by: planning, implementing and assessing innovation, shaping and managing various capabilities; and assembling resources at inter and intra—organizational levels (Zerfass et al. 2004). Communicating by inviting and responding to creative ideas is the first step for engaging employees and defining innovation objectives.

As such, we believe that effective leadership communication is evolving in a VUCAD world as an essential ingredient for successful innovation—whether in tapping innovation resources and investment, contributing to ideation, implementing innovation decisions, or interacting with customers. By communicating effectively leaders can increase their ability to nurture innovation, and translate complex innovations in a way that others comprehend, accept and then embrace. In the innovation workspace, leadership communication is embedded in concrete actions, language, processes and interpersonal relations, and in the depth and breadth of idea dialogue amongst "aware" and motivated participants.

We contend that leadership communication built on the two "tools" ("communication intelligence" and the '*INVOLVE*' communication practices of "fair process"); combined with appropriate innovation based leadership approaches, collaborative behaviors, motivating language and framing; can positively inspire and engage people towards innovation outcomes and support innovation cultures. We believe these should make a substantial contribution to closing the gap between ineffective and effective leadership communication for innovation. These are part of the solution to: ensuring ideation is not cut short; increasing the commitment levels to innovation decisions; clarifying the focus and understanding of innovation implementation issues and priorities; improving articulation of the benefits of the innovation for internal and external stakeholders; and learning from evaluation and knowledge exchange that is objective and constructive.

Furthermore, "communication intelligence" with involvement-based "fair process" will encourage trust, risk-taking, creativity, and collaboration. This supports an environment where people are more likely to contribute and commit to the changes, new directions or initiatives; and see these to fruition. This builds capacity and confidence for innovative and creative work into the future. Effective leadership communication for innovation will be framed to appeal to emotion and rationality. It will be replete with relevant messages, visuality, expressive examples, interactive engagement, listening, motivating language, engagement processes, and "communication intelligence". Such innovation communication will be clear, open and compelling.

Bibliography

Amabile, T. M., Conti, R., Coon, H., Lazenby, J., & Herron, M. (1996). Assessing the work environment for creativity. *Academy of Management Journal, 39*(5), 1154–1184.

Anderson, N., De Dreu, C. K., & Nijstad, B. A. (2004). The routinization of innovation research: A constructively critical review of the state-of-the-science. *Journal of Organizational Behavior, 25*(2), 147–173.

Audretsch, D., Lehmann, E., & Wright, M. (2014). Technology transfer in a global economy. *Journal of Technology Transfer.*

Avery, G. C. (2004). *Understanding leadership: Paradigms and cases*. London: Sage Publications.

Barge, J. (1994). Leadership: Communication skills for organizations and groups.

Basadur, M. (2004). Leading others to think innovatively together: Creative leadership. *The Leadership Quarterly, 15*(1), 103–121.

Bass, B. M. (1985). *Leadership and performance beyond expectations*. New York, NY: Free Press.

Beechler, S., & Woodward, I. C. (2009). The global "war for talent". *Journal of International Management, 15*(3), 273–285. https://doi.org/10.1016/j.intman.2009.01.002.

Bel, R., Smirnov, V., & Wait, A. (2015). Team composition, worker effort and welfare. *International Journal of Industrial Organization, 41,* 1–8.

Butler, J. (1991). Toward understanding and measuring conditions of trust: Evolution of a conditions of trust inventory. *Journal of Management.*

Byrne, C. L., Mumford, M. D., Barrett, J. D., & Vessey, W. B. (2009). Examining the leaders of creative efforts: What do they do, and what do they think about? *Creativity and Innovation Management, 18*(4), 256–268.

Cagliano, R., Chiesa, V., & Manzini, R. (2000). Differences and similarities in managing technological collaborations in research, development, and manufacturing: A case study. *Journal of Engineering Technology Management, 17*(2), 193–224.

Cardinal, L. B., & Hatfield, D. E. (2000). Internal knowledge generation: The research laboratory and innovative productivity in the pharmaceutical industry. *Journal of Engineering and Technology Management, 17*(3), 247–271.

Chen, Y., Tang, G., Jin, J., Xie, Q., & Li, J. (2014). CEOs' transformational leadership and product innovation performance: The roles of corporate entrepreneurship and technology orientation. *Journal of Product Innovation Management, 31*(S1), 2–17.

Collins, M. A., & Amabile, T. M. (1999). Motivation and creativity. In R. J. Sternberg (Ed.), *Handbook of creativity* (pp. 297–312). Cambridge, UK: Cambridge University Press.

Conger, J. A. (1991). Inspiring others: The language of leadership. *The Executive, 5*(1), 31–45.

Conger, J. A., & Kanungo, R. N. (1987). Toward a behavioral theory of charismatic leadership in organizational settings. *Academy of Management Review, 12*(4), 637–647.

D'Aveni, R. A. (1999). Strategic supremacy through disruption and dominance. *Sloan Management Review, 40*(3), 127–135.

Dalkir, K. (2013). *Knowledge management in theory and practice*. London, UK: Routledge.

De Jong, J. P. J., & Den Hartog, D. N. (2007). How leaders influence employees' innovative behavior. *Europe Journal of Innovative Management, 10*(1), 41–64.

Den Hartog, D. N., & Verburg, R. M. (1997). Charisma and Rhetoric: Communicative techniques of international business leaders. *The Leadership Quarterly, 8*(4), 355–391.

Drucker, P. F. (2009). *Managing in a time of great change*. Boston, MA: Harvard Business Press.

Dudek, S. Z., & Hall, W. B. (1991). Personality consistency: Eminent architects 25 years later. *Creativity Research Journal, 4*(3), 213–231.

Elkins, T., & Keller, R. T. (2003). Leadership in research and development organizations: A literature review and conceptual framework. *The Leadership Quarterly, 14,* 587–606.

Fairhurst, G. T. (2011). *The power of framing: Creating the language of leadership*. San Francisco, CA: Jossey Bass.

Fairhurst, G. T., & Sarr, R. A. (1996). *The art of framing: Managing the language of leadership*. San Francisco, CA: Jossey-Bass.

Ford, C. M. (2000). Creative developments in creativity theory. *Academy of Management Review, 25*(2), 284–289.

Goleman, D., Boyatzis, R., & McKee, A. (2001). Primal leadership: The hidden driver of great performance. *Harvard Business Review, 79*(11), 42–53.

Graen, G. B., Hui, C., & Gu, Q. (2004). A new approach to intercultural cooperation. In G. B. Graen (Ed.), *New frontiers of leadership* (pp. 225–246). Greenwich, CT: Information Age Publishing.

Gumusluoglu, L., & Ilsev, A. (2009a). Transformational leadership and organizational innovation: The roles of internal and external support for innovation. *Journal of Product Innovation Management, 26*(3), 264–277.

Gumusluoglu, L., & Ilsev, A. (2009b). Transformational leadership, creativity, and organizational innovation. *Journal of Business Research, 62*(4), 461–473.

Hackman, J. R. (2002). *Leading teams: Setting the stage for great performances*. Boston, MA: Harvard Business School Press.

Heifetz, R. A. (1994). *Leadership without easy answers*. Cambridge, MA: Belknap/Harvard University Press.

Howell, J. M., & Avolio, B. J. (1992). The ethics of charismatic leadership: Submission or liberation? *The Executive, 6*(2), 43–52.

Howell, J. M., & Higgins, C. A. (1990). Champions of technological innovation. *Administrative Science Quarterly, 35*(2), 317–341.

Imai, M. (1986). *Kaizen*. New York: Random House Business Division.

Jablin, F. M. (2001). Organizational entry, assimilation, and disengagement/exit. In F. M. Jablin & L. L. Putnam (Eds.), *The new handbook of organizational communication: Advances in theory, research, and methods* (pp. 732–818). Thousand Oaks, CA: Sage Publications.

Jung, D. I., Chow, C., & Wu, A. (2003). The role of transformational leadership in enhancing organizational innovation: Hypotheses and some preliminary findings. *The Leadership Quarterly, 14*(4), 525–544.

Jung, D. I., & Sosik, J. J. (2002). Transformational leadership in work groups the role of empowerment, cohesiveness, and collective-efficacy on perceived group performance. *Small Group Research, 33*(3), 313–336.

Keller, R. T. (1992). Transformational leadership and the performance of research and development project groups. *Journal of Management, 18*(3), 489–501.

Kim, W. C., & Mauborgne, R. (1997). Fair process: Managing in the knowledge economy. *Harvard Business Review, 75*(4), 65–75.

Koene, B. A., Vogelaar, A. L., & Soeters, J. L. (2002). Leadership effects on organizational climate and financial performance: Local leadership effect in chain organizations. *The Leadership Quarterly, 13*(3), 193–215.

Koryak, O., Mole, K., & Lockett, A. (2015). Entrepreneurial leadership, capabilities and firm growth. *International Small.*

Ling, Y. A. N., Simsek, Z., Lubatkin, M. H., & Veiga, J. F. (2008). Transformational leadership's role in promoting corporate entrepreneurship: Examining the CEO–TMT interface. *Academy of Management Journal, 51*(3), 557–576.

Lonergan, D. C., Scott, G. M., & Mumford, M. D. (2004). Evaluative aspects of creative thought: Effects of idea appraisal and revision standards. *Creativity Research Journal, 16,* 231–246.

Macik-Frey, M. (2007). *A communication-centered approach to leadership: The relationship of interpersonal communication competence to transformational leadership and emotional intelligence*. University of Texas at Arlington.

Mayfield, J., & Mayfield, M. (2002). Leader communication strategies: Critical paths to improving employee commitment. *American Business Review, 20*(2), 89–94.

Mayfield, J., Mayfield, M., & Kopf, J. (1995). Motivating language: Exploring theory with scale development. *Journal of Business Communication, 32*(4), 329–344.

Mayfield, M., & Mayfield, J. (2004). The effects of leader communication on worker innovation. *American Business Review, 22*(2), 46–51.

Mayfield, M., & Mayfield, J. (2008). Leadership techniques for nurturing worker garden variety creativity. *Journal of Management Development, 27*(9), 976–986.

McCauley, D., & Kuhnert, K. (1992). A theoretical review and empirical investigation of employee trust in management. *Public Administration Quarterly.*

McIntyre, R. M., & Salas, E. (1995). Measuring and managing for team performance: Emerging principles from complex environments. In R. Guzzo & E. Salas (Eds.), *Team effectiveness and decision making in organizations* (pp. 9–45). San Francisco, CA: Jossey-Bass.

Mumford, M. D., Baughman, W. A., & Sager, C. E. (2003). Picking the right material: Cognition processing skills and their role in creative thought. In M. A. Runco (Ed.), *Critical and creative thinking* (pp. 19–68). Cresskill, NJ: Hampton Press.

Mumford, M. D., Eubanks, D. L., & Murphy, S. T. (2007). Creating the conditions for success: Best practices in leading for innovation. In J. A. Conger & R. E. Riggio (Eds.), *The practice of leadership: Developing the next generation of leaders* (pp. 129–149). San Francisco, CA: Jossey-Bass.

Mumford, M. D., & Gustafson, S. B. (1988). Creativity syndrome: Integration, application, and innovation. *Psychological Bulletin, 103*(1), 27–43.

Mumford, M. D., & Licuanan, B. (2004). Leading for innovation: Conclusions, issues and directions. *The Leadership Quarterly, 15*(1), 163–171.

Mumford, M. D., Scott, G. M., Gaddis, B., & Strange, J. M. (2002). Leading creative people: Orchestrating expertise and relationships. *The Leadership Quarterly, 13*(6), 705–750.

Mumford, M. D., Whetzel, D. C., & Reiter-Palmon, R. (1997). Thinking creatively at work: Organizational influences on creative problem-solving. *Journal of Creative Behavior, 31*(1), 7–17.

Oldham, G. R., & Cummings, A. (1996). Employee creativity: Personal and contextual factors at work. *Academy of Management Journal, 39*(3), 607–634.

Osayawe Ehigie, B., & Clement Akpan, R. (2004). Roles of perceived leadership styles and rewards in the practice of total quality management. *Leadership & Organization Development Journal, 25*(1), 24–40.

Pfeffermann, N. (2011). The scent of innovation: Towards an integrated management concept for visual and scent communication of innovation. In M. Hulsmann & N. Pfeffermann (Eds.), *Strategies and communications for innovations* (pp. 163–181). Berlin, Germany: Springer-Verlag.

Pfeffermann, N., Hüulsmann, M., & Scholz-Reiter, B. (2008). *Framing innovations to grasp stakeholders' attention: A dynamic capability-based conception of innovation communication.*

Qin, Y., & Simon, H. A. (1990). Laboratory replication of the scientific process. *Cognitive Science, 14*(2), 181–312.

Reina, D. S., & Reina, M. L. (1999). *Trust and betrayal in the workplace.* San Francisco, CA: Berrett-Koehler.

Salacuse, J. (2006). *Leading leaders: How to manage smart, talented, rich, and powerful people.* New York, NY: American Management Association.

Senge, P. (1990). *The fifth discipline: The art and practice of the learning organization.* New York: Doubleday.

Snow, D. A. (2004). Framing processes, ideology and discursive fields. In D. A. Snow, S. A. Soule, & H. Kriesi (Eds.), *The Blackwell companion to social movements* (pp. 380–412). Oxford, UK: Blackwell Publishing.

Sosik, J. J., Kahai, S. S., & Avolio, B. J. (1998). Transformational leadership and dimensions of creativity: Motivating idea generation in computer-mediated groups. *Creativity Research Journal, 11*(2), 111–121.

Stoker, J. I., Looise, J. C., Fisscher, O. A. M., & de Jong, R. D. (2001). Leadership and innovation: Relations between leadership, individual characteristics and the functioning of R&D teams. *International Journal of Human Resource Management, 12*(7), 1141–1151.

Tannenbaum, S. I., Salas, E., & Cannon-Bowers, J. A. (1996). Promoting team effectiveness. In M. A. West (Ed.), *Handbook of work group psychology* (pp. 503–529). Sussex, UK: Wiley.

Tierney, P., Farmer, S. M., & Graen, G. B. (1999). An examination of leadership and employee creativity: The relevance of traits and relationships. *Personnel Psychology, 52*(3), 591–620.

Tourish, D., & Jackson, B. (2008). Guest editorial: Communication and leadership: An open invitation to engage. *Leadership, 4*(3), 219–225.

Van Der Heyden, L. (2013). Setting a tone of fairness at the top. *Journal of Business Compliance, 05/2013*, 19–29. Retrieved from https://scholar.google.com.sg/scholar?hl=en&q=setting+a+tone+for+fairness+at+the+top&btnG=&as_sdt=1%2C5&as_sdtp=#0.

Van Der Heyden, L., Blondel, C., & Carlock, R. S. (2005). Fair process: Striving for justice in family business. *Family Business Review, XVIII* (March), 1–21.

Van Der Heyden, L., & Limberg, T. (2007). Why fairness matters. *International Commerce Review, 7*(2), 94–102.

Vincent, A. S., Decker, B. P., & Mumford, M. D. (2002). Divergent thinking, intelligence, and expertise: A test of alternative models. *Creativity Research Journal, 14*(2), 163–178.

Waldman, D. A., & Atwater, L. E. (1994). The nature of effective leadership and championing processes at different levels in an R&D hierarchy. *The Journal of High Technology Management Research, 5*(2), 233–245.

Weisberg, R. W. (1999). Creativity and knowledge: A challenge to theories. In R. J. Sternberg (Ed.), *Handbook of Creativity* (pp. 226–251). Cambridge, UK: Cambridge University Press.

West, M. A., Borrill, C. S., Dawson, J. F., Brodbeck, F. C., Shapiro, D. A., & Haward, R. (2003). Leadership clarity and team innovation in health care. *The Leadership Quarterly, 14*(4–5), 393–410.

Woodward, I. C. (2015). Understanding communication preference styles guidebook (2nd ed.). INSEAD.

Woodward, I. C., More, E. A., & Van der Heyden, L. (2016). *"Involve": The foundation for fair process leadership communication* (INSEAD Working Paper Series No. 2016/17/OBH/TOM/EFE). Retrieved from http://ssrn.com/abstract=2747990.

Yukl, G., Wall, S., & Lepsinger, R. (1990). Preliminary report on validation of the managerial practices survey. In K. E. Glark & M. B. Glark (Eds.), *Measures of leadership* (pp. 223–238). West Orange, NJ: Leadership Library of America.

Zaccaro, S. J., Rittman, A. L., & Marks, M. A. (2001). Team leadership. *The Leadership Quarterly, 12*(4), 451–483.

Zaltman, G., Duncan, R., & Holbeck, J. (1973). *Innovations and organizations.* New York, NY: Wiley.

Zerfass, A., Sandhu, S., & Huck, S. (2004). Innovationskommunikation-Strategisches Hand-
lungsfeldfür Corporate Communications. In G. Bentele, M. Piwinger, & G. Schönborn (Eds.),
Kommunika-tionsmanagement (pp. 1–30). Neuwied, Germany: Luchterhand.
Zhou, J., & Shalley, C. E. (2003). Research on employee creativity: A critical review and direc-
tions for future research. In J. Martocchio (Ed.), *Research in personnel and human resources
management* (pp. 165–217). Oxford, UK: Elsevier.
Zorn, T. E., Jr., & Ruccio, S. E. (1998). The use of communication to motivate college sales teams.
Journal of Business Communication, 35(4), 468–499.

Prof. Ian C. Woodward is a Senior Affiliate Professor in Organisational Behaviour at INSEAD,
directing its flagship "Advanced Management Programme". Teaching and research focuses on
exceptional leadership communication, strategic/high performance leadership, fair process communication, and personal leadership transformation using innovative development techniques
related to human "drivers, blockers and values". He created the "communication intelligence" concept, and also delivers innovative arts-based learning experiences for business executives (he is
Associate Conductor of Singapore's Metropolitan Festival Orchestra). His research is published
in international journals, and he was formerly Associate Faculty Director of Columbia Business
School's Senior Executive Program. Before academia his business career included Board Director-
ships, Chief Executive, Senior Executive and Broadcasting roles in government, financial, energy
and public broadcasting sectors.

Dr. Samah Shaffakat is a Post Doctoral Research Fellow in the Department of Organisational
Behaviour at INSEAD—the Business School for the World. She is a member of INSEAD's Leadership and Communication Research Group. She is also a visiting Professor at IIM-Ahmedabad
where she is teaching courses in leadership and research methods. Her research interests include:
leadership communication, leadership development, psychological contracts, mindfulness, gender
and media. Her research is published in international management journals.

Chapter 12
Customer-Centricity in the Executive Suite: A Taxonomy of Top-Management Customer Interaction Roles

Noel Capon and Christoph Senn

Abstract The quest for customer-centricity drives top-management relationships with customers in business-to-business (B2B) markets. But the impact of executive engagement varies greatly across supplier-customer relationships. Based on exploratory field research, this paper develops a taxonomy of top-management-customer interaction roles. We also provide suggestions for leveraging senior executives for both supplier and customer benefit.

12.1 Introduction

The fundamental core of this paper emphasizes the importance of the marketing and sales functions to firm success. Whereas many firm functions provide significant value, marketing and sales differ qualitatively from most others. Specifically, the purpose of marketing and sales is to attract, retain, and grow customers. If these functions fail, the firm does not make profits, the organization does not survive, and shareholder value atrophies. And from the perspective of employees, no one gets a paycheck!

These assertions are not particularly revolutionary. After all, a half century ago, Peter Drucker stated, "If we want to know what a business is, we have to start with its purpose. There is only one valid definition of business purpose—to create a customer." As just noted, that is the job of the marketing and sales functions. Indeed, Drucker continued, "Because it is [the purpose of a business] to create a customer, [the] business enterprise has two—and only these two—basic functions: marketing and innovation" (Drucker 1954).

In recent years, two factors have highlighted the prescient nature of Drucker's observations. First is the evolution from considering customers as important firm assets in general, to placing a specific monetary value on individual customers. Indeed, during the past decade, both academics and practicing managers alike have

N. Capon (✉)
Columbia Business School, New York, USA

C. Senn
St. Gallen University, St. Gallen, Switzerland

become enamored of the customer lifetime value (CLV) concept. CLV can be complicated mathematically, but essentially equates to the anticipated stream of future profits a customer delivers to the firm, factored by the probability of retaining that customer year-to-year, and the firm's discount rate (Gupta and Lehmann 2004). Hence, management can calculate the value (individually) of current and potential customers, and make more informed resource allocation decisions about customer retention and acquisition.

The second factor is application of Pareto's law to the firm's customer base. Many firms have realized that a disproportionately small number of customers is responsible for a large portion of the firm's sales revenues. Indeed, the 80:20 rule states that 80 percent of firm revenues derive from 20 percent of firm customers. Of course, this ratio is not exactly 80:20; it may be 70:30, 90:20, or even more skewed. The basic point is that all customers are not equal; rather, some customers are more equal than others. Hence, a relatively small number of customers is disproportionately important to firm health.

As a direct result of this realization, during the past quarter century, many firms have developed key/strategic account programs. In these programs, the firm pays special attention to a relatively small number of its most important customers. These programs come in several shapes and sizes, but are typically staffed by specially trained strategic/key account managers (SAMs/KAMs), frequently backed up by a program office, and employing formal systems and processes like annual planning and periodic supplier/customer reviews. More recently, globalization has led firms to initiate global account programs staffed by even more highly trained global account managers (GAMs), to address customers with operations in many countries/continents around the world.

The foregoing places the following question squarely on the table. Given the critical importance for the firm of attracting, retaining, and growing customers; given the firm's new ability to calculate the economic value of individual customers; and given the large investment many firms have made in strategic/key/global account management programs to optimize relationships with their most important customers; what is top management's responsibility to become involved in this endeavor?

In most corporations, the CEO and executive team are busy people. Executive team members direct and manage areas of functional expertise like accounting, finance, human resources, operations, marketing, R&D, sales. The CEO leads and directs the executive team and guides the firm to achieve its objectives in the face of increasingly difficult environmental challenges, and demanding stakeholders. And from time-to-time, top management intervenes at the frontline in attempting to secure major deals. Consider for example the case of Accenture, IBM, and Merck: A few years ago, along with many other large corporations, pharmaceutical giant Merck decided to outsource its data processing system. Middle managers developed an RFP and evaluated submissions from several potential suppliers. The consensus decision: Accenture had earned the contract. Then IBM CEO Sam Palmisano came calling on Merck's CEO: IBM won the contract!

A major reason for this success was IBM's strongly developed focus on *Integrated Accounts*. High-caliber strategic account managers (Account Managing Directors) are responsible for securing revenue and profit growth from IBM's most important customers. Supported by global team resources and a systematic executive sponsorship program, IBM implemented an *Embedded*-type of customer management (Capon and Senn 2010). This program allowed IBM to take action at a critical time when strategic customer Merck was about to award its contract to IBM's competitor.

Perceptive senior executives understand that customers are crucial to organizational survival and growth, and that success with *strategic* customers can have a significant impact on revenues, profits, and shareholder value. Many firms have initiated programs similar to IBM to supplement ongoing sales efforts. But while some research and received industry wisdom suggests that greater top management involvement tends to be positively related to revenue generation and profitability (Homburg et al. 2002; Workman et al. 2003), we suggest three important caveats:

- Using the CEO as a persuader of last resort is not sustainable. Even *rock-star* CEOs will not be able to turn around customers' closed (and almost closed) deals with competitors on a regular basis.
- CEOs and executive team members face ongoing time constraints. They must carefully balance time they allocate to their day-by-day responsibilities with time they direct to addressing customer issues (Lafley 2009). Spending large amounts of time with customers on a regular basis is not sustainable (Quelch 2008).
- Many executive-customer interactions focus overly on securing immediate sales revenues, rather than discuss how the firm and its customer may shape the future together (Senn 2006).

12.2 About the Research

A half-century of research on management time has identified a broad range of general insights and specific recommendations/guidelines on how best to allocate this scarce resource (Drucker 1966; Mintzberg 1973; Kotter 1999). But findings on top management relationships with customers remain mostly anecdotal (Guesalaga and Johnston 2010; Shi et al. 2005, Bossidy and Charan 2002). Nor do these studies offer practical advice for senior executives who care seriously about customer-supplier relationships (Kohli and Jaworski 1990; Jaworski and Kohli 1993; Narver and Slater 1995; Homburg et al. 2000; Day 2006). Regardless, recent research confirms increasing supplier appetites for top-level customer relationships, leading to deeper strategic dialogues and greater influence on customer decisions (IBM Corporation 2013; Senn et al. 2013).

In our research, we set out to answer three related questions:

How, if at all, do senior executives relate to customers?
What are the impacts of various relationship types?
How may firms leverage these relationships in the best way?

The taxonomy we develop result from a multi-year study embracing 30 personal interviews with top managers, and 15 workshops with more than 300 participants. The main preliminary findings are:

Interviews: CEOs and top managers with sales backgrounds interface with customers more broadly, deeply, and frequently than their peers. They also view spending time with customers as a strategic necessity rather than as an agenda-filler.

Workshops: Four direct and one indirect role capture the ways in which senior executives interface with customers; each role offers rewards and risks. For strategically important customers, the *Growth Champion* role provides the most benefits.

Loose cannon	Senior executives make visits to customers without decent prior and/or post meeting interaction with account managers. Often leads to bad impression and/or unrealistic promises
Social visitor	Senior executive meet with customers but interactions focus on relationship building. Firm-customer interactions do not involve substantive business issues, but creates a certain level of trust
Deal maker	Senior executives (including the CEO) get involved in significant revenue opportunities when the customer is about to make (or has just taken) the supplier decision. Strong signal of commitment
Growth champion	The firm builds senior executive involvement into relationships with strategic customers via regular meetings and/or an executive sponsor program

12.3   Top Management Customer-Focused Roles

Our research identified five relationship roles that CEOs and top managers actually play in addressing customers. To be more precise—four customer-focused roles plus an absence of customer interaction. We start with absence and label this role—***not my problem***.

12.3.1   Not My Problem

The CEO of a large manufacturing firm addressed one of the authors: "I typically do not see customers. That's why we have a sales force. Our products and solutions are world-class and we have one of the strongest R&D groups in the industry. If only our account managers would do a better job in selling our value to customers!" A competitor has since acquired this firm and the former CEO is no longer with the organization.

It is not uncommon for senior managers to adopt this perspective. Indeed, the motto, *Let the sales force do its job,* sounds eminently sensible. After all, as noted above, each top manager has his/her own set of responsibilities; competitive pressures continue to increase and most executives have to do more with less. The perspective underlying this role is quite straightforward: The human resource function should do its job—hire the best talent to direct and manage the selling effort. In turn, senior sales executives should use the tools at their disposal to effectively populate the sales force and ensure salespeople reach their goals. If sales force results do not meet firm objectives, then required actions are quite straightforward: Make the appropriate personnel changes. The common mindset is simply stated: *We don't expect top marketing and operations managers to get involved in finance; why should any functional leader, or the CEO, get involved with sales?*

The underlying assumption of ***not my problem*** is that, at a high level, all functions are equally important. Of course, most functional areas are critical to firm success, but as our observations of companies around the world show, and noted above, sales (and marketing) is perhaps a little more equal than others. More than any other function, sales is a boundary role—the critical interface between the firm and its customers. If sales underperforms, revenues do not reach expectations, and everyone's budget suffers. If top managers can assist sales in doing its job, then all functions benefit.

12.3.2 Four Direct Customer-Focused Roles

Not my problem assumes that senior managers cannot add value to the firm's selling effort. This judgment is typically erroneous. Indeed, each executive team member has significant ability to improve firm-customer relationships, but there are pitfalls. Some approaches can be positively damaging to customer relationships. In our interviews and workshops, we found evidence for four separate roles. We organize these roles along two dimensions—*revenue seeking* and *relationship building*. Each dimension has two levels—*low* and *high* (Exhibit 12.1).

Loose Cannon

The ***loose-cannon*** role can be positively damaging for the firm. Top managers must clearly understand the potential negative consequences from ***loose-cannon*** behavior, and implement eradication processes.

A U.S.-based account manager for an Indian technology outsourcing firm complained to one of the authors during an executive education program: "I had been working my account for two years, gaining trust and making steady progress, and then it all fell apart. A senior Hyderabad-based executive made an appointment with my customer's top management without letting me know, and without any briefing or debriefing. He had no idea what was going on with the customer, and we just could not honor he promises he made. The meeting was a disaster and set us back at least one year."

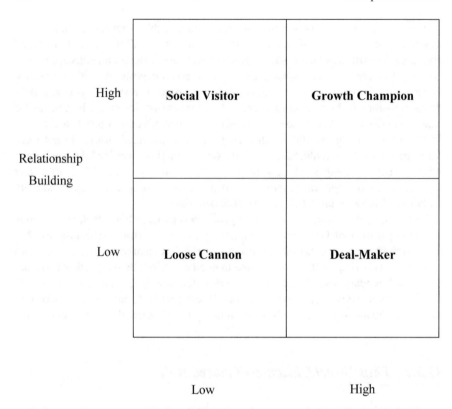

Exhibit 12.1 Top management customer relationship roles

It was later revealed that the Indian executive's daughter was at college in the U.S., and the customer visits were an excuse to charge the trip to his company expense account.

A top-level executive at a European chemical firm met with a major customer. The executive was unfamiliar with both customer challenges and his firm's initiatives—he left a negative impression. This executive also made unrealistic promises the firm could not fulfill; he did not tell the account manager about the meeting nor about the promises. The result: A badly damaged relationship that remains unrepaired despite heroic efforts by the account manager and her team.

The critical question is straightforward: Are the CEO and/or senior executives acting as *loose cannons*? *Loose-cannon* behavior typically occurs when a senior executive meets with a major customer but has no (or minimal) prior interaction with the account manager. Sometimes the account manager only finds out long afterwards—often from customer executives! Not only was the senior executive not (or improperly) briefed, the executive commits the firm to actions without full awareness of intricacies in the firm-customer relationship. And the senior executive may fail to advise the account manager of these agreements.

Loose-cannon interactions with customers are typically short-lived and not particularly focused on revenue generation, at least not in the short term. But they seem to be ubiquitous. We have met few account managers who could not identify *loose-cannon* behavior and its negative consequences from one or more of their senior executives.

Account managers have another name for *loose cannons—seagulls*. The *seagull* flies in, leaves a deposit, and flies off, rarely returning to the same spot! Regrettably, many readers probably have *seagulls* in their firms. Consider what the senior sales executive at a major Canadian financial institution told us when we discussed our evolving framework with him: "We don't have isolated senior executives acting as *seagulls*; we have a whole flock of them!"

If such interactions are so damaging, why do they occur? Our research revealed that, in many cases, senior executives thought they were being helpful. They saw their organizational positions as door openers to customer executive suites. Typically, they were not mistaken. But lack of preparation and failure to develop the appropriate working relationships with relevant account managers led to disaster.

Fortunately, most sample firms had sufficiently strong processes to avoid *loose-cannon* behavior most of the time. They employed some form of strategic account management and executive-sponsor program to interface with major customers. These firms clearly defined roles and responsibilities as a central program element.

Social Visitor

Social-visitors also have a low revenue focus, but a long-term relationship focus. The role's purpose is to demonstrate supplier commitment, and to create trust through long-term personal relationships. Though less negative than *loose cannons*, the *social visitor* has its own set of risks and rewards.

The CEO of a European-based engineering firm frequently met with customer CEOs at social events and trade shows. Attempting to take these conversations to a more business-oriented level, the CEO accepted an invitation to visit a major customer's U.S.-headquarters. The CEO advised the account manager of the visit but decided to go alone. The customer CEO, used to high-level visits from high-performing suppliers, awaited the supplier CEO with a senior executive entourage. After a moment's surprise at seeing the supplier CEO alone, the customer CEO asked: "Good to see you, but where on earth is your account manager and his team?" The supplier CEO responded: "Well, I thought we were going to discuss confidential issues." "Sure," was the response, "but who's going to take the notes?"

Whereas *loose cannons* are positively destructive for the firm-customer relationship, the *social visitor's* impact ranges from mildly positive to mildly negative. The *social visitor* specializes in relationship building via meeting and greeting. Typically, the firm arranges meetings for senior customer executives—educational events on company premises, cocktail parties at trade shows, and/or hosting trips to sporting events. The *social visitor* works the crowd, engages in conversation about industry issues, and builds personal relationships with individual customer executives.

These interactions rarely involve deep discussion of business issues; indeed, spouses may attend these events. Relationships are often long-term in nature and occur at regular intervals—the annual duck hunt, U.S. Open Tennis Championships, U.S. Golf Masters, or Formula 1 Grand Prix races in Europe or Asia. Customer executives often look forward to these events.

There is nothing particularly undesirable with social interactions between senior firm and customer executives—everyone has a good time, and leaves with pleasant feelings; these meetings enhance personal relationships. Rarely do such social interactions cause positive harm. But customer executives may wonder why these interactions are all they see of senior supplier executives. Indeed, they may feel mildly negative about lack of relationship depth, and be frustrated about missed opportunities for meaningful discussion. Negative feelings maybe especially strong if other suppliers make concerted efforts to engage in deeper business relationships.

Customer expectations can be easily misconstrued. The *lonely* CEO and his firm described in the opening vignette learned a big lesson, and learned it quickly. Today, the firm prohibits executives from visiting customers alone, except under special circumstances, and only with proper briefing and de-briefing. A centrally managed tracking system ensures no top-management visit occurs without evaluating potential alternatives. Results from recent meetings are built into briefing packages for upcoming customer discussions.

Deal-Maker

Deal Makers have high revenue focus but low relationship focus. The purpose of this role is to stabilize shaky relationships or secure deals. Sam Palmisano's actions as described earlier exemplify a senior executive fulfilling the **deal-maker** role.

A mid-sized coating systems manufacturer was trying to supply eco-friendly, cost-saving production technology to a leading German automotive firm. Contrary to the experiences during lab- and field-tests, a stable operational solution at the customer's main production line could not be reached within the contractually agreed timeframe. The supplier CEO personally intervened. The purchase-decision delay he secured enabled the firm to demonstrate technical ability to solve customer problems in a full-scale production environment.

Senior executives can have a major positive impact as **deal makers**. Sophisticated suppliers with good sales pipeline systems know what deals customers are deciding on a week-by-week basis. For many opportunities, the critical deciding issue is not so much the value propositions that supplier account managers offer, but customer beliefs that the supplier will/or will not live up to its commitments. Account manager promises go so far, but only a senior executive, often the CEO, can fully commit firm resources. The **deal maker** is very revenue focused.

In many customer organizations, middle managers negotiate deals and make purchase recommendations, but senior managers have the final word. As our opening vignette showed, Merck's CEO overruled internal advice to outsource its data-processing system to Accenture when IBM's CEO came calling. Indeed, we have observed several cases where top management intervened in the buying process has reversed previously made purchase decisions. On some occasions, customers even assumed penalties to void signed contracts.

Senior executives, especially CEOs, are uniquely placed to fulfill the **deal-maker** role. Only senior executives can truly commit the firm to allocate necessary resources to serve a customer. But there are dangers: Too many deals escalated to the executive suite creates a culture of upward delegation. The senior executive team can certainly help the sales force close deals, but too frequent rush meetings on customer schedules may be unsustainable. And although customers may be happy to see senior supplier executives engaged in real business issues (versus meet-and-greet **social visitors**), they may also wonder where senior executives are the rest of the time. Could they not have been involved earlier in the process?

Notwithstanding potential success, **deal makers** are no panacea. Consider what one account manager told us: "The division president decided to accompany me to my customer to close an important deal. Shortly after the meeting started, the customer pushed for significant additional price concessions. The division president was so focused on winning the deal, he was about to agree on a massive discount that would have cost us more than $2 million. I didn't know how to stop him, so I faked a heart attack. The meeting ended and the negotiations were put on hold. A few weeks later I negotiated the deal at a much better price!"

Growth Champion

Growth champions have a strong revenue focus and a long-term relationship focus. Unlocking new growth opportunities and serving as a role model to inspire others are the main purposes of this role. Thus, growth champions represent the most positive customer-facing behavior for senior executives.

Cisco's former long-time CEO John Chambers fit the profile of a truly customer-centric top manager. Chambers accompanied strategic account managers on business trips and demanded feedback on how to improve his participation. And Chambers regularly interacted with customer CEOs by leveraging Cisco's tele-presence technology. eBay's CEO John Donahoe also plays a major role in developing and sustaining retailer partnerships that drive substantial revenue growth for eBay.

Senior executives acting as **growth champions** play a critical role in driving revenue and profit growth. Our research suggests that well-planned, well-organized, and well-executed executive involvement at customers can pay huge dividends. Findings from our preliminary research indicates that executives who combine revenue seeking and relationship-building create significantly more favorable conditions for customer-centricity than other approaches.

Two items merit further consideration: First, some of the most successful executives we identified commenced their customer-facing behavior as **social visitors** or **deal makers**; later, they evolved into **growth champions**. Second, senior executives should take customer behavior into account when planning customer relationship efforts. When customers behave in a transactional manner, **growth- champion** behavior is often a waste of resources. Conversely, if customers desire long-term collaborative relationships, behaving as a **deal maker** or **social visitor** delivers suboptimal results.

Perhaps the most significant corporate initiative that firms take to implement *growth-champion* behavior is installing and executive sponsor program.

Executive Sponsor Programs. Over and above periodic and episodic interactions between top supplier and customer executives, many suppliers are appointing executive sponsors for important customers (Capon and Wallner 2012). Executive sponsors are senior managers who work with individual strategic account managers at specific customers. The executive sponsor is a *permanent* senior executive who interacts with senior customer executives. Executive sponsors can add significant value to the supplier-customer relationship.

Suppliers with these programs typically draw executive sponsors from the C-suite. These firms work through a serious matching process to ensure positive chemistry between and among the strategic account manager, senior customer executives, and the executive sponsor. Strategic account managers know that some senior firm executives make good executive sponsors—others do not; sometimes divorce is necessary. DHL gives strategic account managers the right to suggest changing executive sponsors without penalty, provided they back up proposals with solid evidence. The strategic account manager—executive sponsor relationship is particularly tricky because, at some level, the executive sponsor *works for* the account manager. Frequently, the CEO is executive sponsor for the firm's most important customers: J. W. Bill Marriott Jr. was executive sponsor for major Marriott customer Accenture; Oracle's executive sponsor at GE is Larry Ellison. At Siemens, the entire Executive Board participates in the executive sponsor program—members, including the CEO, average seven customer sponsorships.

Most firms introduce executive sponsor programs for customer-facing reasons. But executive sponsors may offer significant benefits within the firm. It's one thing for the account manager to discover customer needs and propose value propositions; it's quite another to actually deliver promised value. Powerful executive sponsors can cut through supplier bureaucracies when account managers face internal roadblocks. Some executive sponsors offer direct support to account managers:

In a significant downsizing, IBM released an engineer with considerable expertise in transaction processing. IBM's strategic account manager at Merrill Lynch realized IBM's position would be severely weakened by this loss; he contacted his executive sponsor. The sponsor said he could not reverse the layoff, but authorized hiring the engineer on a consulting contract and agreeing to pay on his budget. The engineer remained as a consultant for the next dozen years, helping IBM to secure business with a strategic customer during economically difficult times.

Senior executives acting as *growth champions* play a critical role in driving revenue and profit growth. Our research suggests that well-planned, well-organized, and well-executed executive involvement at customers can pay huge dividends. Findings from our preliminary research indicate that executives who combine revenue seeking and relationship-building create significantly more favorable conditions for customer-centricity than other approaches.

12.4   Conclusion

Leveraging top management-customer interactions requires a clear view of the firm's strategic direction. Greater customer focus by top management and appropriate executive relationships with customers are key differentiators for many firms. Senior executives have critical functional roles, but they may also play an important additional role in helping the firm reach sales and profit objectives. Forming meaningful relationships and becoming more engaged in customer affairs allows senior executives to provide higher degrees of customer-centric behavior, and hence greater value for both the supplier and its customers. Peter Drucker was very clear about the purpose of a business. Active and well-managed participation in fulfilling that purpose should become an increasingly important role for senior managers, led by the CEO.

Bibliography

Bossidy, L., & Charan, R. (2002). *Execution: The discipline of getting things done*. New York: Crown Business.

Capon, N., & Senn, Ch. (2010). Global customer programs: How to make them really work. *California Management Review, 52*(2), 32–55.

Capon, N., & Wallner, C. (2012). Key account management at Siemens: The executive relationship program. In N. Capon & C. Senn (Eds.), *Case studies in managing key, strategic, and global customers*. Wessex: Bronxville, NY.

Day, G. S. (2006). Aligning the organization with the market. *Sloan Management Review, 48*(1), 41–49.

Drucker, P. F. (1954). *The practice of management* (pp. 37–38). New York: Harper and Row.

Drucker, P. F. (1966). *The effective executive*. New York: Harper & Row.

Guesalaga, R., & Johnston, W. (2010). What's next in key account management research? Building the bridge between the academic literature and the practitioners' priorities. *Industrial Marketing Management, 39*(7), 1063–1068.

Gupta, S., & Lehmann, D. R. (2004). *Managing customers as investments*. Philadelphia, PA: Wharton.

Homburg, Ch., Workman, J., & Jensen, O. (2000). Fundamental changes in marketing organization: The movement toward a customer-focused organizational structure. *Journal of the Academy of Marketing Science, 28*(4), 459–478.

Homburg, C., Workman, J. P., & Jensen, O. (2002). A configurational perspective on key account management. *Journal of Marketing, 66*, 38–60.

IBM Corporation. (2013). *The customer-activated enterprise—Insights from the global c-suite study*. NY: Somers.

Jaworski, B., & Kohli, A. K. (1993). Market orientation: Antecedents and consequences. *Journal of Marketing, 57*(3), 53–70.

Kohli, A. K., & Jaworski, B. J. (1990). Market orientation: the construct, research propositions and managerial implications. *Journal of Marketing, 54*(2), 1–18.

Kotter, J. P. (1999). What executive general managers really do. *Harvard Business Review, 77*, 145–159.

Lafley, A. G. (2009). What only the CEO can do. *Harvard Business Review, 87*, 54–62.

Mintzberg, H. (1973). *The nature of managerial work*. New York: Harper & Row.

Narver, J. C., & Slater, S. F. (1995). Market orientation and the learning organization. *Journal of Marketing, 59*(3), 63–74.

Quelch, J. A. (2008). How CEO's should work with customers. *Harvard Business Review Blog Network*. Retrieved at http://blogs.hbr.org/2008/09/how-ceos-should-work-with-cust/ [September 22, 2008].

Senn, Ch. (2006). The executive growth factor—How Siemens invigorated its customer relationships. *Journal of Business Strategy, 27*(1), 27–34.

Senn, Ch., Thoma, A., & Yip, G. S. (2013). Customer-centric leadership: How to manage strategic customers as assets in B2B markets. *California Management Review, 55*(3), 27–59.

Shi, L. H., Zou, S., White, J. C., McNally, R. C., & Cavusgil, S. T. (2005). Executive insights: Global account management capability: Insights from leading suppliers. *Journal of International Marketing, 13*(2), 93–113.

Workman, J. P., Homburg, C., & Jensen, O. (2003). Intra-organizational determinants of key account management effectiveness. *Journal of the Academy of Marketing Science, 31*(1), 3–21.

Noel Capon is the R.C. Kopf Professor of International Marketing and Chair of the Managerial Marketing Division, Columbia University, Graduate School of Business, New York.

Christoph Senn is Director of the Competence Center for Global Account Management and Senior Lecturer of International Management at St. Gallen University, Switzerland.

Part III
New Leadership in Education

Chapter 13
A Competency Based Approach to Leadership Development: Growth Mindset in the Workplace

Thomas Sullivan and Nadine Page

Abstract This chapter, examines the impact of fostering pro-development self-theories on early career success for a cohort of MBA graduates at an international business school in Boston, Massachusetts. Day (Leadersh Q 11:581–613, 2001, p. 601) proposes that "lecture-based, classroom training found in most formal leadership development programs is at best only partially effective," citing short-lived behavioral change as a source of frustration. The chapter describes the early results of a longitudinal study of lasting change produced by incorporating corporate best practices in leader development identified by Day (Leadersh Q 11:581–613, 2001) in a business school setting. This the chapter includes the research question, theoretical background, research context, methodology, and the high-level findings.

13.1 Introduction

> So, it was remembering to always take that deep breath, relax, just listen to what they have to say, you don't have to be right today, I don't have to necessarily have an opinion today… (Development S20)

> I remember that quite often and I am like, "oh yeah, time to listen and I do that a lot and it works, it really works". (Development S16)

The quotes above illustrate two managers in our study describing challenging situations they encountered in their workplace. They show an awareness of themselves in a social situation (leadership awareness), describe a way of framing their experience of the situation (leadership strategy), and reveal the process of making a deliberate choice of action (leadership act), rather than defaulting to a habitual pattern of behavior. These early-career managers show a level of cognitive competence in a

T. Sullivan (✉)
Hult International Business School, 1 Education St. , Cambridge, MA 02141, USA
e-mail: thomas.sullivan@faculty.hult.edu

N. Page
Hult International Business School, Ashridge House, Ashridge, Berkhamsted, Hertfordshire HP4 1NS, UK
e-mail: nadine.page@ashridge.hult.edu

© Springer Nature Switzerland AG 2020   179
N. Pfeffermann (ed.), *New Leadership in Strategy and Communication*,
https://doi.org/10.1007/978-3-030-19681-3_13

leadership context that, we believe, can be attributed to their preparation and development during a 12-month program on leadership, which they completed as part of their MBA in a business school in Boston, MA. At the core of this development program was the learning and development of a "growth mindset." Dweck (2006) proposed that a growth mindset is a latent variable underlying durable skills acquisition of developing leaders. Developing a growth mindset is fundamental to leaders and leadership development because it allows an individual to be aware of; their situation and desired outcomes, recognize, choose, and sometimes invent from among a variety of leadership strategies, and effectively employ a variety of potential leadership actions.

This chapter examines the impact of that leadership development curriculum for MBA students at the early stages of their career. We explore their ability to demonstrate competence in a defined set of key leadership behaviors. The students were measured and identified according to their increase in skill during their 12-month course, and in the interviews we conducted with them 2-years post-graduation, they described their ability to show growth mindset competence in the workplace.

The promise of using a competency-based approach to learning in higher education is to both provide learners with a more efficient and less expensive path to success in their field, while at the same time allowing employers a closer link between graduates' abilities and their application in the workplace. By focusing on growth mindset, students were encouraged to see their development as leaders as more than skill building. They actively worked on an integrated approach that meshed awareness and frameworks with actions. We sought to understand the impact and longevity of this growth mindset exposure on leader behavior in the workplace.

Why are competencies important?

Given the demands that today's volatile economic, technological, and political environments place on formal and informal leaders at all levels, the ability to continually adapt and respond effectively to ever-changing and unpredictable circumstances is a critical leadership competence. A growth mindset may be the key enabling variable for leaders to succeed. While competencies and skills are interrelated, Klein-Collins (2012) places competencies at a higher level than skills and knowledge. She states that a learner demonstrates competence through application of skills and knowledge in a variety of circumstances and situations. A competence therefore requires an awareness of desired outcomes in any situation, an implicit or explicit model or frame to diagnose and develop a strategy in the situation, and a theory of action and practice to engage in an appropriate response to the needs of the situation.

In any field, not just leadership skills development, competency-based education requires a rethinking of the approach to teaching, as well as to the measurement of student achievement. Competence means learners can do something not just know about something. Demonstrating this development can, however, be very challenging.

For leadership, the challenge begins with "leadership development", a term that is used loosely, as if the meaning of leadership and the meaning of development are assumed to be shared. Brief reflection on one's own experiences shows that they're not. People's expectations of leadership vary widely and often are contradictory. This is complicated by the understanding that while people can develop as leaders, trying to identify the "impact" of leadership development is difficult to achieve. It is rarely measured in a way that allows us to understand the leadership actions and the unique social dynamic of the context at a given time that leads to impact. When it is measured, it is often found to be lacking. Day (2001, p. 601) proposes that "lecture-based, classroom training found in most formal leadership development programs is at best only partially effective," citing short-lived behavioral change as a source of frustration.

In addition, leadership thinkers and practitioners have segmented and chunked leadership into buckets that can be limiting and regimented for a user. No single practice can be effective at any time when we understand that the context is variable. Even a situational leadership approach can be crude when employed to dynamic social situations rather than one-on-one interactions. What might be most helpful to users is to see leadership competence as the application of leadership awareness, a set of leadership models and strategies, and leadership actions employed effectively in a given situation to produce the desired impact.

Having a growth mindset appears to enable the development of leadership competence as it requires that learners engage in a dynamic feedback loop between leadership awareness, strategies and effective action. Growth mindset is a cultivated approach that helps individuals to cope and manage change, adversity, and uncertainty—all circumstances that require new learning. This research explores the characteristics of leader development initiatives with the power to cultivate growth mindset among young leaders.

In follow up to the insight from Day and Sin (2011, p. 546) that development "inherently involves… the consideration of time in underlying theory and research," the research investigates whether a growth mindset developed in MBA graduates during such management education can propel leadership trajectories over time and across contexts (Day and Sin 2011), and, if so, how this associates with specific behaviors in the workplace.

Theoretical Background

The implicit theories we have of ourselves shape our perception, motivations, and achievements. Mindset theory, as developed and articulated by Dweck (2006) and others (e.g., Burnette et al. 2013), acknowledges that individuals differ in their beliefs about whether human attributes, such as intelligence, talents, and abilities, are stable or malleable, and what they can do to change them over time.

People tend to have a natural predisposition towards a certain type of mindset. At the 'fixed' end of the continuum, individuals hold the implicit belief that attributes, talents, and abilities are defined at birth and are static, stable, and not amenable to change. At the 'growth' end of the continuum, individuals hold the implicit belief that attributes, talents, and abilities are malleable and can be developed and strengthened with effort and practice (Dweck 2006; Burnette et al. 2013; Keating and Heslin 2015). These beliefs create mental frameworks that encourage deliberate engagement in different behaviors, and ultimately impact outcomes such as achievement and performance (Heslin and VandeWalle 2008). Decades of research show that when people have a growth mindset, they tend to be more successful (Rae-Dupree 2008).

Mindset theory has often been applied in the context of learning and education (see Rattan et al. 2015; Asbury et al. 2016). Research exploring the attributes of a growth mindset in an educational setting has identified a distinctive behavioral profile. Individuals pursue challenging development opportunities even in the face of frustrations or setbacks (Hong et al. 1999), they perceive effort as essential for development (Blackwell et al. 2007), and they seek feedback to garner useful insights and view such feedback constructively (Mangels et al. 2006). Not surprisingly, therefore, a growth mindset has been linked to several positive outcomes associated with learning and development, including motivation and achievement (Dweck 2008; Boyd et al. 2014), especially for students facing situational challenges (Yeager and Dweck 2012). In this chapter we explore growth mindset in relation to a different outcome—leadership competence.

Mindset is different from other personal resources (e.g., proactive personality, for example) in that it is not a stable characteristic of an individual; it reflects a belief that can be induced, changed and developed intentionally by interventions (Keating and Heslin 2015; Mueller and Dweck 1998). It is possible to get people to shift their operating mindset from growth to fixed, or from fixed to growth (Dweck 2006). This presents an opportunity for business schools to 'intervene' and focus on nurturing the development of a growth mindset as part of the business school curricula. At the same time, it presents an opportunity for graduates to accelerate their career progression, and for businesses to differentiate and develop critical skills that potentially produce a jump-shift in individual-level performance.

Despite the recognized benefits of a growth mindset, to date, very little research has explored whether, once developed, a growth mindset becomes embedded into an individual's psyche and integral to who they are, what they become, and the impact they have on their organization. In light of this, this research explores to what extent a growth mindset, developed during management education, is retained in graduates during their early career, and if so, what this transfer of learning looks like in the workplace, and how it associates with workplace behavior.

Research Context: The leadership development program

In 2014 students in a one-year, intensive, international MBA completed a leader development initiative designed to shape leader identities (Day and Sin 2011), build skills, and cultivate a learning orientation with the power to produce further development over time. The leader development program was a series of dedicated courses encompassing a variety of non-traditional design elements supporting the development of competencies and skills over an eight-month period. The program was recognized as unique and innovative by accrediting bodies (AMBA 2014).

The program was not designed to teach students about leadership. Rather, it was an intervention directly targeting leader competency development, in itself "comprehensive in terms of integrating assessment, challenge, and support in the name of development... when linked with... feedback (Day 2001, p. 591)." It was designed using elements of intentional change theory (Boyatzis 2006), with a substantial focus on experiential learning involving experimentation and practice. In practice, students were encouraged to facilitate competency and skill building by developing a growth mindset, taking ownership and responsibility for their learning, noting and improving their own relationship with failure and rejection, and seeking and absorbing help from peers and faculty coaches.

The program was modeled explicitly after corporate best practices and taught as part of the MBA curriculum by dedicated professionals who had specific industry experience with leader development. Day (2001) identifies a number of best practices for leader development that were incorporated into the program. These include quantitative and qualitative feedback against a research-based competency model (collectively referred to as the graduate "DNA"), "executive" coaching support (peer and faculty coaches), the development of supportive peer networks for collaborative learning and social support, and simulated action learning through substantial team project work on real corporate or entrepreneurial ventures.

Mindset change and leader competency development was emphasized within this framework. Students were coached to understand a growth mindset and demonstrate associated practices, develop self-awareness of current capabilities, and put into practice a self-designed development plan to strengthen competencies. Progress against development goals was frequently evaluated through regular written and observation-based assessments by faculty, self, and peers. The idea of trajectories was incorporated explicitly. As the program progressed, students were expected to improve their ability to frame common business problems as requiring leader intervention, demonstrate increasing capability to address team-based challenges over baseline levels, and address and master challenges of increasing complexity.

Methodology

Findings result from a two-staged process of subject screening and qualitative data collection. First, a sample of 17 students was selected from a cohort of roughly 190 graduates of the class of 2014–2015. The cohort was the first to experience the leadership program and has produced young leaders with enough experience to evaluate its effects over time. The sample was roughly balanced between males and females, reflected a global population from over 40 different countries, had an average age of 30.8 years, and had performed in what could be described as early career team leadership positions from 1 to 3 years.

The sample was selected based upon an analysis of data collected during the 12-month program, designed to identify changes in competence and demonstration of growth mindset in students while on campus. The assessments included self and peer-feedback (based on shared team experience), faculty observation of coaching competences in fixed peer triads, diagnostic assessments which were double graded for evidence of application of the "DNA", written self-reflection against progress in prioritized skills from the "DNA", and participation levels captured in team-based simulations at three points during the course.

The change in "DNA" profile from the start to the end of the program was used as an indicator of the development of a growth mindset, and as an identifier for suitable participants. The aggregate data for the cohort showed improvement over the 12 months. The sample was chosen from two groups of students, those with demonstrated significant increases in leader competence over 12 months ("the development sample") and those with no demonstration of improvement "the plateau sample"). To further isolate the effects under study, a control group of 7 students from the graduating class of 2013 was solicited.

Data were collected from identified participants using semi-structured interviews and explored the sustained impact of the leadership development program on graduates' cognition, affect, and behavior. The interviews explored how graduates approached their work and future career, specifically focusing on aspects of challenge and motivation, learning, achievement, and aspiration. In essence, the research explored whether students had sustained their growth mindset and transferred this to the workplace, and how this associated with certain kinds of workplace behaviors. To accommodate the global distribution of participants, the interviews were conducted and recorded using an online meeting platform.

Growth Mindset and competencies in the workplace findings
Patterns and competencies formed a taxonomy of Growth Mindset

Having a growth mindset helps to equip people with the affective approach, cognitive competencies, and behavioral skills they need for work. Participants in the development sample described an appropriate and extensive repertoire of behaviors and competencies that supported their professional performance. Analysis of the data led to a derived taxonomy of cognition, affect, and action. The taxonomy is broadly categorized in terms of as purpose, passion, and practice. It is summarized in Table 13.1 along with representative participant quotes.

Purpose—Self-awareness and self-management

...because those soft skills are very difficult to develop. Sometimes because it's a part

of your personality. So if you don't have it you have it sometimes you get it, but you have this self-awareness and say ok now I need to improve these kinds of situations. Maybe I'm not a good listener, maybe I can communicate how I really want to, so if you know that you have that self- awareness you can work towards those things and it's going to be the difference. (Development S5)

So when I started to understand my character in a lot more depth I was able to adapt in surprising ways, and by surprising ways, I mean not just dealing with customers or dealing with like an extra set of analysis

I have to go through, it's also dealing with authority figures and that was always an area I had an issue with. (S9)

Passion—Motivation

I had no transition. I had no training. I had no idea what to do and then when I was introduced to the client and they seemed to be all high-risk clients and here I am, I mean imagine like on the sixth day of your job you just see somebody just got fired because obviously they didn't do well. The anxiety that you get from that, that's more than enough, you know, but what I did though, I stood calm. I decided no matter how much pressure this is I'm going to do a good job. (Development S23)

I know there is struggle, like how am I going to change it so that I am excited to go in everyday because when I first started I wasn't excited, I wasn't ready and I came up with a game plan of how I was going to change what I was doing every day. So, I really thought oh I can get through this, I didn't really know what I was going to be doing, I liked it, but it was a struggle adapting. (Development S20)

Practice—Listens to learn

So, I just breathe, you don't have to necessarily know everything but if you sit back and you listen to everything they have to say, you can actually learn something from them and I think that helped me. (Development S20)

The profile of the "development sample" was distinct as compared with the "plateau sample" and the control group.

Those students with an increasing ability to demonstrate a growth mindset during the program continued to distinguish themselves two years past graduation. As compared to the plateau sample and the control group, students in the development sample had achieved higher levels of self-perceived success, described their mindset and related behaviors with more granularity and positivity, exhibited strengthened leader identities (Day and Sin 2011), and articulated fewer beliefs in terms that would be characteristic of a more fixed mindset. While all three groups described determination and diligence in describing their work, the control group and the plateau subjects

Table 13.1 Taxonomy of workplace competencies associated with a growth mindset	Purpose	Clarity of thought	Self-awareness and self-management
			Active approach to learning
			Strong networks and relationships
			Clear vision
	Passion	Positive energy used productively	Motivation
			Aspiration
	Practice	Purposeful action	Listens to learn
			Communicates to connect
			Cognitively curious
			Feeds back to feed forward
			Agile and adaptive

both described their work in terms that were less affective and did not show the same levels of awareness (self and contextual) as the development sample. (Though the plateau subjects showed more explicit awareness of the importance of learning.) Those students in the development sample showed more awareness (self-awareness and contextual awareness), more deliberate choice among strategies and more persistence in application of effective actions.

e.g.

Sample quotes showing differences between groups

On seeking learning at work after MBA

That's the problem, I'm not sure I keep learning, [okay] To be frankly - honest, I'm not sure…is a little bit like working for a startup company but it's still too big for you to really– it takes time…know which…. So I'm not a hundred percent sure, it's not where I'm gonna grow this job.

(Control Group S31)

I think I waited too long to speak up to my Manager and yeah, the management team about that because I was…. I guess you're hesitant about is it my fault or if I do something why can't I make this work now I think I know better that you know this is just like the right way to work together or you know this person is too good at this thing we need to have to do something else.

(Control Group S27)

I would say when it came to learning at work, I was more like, I want to learn but I'm going to wait till someone comes and you know teaches me something. Sometimes and more I just sit there and I think it was……. Maybe it was fear of not wanting to sound stupid or just you know any……. I was just not comfortable asking for help and that was pretty much it.

(Now) I found that I am now a lot more comfortable just going up somebody and say I need help. so I find something to something I don't know I just say you know what, I don't know what let me ask someone, what this is all about and I find that I'm a lot more comfortable doing that.

(Plateau S3)

I think there's one junior person who was an intern and recently became a permanent employee. The way he puts up proposals and other things, he didn't have any experience with most of the consulting, (proposals) is one thing that has to be mentioned…

(Plateau S24)

I would see one of my reportee who was probably 64 or 65 because he had experience and I had knowledge sometimes I feel a I don't understand things I just go and discuss it with him and then get an honest opinion from him.

(Plateau S25)

A growth mindset can be cultivated, retained, and embedded

The results suggest that development of a growth mindset is possible with a well-structured and dedicated experiential learning program. Repeatedly, participants in the development sample and the plateau sample distinctively referred back to their management education and described how this helped them to develop a growth mindset. Students who reported a predisposition towards a fixed mindset changed their perspective towards a more positive disposition, and they recognized that this was a result of their learning on the MBA. In both the comments about growth mindset, and in their descriptions of how thinking influenced their workplace behavior, participants described this as a learning mindset that afforded a constructive approach to challenge, involving concerted effort, and perseverance.

The results suggest that the learning participants undertook as part of their MBA was retained over time, transferred, and applied to a workplace setting with beneficial effects. Participants put into practice the tools and techniques they used as part of their management education to encourage further learning and development in the workplace, including encouraging a growth mindset in others. It can be inferred that once developed, a growth mindset is significant in its persistence over an extended period of time and in a variety of situations that are distinct from the original learning environment. Mindset gets embedded into an individual's psyche and becomes integral to who they are, what they become professionally, and how, when given the opportunity, they shape their organization.

13.2 Conclusion

We examine an under-studied phenomenon, which is the impact of best practice development methods on cultivating pro-development leader attitudes and the translation of these attitudes into specific workplace practices over time. Developing a growth mindset requires an ability and practice of self-reflection, deliberate choice of mindset and strategy to achieve an outcome and a persistence in action and incorporating feedback from action into an awareness of oneself. Effective leadership requires a leader to engage in the same process, where leadership effectiveness is a consequence of leader awareness of self, outcomes and context, options among leadership strategies to choose and a repertoire of actions to achieve a desired outcome. This early-stage longitudinal study finds evidence for the development and persistence of growth mindset in a program that helps students develop leadership awareness, leadership strategies and effective leadership action. As such, this chapter contributes to the limited literature that has investigated growth mindset in the workplace, and its role in supporting the development of leadership competencies. We contribute a taxonomy of behavior to define the characteristics of growth mindset at work. We know of no other study that links these aspects of leader development to a specific developmental approach in a formal school setting.

Bibliography

Asbury, K., Klassen, R., Bowyer-Crane, C., Kyriacou, C., & Nash, P. (2016). National differences in mindset among students who plan to be teachers. *International Journal of School & Educational Psychology, 4*(3), 158–164.

Association of MBAs (AMBA). (2014). *XXX Wins AMBAs 2014 Innovation Award*. Retrieved November 25, 2018 from https://www.mbaworld.com/en/news/amba-in-the-news/2014/november/hult-wins-ambas-2014-mba-innovation-award.

Blackwell, L. S., Trzesniewski, K. H., & Dweck, C. S. (2007). Implicit theories of intelligence predict achievement across an adolescent transition: A longitudinal study and an intervention. *Child Development, 78*(1), 246–263.

Boyatzis, R. E. (2006). An overview of intentional change from a complexity perspective. *Journal of Management Development, 25*(7), 607–623.

Boyd, M., Kim, M. S., Ensari, N., & Yin, Z. (2014). Perceived motivational team climate in relation to task and social cohesion among male college athletes. *Journal of Applied Social Psychology, 44*(2), 115–123.

Burnette, J. L., O'Boyle, E. H., VanEpps, E. M., Pollack, J. M., & Finkel, E. J. (2013). Mind-sets matter: A meta-analytic review of implicit theories and self-regulation. *Psychological Bulletin, 139*(3), 655.

Day, D. V. (2001). Leadership development: A review in context. *The Leadership Quarterly, 11*(4), 581–613.

Day, D. V., & Sin, H. P. (2011). Longitudinal tests of an integrative model of leader development: Charting and understanding developmental trajectories. *The Leadership Quarterly, 22*(3), 545–560.

Dweck, C. S. (2006). *Mindset: The new psychology of success*. New York City: Random House Incorporated.

Dweck, C. S. (2008). *Self-theories and lessons for giftedness: A reflective conversation*. London: Routledge.

Heslin, P. A., & VandeWalle, D. (2008). Managers' implicit assumptions about personnel. *Current Directions in Psychological Science, 17*(3), 219–223.

Hong, Y. Y., Chiu, C. Y., Dweck, C. S., Lin, D. M. S., & Wan, W. (1999). Implicit theories, attributions, and coping: A meaning system approach. *Journal of Personality and Social Psychology, 77*(3), 588.

Keating, L. A., & Heslin, P. A. (2015). The potential role of mindsets in unleashing employee engagement. *Human Resource Management Review, 25*(4), 329–341.

Klein-Collins, R. (2012). *Competency-Based Degree Programs in the U.S. Council for Adult and Experiential Learning*. http://cdn2.hubspot.net/hubfs/617695/CAEL_Reports/2012_CompetencyBasedPrograms.pdf.

Mangels, J. A., Butterfield, B., Lamb, J., Good, C., & Dweck, C. S. (2006). Why do beliefs about intelligence influence learning success? A social cognitive neuroscience model. *Social Cognitive and Affective Neuroscience, 1*(2), 75–86.

Mueller, C. M., & Dweck, C. S. (1998). Praise for intelligence can undermine children's motivation and performance. *Journal of Personality and Social Psychology, 75*(1), 33.

Rae-Dupree, J. (2008, July 6). If you're open to growth, you tend to grow, *The New York Times*. Available at http://www.nytimes.com/2008/07/06/business/06unbox.html?_r=0.

Rattan, A., Savani, K., Chugh, D., & Dweck, C. S. (2015). Leveraging mindsets to promote academic achievement: Policy recommendations. *Perspectives on Psychological Science, 10*(6), 721–726.

Yeager, D. S., & Dweck, C. S. (2012). Mindsets that promote resilience: When students believe that personal characteristics can be developed. *Educational Psychologist, 47*(4), 302–314.

Thomas Sullivan holds the position of Prof. of Leadership Development at Hult International Business School where he teaches MBA and EMBA students leadership and coaching skills based on a competency framework. He is the co-founder of My City At Peace a social venture dedicated to reducing gang and youth violence in cities around the USA and Season of Peace a non-profit national program for coordinated cease fires in cities across the US during the Thanksgiving to

New Year period. Thomas has developed, designed and lead teams to deliver leadership development programs and coach teams and managers at all levels of leadership with the World Bank Group, the International Monetary Fund (IMF) and the Inter-American Development Bank (IDB).

Dr. Nadine Page is a member of Faculty and Senior Research Fellow at Hult International Business School. Her expertise and research interests are in contemporary issues in organisational behaviour, individual differences, leadership, sustainability, and change. Nadine has worked with both public and private sector clients and explored individual differences and behaviour change in a variety of contexts including health, transport, and education. Her research has been published worldwide including in Frontiers in Psychology, Harvard Business Review, and Forbes.com. Nadine has an honours degree in Psychology, a Master's degree in Research Methods and Statistics, and a Ph.D. in Behavioural Science. She is an Associate Fellow of the British Psychological Society (BPS) and is a qualified FIRO practitioner.

Chapter 14
Leadership Disrupted—No Time for Egoism!

Hanane Bouzidi

Abstract This book chapter claims the outcome of the digital transformation and its implications for leaders in organizations. When organizations change their strategy in accordance to the digital transformation to be not outperformed by the pioneering companies on the market, the inevitable consequence is to translate this shift thoroughly into the leadership levels. The more important it becomes for the organizations to develop initiatives and upskilling measures to contribute to and simplify innovative changes in leadership. I argue that leaders who will not invest in a digital mindset, instead strive for egoism, will cause the company's digital transformation failure.

14.1 Introduction

The role of a leader becomes increasingly important in times of acceleration through digitalization and its implications for the organization. Digitalization allows innovation and new business models to rapidly emerge, making new products and services available to customers anytime and anywhere. The challenges that come alongside with it, require a lot of attention and should not be ignored. Digitalization puts the time factor in a pole position: Who learns to adapt quickly and maintains flexibility puts other companies in the shade, secures a competitive edge and counts as an innovation carrier. But who can be the driver of such organizational adaptability? One for sure should put this on the agenda: The Executive. The way they deal with the digitalization, the willingness to embrace changes while becoming more creative, collaborative and adaptive to scale up the growth of the company can be a gamechanger for any business. In times of the Digital Economy and emerging startups that are willing to raise the bar in customer centricity and perfect user experience to bond with and bind their customers, organizations and big corporates would be well advised to do the same and to proactively shape their digital future. Setting this on their top priorities list, implies also finding the right set of activities to lead

H. Bouzidi (✉)
Deutsche Telekom AG, Bonn, Germany
e-mail: hananebouzidi@icloud.com

© Springer Nature Switzerland AG 2020

191

N. Pfeffermann (ed.), *New Leadership in Strategy and Communication*,
https://doi.org/10.1007/978-3-030-19681-3_14

the digital transformation within the organization. Fostering an innovative corporate culture and mindset while maintaining stability and constantly scaling up the abilities of the business leaders and employees should be one amongst those building blocks of embracing the transformation. Leadership development is taking into consideration the factor "digital" and its impact on how to support the development of future oriented and customer centric leaders. The most important fact is, that it is also the responsibility of the leader to foster a digital mindset and start using the power of the digital age to be better at innovation, team collaboration and adding value to the business.

The one risk a leader may fall for while navigating the business through the digital transformation is becoming egoistic.

"Egoism" is a driver of so many things harming the organization and being a show stopper for innovation and a solid failure culture within the company. Egoism withholds a leader becoming a "good" leader. Breaking free of the ego that a leader might be at risk of developing in so many hardworking years achieving their position, determined to be the number one, is vital for rethinking their leadership role and their overall responsibilities. In times of virtual collaboration and flexible working has become every day, letting go of command and control to empower the employees giving their best performance is becoming indispensable. A leader in times of uncertainty should be the solid backbone in forming and transforming the corporate culture with the right set of values. Jennifer Woo, CEO and chair of The Lane Crawford Joyce Group, Asia's largest luxury retailer ones said: "Managing our ego's craving for fortune, fame, and influence is the prime responsibility of any leader" (Carter and Harvard Business Review 2018). A leader investing in the ego by fame and people's charm might ignore that it leads to a clouded vision, making false decisions and attracting and relying on to the wrong people and opinions. A false ego corrupts the behavior, that might be a major cost factor in times of setbacks. Building strong reliable connections to the people that are led, customers and stakeholders and keeping always track of the critical voices, makes the leader vigilant for changes and sharpens the perception for risks. A "good" Leader is the driver of the change and enables the environment to adapt flexible to the challenges that come along with the uncertain times of digitalization, by reducing complexity and enhancing stability within the organization" (Carter and Harvard Business Review 2018).

This book chapter will discuss the effects of digitalization on leadership and the resulting claims to leadership. The described factors are the practical subject of the discussion and will be illustrated below in more detail under the questions: What role does digitalization play in terms of organizational changes and leadership? What role does leadership development play to guide executives through the challenges of acceleration in times of digitalization? How can a leader contribute effectively to the corporate culture enhancing degrees of freedom to explore and adapt to rapid change while maintaining stability?

14.1.1 The Sound of "Digital" Speed

67 percent of all specialists believe that their supervisor is not well prepared for the future. Are executives standing in the way of innovations? The fact is, that leadership of tomorrow needs more than just assertiveness (Dettmers 2018).

A strong statement towards leadership and its importance in being a driver for the future. To better understand the interrelations between leadership and its influence on the organization, let's have a look at the digital economy first to understand the challenges for the business and for leaders that arise in terms of the digital transformation.

It only took roughly 20 years for the digital economy to build up. It continues to expand and it has a strong influence on the market in vast and diverse sectors. It has disrupted old industries and made it possible for completely new ones to rise. Looking at Online-Bloggers or Artist (non-famous) using Social Media Tools to promote their latest fashion trends made them more then fortunate out of the blue. Even more it seems only logical that sighting the benefits of the digital economy is the most prominent topic in the highest ranks of the companies (Gada 2016). Statista released the Digital Economy Compass in 2018a, which shows very clearly what is happening online in only one minute.

Figure 14.1 describes the online activities of users on different online platforms in only 60 s.

With this image in mind, it is hard to imagine that there will be an economy, a market without digitalization. Like an electric current, digitalization flows through the most diverse areas of life within a very short space of time and ensures that these are interlinked in a variety of forms and integrated into the structures of our society

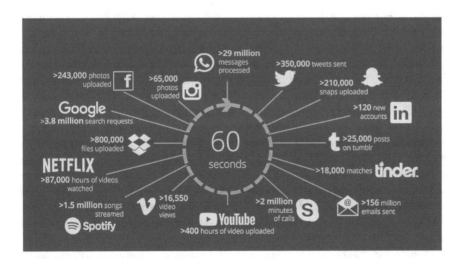

Fig. 14.1 Things that happened online in 2017 within 60 s (*Source* Go-Globe.com, Company Information, Statista Research, 2018a)

to connect people, organizations and things. In 2018 almost 4.2 billion people used the Internet, this is more than a half of the world's population. One medium that has evidently given a strong boost to achieving this development to a special degree is the Smartphone. Since its launch 2007 3.9 billion people were connected to the internet by mobile devices. The manifold use of the internet has made it clear that digitalization is making it's contribution to shaping the societal change. Looking at Europe the digital transformation is driven by fast broadband connections that are more and more expanded worldwide, social media platforms that mushroom with high speed and two million mobile applications that are available on the app stores (European Commission, DESI Report, 2018: 7; Statista 2018b). The use and interaction of people with new media, born of the new possibilities of communication and the knowledge exchange in the digital age, yield also a new form of online platform networks, the Social Networks. Today 3.4 billion people use Social Networks to interact with others and to share information of all kinds. China, India and the United States rank ahead all other countries in terms of internet users (Statista 2018c). No wonder that this also attracts a special attention among the companies to win over and retain customers for their brand and their products, because half of them are online and almost permanently attached to their mobile device. With these developments, businesses and organizations are under pressure, in the face of the *digital haze* that opens to the multiple possible use-cases, to clearly identify the growth potential of digitalization for their business and to use it for their own competitive advantage. The question that arises in the wake of this flood of information, availability of knowledge and digital possibilities: "What is the difficult thing to access and control that firms will base their future competitive advantage on? […] it is our attention, our capacity to focus on and respond in an effective way to the stimuli we receive, that we need to worry about…" (Birkinshaw and Ridderstråle 2017: p. 8*)*. Birkinshaw and Ridderstråle have put it with their statement straight to the point. The ability to focus the *attention* on the changes that move entire societies and translate them into actions that are good for the business will be one of the key aspects of top management in striving for becoming the market leader. In this sense the Digital Economy challenges all industries to accelerate, adapt and look for the latest trends. Those who want to be ahead in terms of customers and trends must learn to raise their *attention* and even foresee rapid changes or will be left behind. An even more hard fate for companies that already struggle with rigid structures and processes that make them *late adopters* to new market demands and shaping the future with innovative strategies. The adoption of digital technologies varies strongly with the company size. Especially for large companies there is a chance in the digital economy. Through their ability to invest and build new areas and attract new talents in IT, they empower themselves to engage in digital transformation and using new technology (see Fig. 14.2). However, the larger the company the greater the complexity in structure, responsibilities and processes. Complexity is the antagonist of adaptation. The more complex the structure of an organization the more complicated the handling of organizational adjustments. A flexible organization at different levels learns from mistakes and finds answers to current challenges without increasing the level of complexity but rather producing simple, fast and adaptive solutions.

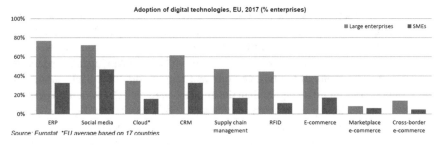

Fig. 14.2 Adoption of digital technologies, EU, 2017 (European Commission based on Eurostat Data, 2017)

Figure 14.2 shows that company size is a major factor enabling companies to digital transform. Small and Medium Sized Companies (SMEs) are closing the gap with large companies but there are a lot of opportunities still to be exploited.

What kind of instincts do companies rely on with so much uncertainty from the sheer number of unfiltered information that come across an industry while diving into the depths of digitalization and technological potentials? Who has the right "*nose*" for the selection of profitable measures and serves a pioneering corporate strategy? An essential role that leadership should be in charge with. Being vigilant and ready to act in constant change and uncertainty is a strategic maneuver any business can benefit from. Companies that do not promote and encourage their own source of transformational support are quickly being outperformed by organizations that favor a flexible organization with vigilant, bolt and innovative leaders.

14.1.1.1 Leading the Change

In times of the digital economy new promising technologies, products and businesses are growing fast and uncontrolled. The time span in which organizations and companies can focus on new profitable technologies and products to use them for their competitive advantage is relatively short. Gartner presented in 2018 a collection of promising new emerging technologies. Gartner's "Hype Cycle" (see Fig. 14.3) is a cross-industry perspective on the technologies and trends that business leaders should consider in developing emerging-technology portfolios. It presents the next generation of technologies that will make a difference in the way we work and collaborate within the workforce, create new products and services and in this sense also promote a highly fashionable customer experience. Just to mention a few groundbreaking technologies of the report that should be placed on the watchlist: In the next 10 years *Artificial Intelligence* technologies, like autonomous driving vehicles, smart robots, deep neural nets and virtual assistants will be almost available for everybody. Advanced compute power and Blockchain Technology, Digital Twin, IoT Platform and Knowledge Graphs will pave the way to *Digitalized Ecosystems*, making new business models possible. Over the next decade, humanity will begin its *"transhuman"* era using *Biohacking* tools like Biochips, Cultured or Artificial Tissue, Brain-Computer Interface, Augmented Reality, Mixed Reality and Smart Fabrics to hack their Biology, depending on lifestyle, interests and health needs. However, in this case the question remains open how unbiased and ready mankind is to make these artificial improvements of the human body and mind to optimize itself. Above all, the ethical barrier might be higher than the human benefit and the economic purpose. First steps have already been taken in biohacking by hypodermic digital implants. Adding to more human-centric technologies, *transparently immersive experiences* might help building a bridge between people, business and things and how we perceive the digital world. Looking at Smart spaces or—working Technologies this might be a milestone for innovative working and collaboration frameworks. Technological progression is in fact unstoppable. It is inevitable that top management and executives need to understand and determine on the efforts of technology to create new opportunities for the business with high strategic relevance and impact and inspire customers and employees being a forward thinker. Furthermore, leaders will benefit from assessing, using and piloting emerging technologies to engage with their workforce and customers and creating new ways of working, communication and product development and customer experience (Gartner 2018).

Figure 14.3 shows Gartner's Hype Cycle of 2018 with a specific focus on the set of technologies that show promise in delivering a high degree of competitive advantage over the next five to 10 years.

What should executives now pay *attention* to? Is it just hunting for and piloting technologies, because the *classical* management tasks remain on track while dealing

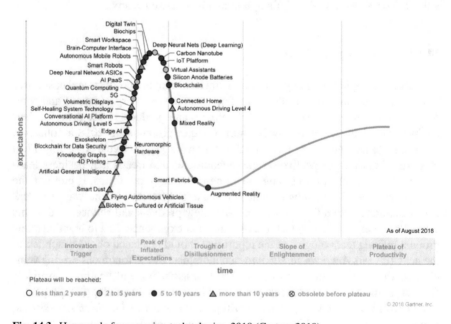

Fig. 14.3 Hype cycle for emerging technologies, 2018 (Gartner 2018)

with the challenges of the digital transformation? In this context, managers can feel quite torn in their role promoting efficiency in a highly competitive environment that puts pressure on reducing costs, while promoting digital transformation and the accompanying change, which also means watchingout for the latest trends to promote innovation in the company. One approach that connects both worlds is *ambidexterity*. "In organizational terms, dynamic capabilities are at the heart of the ability of a business to be ambidextrous—to compete simultaneously in both mature and emerging markets–to explore and exploit" (O'Reilly and Tushman 2007: 12). Diverse and complex environments often demand diverse strategies even simultaneously. "Companies need to be ambidextrous when bringing new products and technologies to market while exploiting existing ones […] Exploration and exploitation require different ways of organizing and managing. Exploration is facilitated by long-term targets, a flexible and decentralized structure, and a culture of autonomy and risk taking, while exploitation typically requires short-term targets, centralization, standardization, and discipline in execution" (Reeves et al. and Boston Consulting Group, February 2013). To go for different strategic styles fostering either *exploration* or *exploitation* a *digital culture* and *mindset* within the organization can be a major support in paving the way towards a holistic corporate strategy. Digital transformation is more then just an arms race of new products and services, it is about changing a company's attitude. This will not happen by accident. Culture reflects the organization's reaction towards change by behavior that results in their values, purpose and goals. Without changing the corporate culture, there will be no openness to innovation. Behavioral forms that support establishing openness for failure and mistakes, breaking-up internal and external silos and an outward perspective, really pay for a digital culture. Culture promotes the achievement of desired goals and helps to create a common understanding of corporate purposes. If companies foster a culture that embraces the changes, a future-oriented organization can be created, and new challenges can be better addressed. In concrete that means for the behavior that the digital corporate culture improves the way we work and think; the way we perform our tasks, decide and collaborate within and outside the organization. A corporate culture can also help getting rid of complexity and simplifying tasks and processes. Transforming the company successfully means transforming the culture as a driver for corporate performance. According to a study of Hemerling et al. and Boston Consulting Group [BCG] in April 2018 nearly 80% of the companies that focused on culture sustained strong performance and 90% setting culture under their high priority reported breakthrough or strong financial performance. A digital culture can add on the speed of a corporate transformation by dismantling hierarchies and allowing faster decision making. There is no *one fits all approach*; therefore it is undeniable that each company must define their own intricacies in the implementation of a digital culture. Even in the company itself, there will be differences from one entity to another, for example in aspects such as risk taking and decision making. Nevertheless in 2018, [BCG] has very well described the "*typical five digital culture defining elements*" a company should take into consideration transforming a corporate culture in times of digitalization:

- **"Promoting an external, rather than an internal, orientation**. A digital culture encourages employees to look outward and engage with customers and partners to create new solutions. A prime example of external orientation is the focus on the customer journey; employees shape product development and improve the customer experience by putting themselves in the customer's shoes."

- **"Prizing delegation over control**. A digital culture diffuses decision making deep into the organization. Instead of receiving explicit instructions on how to perform their work, employees follow guiding principles so that their judgment can be trusted."

- **"Encouraging boldness over caution**. In a digital culture, people are encouraged to take risks, fail fast, and learn, and they are discouraged from preserving the status quo out of habit or caution."

- **"Emphasizing more action and less planning**. In the fast-changing digital world, planning and decision making must shift from having a long-term focus to having a short-term one. A digital culture supports the need for speed and promotes continuous iteration rather than perfecting a product or idea before launching it."

- **"Valuing collaboration more than individual effort**. Success in a digital culture comes through collective work and information sharing across divisions, units, and functions. The iterative and fast pace of digital work requires a far greater level of transparency and interaction than that found in the traditional organization" ([BCG], April 2018).

Leaders play an important role in the cultural transformation of an organization or a company by making a point of fostering the new organizational values, purposes and goals. Dealing with the transformation of culture and the accompanying changes, such as the dismantling of hierarchies, new forms of collaboration, in that context also the use of new ways of working and technologies, will also reveal limits of employee's acceptance towards these changes. It is very important for leaders to engage in open dialogues with their workforce and peers, to talk about the opportunities and limitations of transformation and to be open to change in their own behaviors to act as a role model. When "walking the talk" leaders can serve as ambassadors for the transformation. They engage in open exchanges with their workforce to discuss questions about the new developments or tackle challenges together with them to manage and implement the transformation properly. Being a role model of the new determined behaviours that stand for the digital culture, means, in the best case, embodying the values and exemplifying them. That this is not always as easy as it sounds is for sure, but a clear task of leading the change and being accountable for the corporate transformation and being considered as the driving force of the overall digital transformation. If being successful, this also means highly contributing to the competitive advantage and a long-term sustainable performance of the company.

14.2 A New Star Is Born—The "Digital" Leader

During the digital transformation and the emergence and expansion of the digital economy, leadership itself is being disrupted. Leadership is rethought and innovated at different levels. While technological progress paves its way into the corporate strategy, it innovates the way people work and refines the tasks of leadership. New opportunities and challenges in collaboration arise. Work is no longer organized just in a *9 to 5 model* aiming at recurring routines and tasks to be done at a daily basis. Work is getting more and more complex due to the speed of technological progress, new challenges must be solved. This often requires subject matter experts from a variety of fields with different background knowledge. Project teams are set up to work on specific issues that are both location-independent and interdisciplinary. More and more virtual teams emerge in which the project members from different nations, expertise and locations work together. Due to the mass of information and information exchange, from emails to cloud-based data services such as sharepoints and project management tools, time also plays a key role. Project-organized collaboration turns working times upsidedown. There are different performance curves and peaks that require flexible working hours where employees are free to work remote on their tasks. These challenges demand a new leadership style. *Digital Leaders* who are willing to embrace the technological progress, are committed to implement and leverage the digital strategy of the company and drive business success. In doing so, they will have the task to translate the digital strategy into the workforce and put an emphasis on new collaboration frameworks while being a source of inspiration and a role model. Jim Link Community Voice at Forbes Human Resources Council and Chief Human Resources Officer of Randstad North America, one of the world's largest HR services providers and staffing firms, has performed a study with 3000 workers and 900 hiring decision-makers across the U.S, investigating the question: Why organizations need *Digital Leaders*. He found that: "95% of the surveyed organizations agreed that a different, new type of leadership is required to effectively address changes in organizational structures and operating models due to digitalization." Only 37% "believe they have a strong *Digital Leader* in place today." "The absence of effective leadership can even negatively impact an organization's ability to attract and retain top talent. Innovative *Digital Leaders* are often challenging to identify within a company, and they are even more difficult to cultivate." Furthermore, he assessed the five key competencies of successful leaders in the *post-digital workplace* (Link 2018). Link's core findings are that *Digital Leadership* needs executives who are willing to "*Inspire others*" being able to personally affect others' behaviours is "key to securing genuine commitment to objectives, cultivating satisfying working relationships, developing careers and ultimately driving higher productivity and performance." "This includes the ability to articulate the organization's strategic vision, foster an atmosphere where people can freely discuss the company's goals and outwardly display optimism about the long-term mission for which everyone is striving" (Link 2018). "*Leveraging Technology*" means for leaders embracing the use of digital tools and innovations. "94% of workers agree that technological innovations are influencing what is required for a leader to be successful." The *digital readiness* of executives is crucial in translating the digital strategy of the business into the actions and behavior of their workforce and their own. Seeking greater expertise and a reputation for innovativeness, digital tools and technology can increase the commitment of their peers and workforce and support their authenticity (Link 2018). According to a study by SAP and Oxford Economics who surveyed over 1500 CFOs and other senior executives globally on their strengths, weaknesses, challenges, and opportunities for growth, the most valuable and effective factor described was "*Encouraging Collaboration*". Encouraging collaboration can mean higher business performance and revenue growth. According to the study "46% of companies with zero or negative revenue and profit growth say an isolated finance function is keeping them from achieving their business goals. That percentage shrinks to 28% among respondents whose revenues are growing by 5.1–10% a year" (Brown 2017). Encouraging people to collaborate cross departments and functions is a competitive advantage for the whole company and builds new valuable and trustful relationships. "*Driving Innovation*" is also one of the most important skills for a Leader becoming a *Digital Leader*. It means being open to new ideas and testing new innovative methods and using new tools, also digital ones, to increase more acceptance for new technologies. In Link's study only "37% of employees describe their employer as a leader in digital innovation." "Innovation remains a difficult quality to cultivate, both in leaders and in organizations" (Link 2018). The role of the *Digital Leader* requires courage, even in times of setbacks and risks, not to relinquish the benefits of new technologies and to be willing to break new ground, even if it appears risky and uncertain. This is the only way an executive can support the belief in their intentions and promote a culture of digitalization. This requires not only new patterns of behavior that benefit the digital transformation, but also the control of the Leader's own ego. The vigilance for new opportunities through technological advances and the consistent handling of new forms of work require a high concentration and focus on the internal and external environment of a leader (Dettmers 2018). Which often means adapting one's own behaviors to allow space for other opinions and new ideas. An executive who is self-absorbed and dedicated to one's own success and ego has a clouded view of others' opinions and the opportunities that may rise with new ideas and even more finds it hard to break free of business as usual habits. Self-overestimation also promotes silos in the company and prevents collaboration by protecting too much what is perceived as *good practices* (Carter and Harvard Business Review 2018). There seems to be a high level of responsibility for the leaders in the digital age. Technology is being disrupted and so the role of the leader is also being disrupted by the acceleration of digitalization and it's challenges that come across a leader's daily life. It is therefore crucial to meet the pressure on the executives by future-oriented and easy-to-implement skills development to get them ready to lead the change,meet the businesses needs and to add on the corporate's success.

14.3 Setting the Path for Leadership in the Digital Age—A Guidance by Experience

In the wake of these overall changes and challenges that digital transformation causes, it is imperative to foster leaders in their development and adaptation to the role of a *Digital Leader* who incorporates a *digital mindset* and adds value to a *digital culture* in the company. Leadership development plays a key role in implementing and enhancing new behaviors and principles of leadership in the context of the company's digital transformation. It is a driver and multiplicator of the digital corporate culture and through its contribution to the development of future-oriented and digital ready executives, is effectively supporting the organization's digital corporate strategy and performance (e.g. Knies et al. 2016). Leadership development enhances leadership skills by inspiring to break fresh ground by applying newly learned practical methods and knowledge that are easily applicable in leader's daily business life and foster sustainable outputs in leadership behavior.

As described in the previous chapters, skills of *new ways of working*, *technological knowledge* and *digital literacy* are becoming increasingly important and can be promoted through the purposeful use of digital tools and media. But also, so called *Soft skills* like creativity and *empathy* are becoming increasingly important in times of digitalization. There are several key factors in the provision of trainings aiming at digital skills. It is not just about the offer of leadership development, but especially at the beginning of the change the most important role plays the *Storytelling of Transformation* and therefore the communication of: The intention or the sense *(Why?)*, the goal *(What?)* and the implementation *(How?)*. If executives recognize no clear intention behind the changes, they will follow the same patterns of behavior that have proven their worth over the past years and the transformation will be at risk to fail. As explained in Sect. 14.3, it is beneficial to draft a guidance based on the daily needs and requirements of the executives. Involving executives and important stakeholders, like board members, in the development of new programs and formats is an important lever to align with the business strategy and needs of the business to achieve greater commitment *(buy-in)* for the companies transformation. At the same time, the offer of executive development must always be challenged and adapted to the coherence and compatibility with the corporate mission and the needs of the executives. Not every training or seminar is purposeful. In the offer itself a variation of different combinations of digital as well as presence formats is an advantage. Different learning types will have the chance to access and expand learning possibilities to enhance their abilities.

The acceleration of digitalization has caused information over flow and the working days of managers are also fully packed with tasks, meetings and important decisions, even better if they can benefit from learning contents that are consumable anytime and anywhere without being bound to time or location restrictions. The learning content can be designed accessible via various media and technology. *Online* through diverse platforms made available on all mobile devices and in *presence* in various inspirational locations that promote open communication, networking and reflection. In designing the digital or non-digital content, it should be acted on the maxim: learning contents are *inspirational, fun, appealing, easy to consume, quickly available, easy to use* and *instantly applicable* in an executive's daily business life! *Videos, online readings, cases* and inspirational *self-assessment tests* or *quizzes* offer a great variety and keep the executives motivated and interested on their learning journey. Also, so-called reward strategies can be considered. In many *gamification* approaches, scores for a learning journey can be used in such a way that special or additional material or formats release when certain *Scores* are reached. However, it may be handled, it is important to design the content being useful and informative. Shaping a mix of different digital and non-digital formats favours the increase of interest and motivation. In the *presence formats*, special attention should be paid to encouraging learning from and with others in small groups when contents are intensive and in bigger groups when focusing more on inspiration. Explaining the use of new tools and methods in a vivid way encourages leaders to try out, test and discuss, breaks silos and counteracts egoism by emphasizing on openness. In an environment where mistakes are not a problem but are there to highlight new opportunities through reflection, executives are more likely to implement new ideas and methods. The use of specific technology, like virtual collaboration tools, virtual- and augmented reality can be taken into consideration when developing digital and technological literacy, nevertheless it remains important to check the deployment and the educational benefit of them. Creating space for curiosity by inserting playful and fun elements will give a boost to the creative and innovative power of leaders.

In leadership development more and more, special topics have emerged and have become important in up skilling leaders with regards to the requirements of the digital transformation. They address different skills of a *Digital Leader* promoting *Communication, Innovation, Collaboration and Resilience.* In the following, not all but some highlights of the formats that foster digital leadership skills are described:

- **Storytelling** If you cannot tell stories today nobody will listen. Boring powerpoint battles and long boring talks are not a good basis for convincing and inspiring people. Pictorial, vivid and simple language that tells a powerful story creates a human connection and a higher identification for the speaker's intention. Ever since the TED Talks this kind of verbal content mediation has received approval and worldwide recognition and a high attention to all kinds of content using it to describe the strategic relevance of a transformation can be a powerful tool. (e.g. Duncan 2014).

- **(Virtual) Collaboration** Having different people with different background and even different cultures sitting in different locations and working together on a common task is with no doubt not as easy as it seems. People must build a trustful relationship, set aside their egos and share their knowledge and expertise. Building a common ground for the virtual or non-virtual operating team is necessary to have a good start and nurtures the collaboration. Using games or social media, blogs, wikis, online collaboration tools in which they can share their best practices or workarounds for specific problems can help. Being open for brainstormings and giving time for reflections gives freedom for innovativeness and brings in new ideas (e.g. Ferrazzi 2018). Setting a clear structure and targets for meetings will remove barriers of collaboration. Maintaining a role clarity while aming for clear task accountabilities will increase collaboration and efficiency.

- **Design Thinking** Is a process and a mindset that has the power of innovatively solving problems by creating customer centric and userfriendly solutions. Following an iterative 6-step process unfolding a lively culture of innovation and fostering flexibility and creativity in the organization. An inspiring method that can be performed in small working groups to work on customer- or user specific challenges, test and pilot new ideas and build on the ability of innovativeness (e.g. Gibbons 2016).
  Figure 14.4 describes the six steps of the iterative Design Thinking Process by Nielsen Norman Group

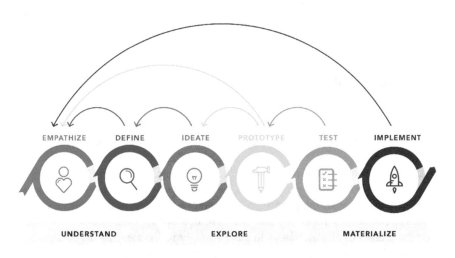

Fig. 14.4 The design thinking process (Nielsen Norman Group, Gibbons 2016)

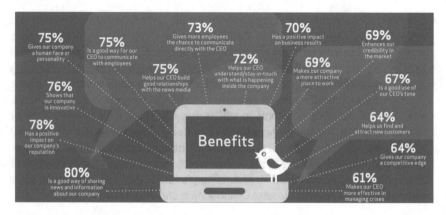

Fig. 14.5 CEO sociability yields multiple dividents (Holmes 2016)

- **Social Media** Used rightly, can be a productive companion for executives with a broad set of benefits giving executives a direct pipeline into what their customers are thinking and doing—in real time. It even requires little time and effort. Using Social Media in an authentic way, opens vast possibilities staying in touch with customers and employees and brings insights on what is going on the market. Using Social Media as an executive can benefit their outward perception, business reputation and increase company attractiveness. Being present on social networks can create a closeness to employees that would otherwise not be possible and flattens hierarchies that normally would not be possible. 3.4 billion people are on social media, what a pity it would be not to take advantage off so much marketing potential. Let it be sales, customer service or internal communication (e.g. Holmes 2016).
  Figure 14.5 describes CEOs use of social media and their perception by employees Image: webershandwick.com

- **Resilience and Mindfulness** Always being alert, communicating and reachable due to the vast possibilities of digitalization can make one loosing ground, calmness, focus and diminishes the creative spirit that is important for fresh new ideas and maintaining mental and physical health. Being able to turn of the "noise" and regulating negative emotions due to high and permanent stress can be healing and relieving. Developing strategies to improve our responses to stress can set stress free and help regaining energy and health. Some of the practices can focus on meditation or characteristics to build stronger resilience. The importance is to practice the methods as a habit to gain a sustainable effect (e.g. Swart et al. 2015: 162–165).

14.4 Conclusion

The digital transformation demands a lot of attention from the organization and the executives. Attention for fast emerging technologies, which can be a decisive competitive advantage for the entire company. Attention to changing working conditions and working forms of collaboration that are redesigned and become more and more virtual. Project work exchanges with 9 to 5 models and working routines, and the acceleration of working processes calls for more flexible work. In doing so, new forms of communication incorporating a digital nature, are being introduced into the working life. All these changes require an immense attention, which should also pay for the digital corporate culture. There is *no time for egoism* and the polishing of one's ego. Leaders can no longer isolate themselves. The digital transformation demands it's tribute. Keeping the finger on the pulse of change and sharpening the perception is imperative to engage with employees, customers and peers and to foster an open dialogue and exchange knowledge through collaboration. Outdated principles and behaviors must be discarded to adapt to the new challenges to not come in last in terms of the digital transformation. In times of digitalization it is especially important to be vigilant, mindful and attentive with oneself and the internal and external environment that surrounds a leader. With the external view towards the customer and the market, towards the trends of digitalization and innovation, chances can be better recognized and used for the company to spark a competitive advantage is generated. In the sum of the challenges that digital transformation brings to leadership and businesses, it is essential to approach executives with innovative and smart leadership development. Good leadership development enables the growth of a digital corporate culture and the associated values that result in leadership behaviour. The Implementation of future-oriented leadership development should also focus on the education and training of technological knowledge and digital literacy, *Communication, Innovation, Collaboration and Resilience*. Communication for instance will always be an essential core aspect of good leadership, investing in the development of communication skills that contribute to openness, appreciation and the establishment of a failure culture build a solid ground for digital leadership. Promoting an open dialogue and exchange with employees, peers, and customers has a lot of advantages and can, for example, support the ability of risk perception. One thing should not be forgotten: Acceleration, the *digital speed*, as a consequence of the *digital age*, has made it even more difficult for executives to find the right decisions while keeping a cool head with all the overflowing information and stimuli that are around. Therefore, the topic of resilience and mindfulness becomes more important. Although dealing with stress and negative emotions is genetically hardwired (e.g. Swart et al. 2015), it is still essential to develop mechanisms that can better manage high stress. Whatever the focus of a leader may be, it should not be forgotten that leadership has its limitations the more important it becomes to trust others. Leadership should be a source of inspiration whether for employees, customers or peers. Motivating and inspiring others is a skill that benefits all and should be the basline for changing behavior and creating an openminded, diverse and innovative corporate culture.

When executives embark on the development of their skills, it will benefit their abilities to deal with the challenges in their daily business life and enhances the company's performance. Leadership then plays a powerful role as the driver of a successful digital transformation and really is in charge of *leading the change*.

Bibliography

Birkinshaw, J. M., & Ridderstråle, J. (2017). *Fast/forward: Make your company fit for the future* (p. 8). Stanford, CA: Stanford Business Books, an imprint of Stanford University Press.

Brown, T. (2017, July 14). *The traits of successful CFOs.* Retrieved December 25, 2018 from https://www.financialdirector.co.uk/2017/07/14/successful-cfos-collaborate-not-isolate/.

Carter, R. H., & Harvard Business Review. (2018, November 07). *Ego is the enemy of good leadership.* Retrieved December 25, 2018, from https://hbr.org/2018/11/ego-is-the-enemy-of-good-leadership.

Dettmers, S. (2018, May 2). *Der alte Führungstyp hat ausgedient.* Retrieved December 25, 2018 from https://www.wiwo.de/erfolg/management/digitalisierung-der-alte-fuehrungstyp-hat-ausgedient/21235074.html.

Duncan, R. D. (2014, January 4). *Tap the Power of Storytelling.* Retrieved December 25, 2018, from https://www.forbes.com/sites/rodgerdeanduncan/2014/01/04/tap-the-power-of-storytelling/.

European Commission. (2018). *The digital economy and society index (DESI)* (p. 7). Integration of Digital Technology (n.d.). Doi: file:///C:/Users/hb187/Downloads/4DESIReportIntegrationofDigitalTechnologypdf%20(1).pdf.

Ferrazzi, K. (2018, February 01). *How successful virtual teams collaborate.* Retrieved December 25, 2019, from https://hbr.org/2012/10/how-to-collaborate-in-a-virtua.

Gada, K. (2016, June 16). *The digital economy in 5 minutes.* Retrieved December 25, 2018, from https://www.forbes.com/sites/koshagada/2016/06/16/what-is-the-digital-economy/.

Gartner Hype Cycle for Emerging Technologies, 2018. (2018, August 06). Retrieved December 25, 2018, from https://www.gartner.com/doc/3885468/hype-cycle-emerging-technologies.

Gibbons, S. (2016, July 31). *Design thinking 101.* Retrieved December 25, 2018, from https://www.nngroup.com/articles/design-thinking/.

Hemerling, J., Kilmann, J., Danoesastro, M., Stutts, L., & Ahern, C. (2018, April). It's not a digital transformation without a digital culture. Boston Consulting Group. Retrieved December 25, 2018 from https://www.bcg.com/publications/2018/not-digital-transformation-without-digital-culture.aspx.

Holmes, R. (2016, April 20). *The most important digital skill for tomorrow's CEOs.* Retrieved December 25, 2018, from https://www.weforum.org/agenda/2016/04/the-most-important-digital-skill-for-tomorrows-ceos.

Knies, E., Jacobsen, C., & Tummers, L. G. (2016). Leadership and organizational performance: State of the art and research agenda. In: J. Storey, J. L. Denis, J. Hartley, & P. 't Hart (Eds.), *Routledge companion to leadership* (pp. 404–418). London: Routledge.

Link, J. (2018, October 04). *Why Organizations need digital leaders with these five key strengths.* Retrieved December 25, 2018, from https://www.forbes.com/sites/forbeshumanresourcescouncil/2018/10/04/why-organizations-need-digital-leaders-with-these-five-key-strengths/.

O'Reilly, III, C. A., Tushman, M. L., (2007). *Ambidexterity as a dynamic capability. Resolving the innovator's dilemma* (p. 12). http://www.hbs.edu/faculty/Publication%20Files/07-088.pdf.

Reeves, M., Haanæs, K., Hollingsworth, J., & Scognamiglio Pasini, F. (2013, February). *Ambidexterity: The art of thriving in complex environments.* Boston Consulting Group. Retrieved December 25, 2018 from https://www.bcg.com/de-de/publications/2013/strategy-growth-ambidexterity-art-thriving-complex-environments.aspx.

Swart, T., Chisholm, K., & Brown, P. (2015). *Neuroscience for leadership: Harnessing the brain gain advantage* (pp. 162–165). Hampshire, UK: Palgrave Macmillan. Palgrave's The Neuroscience of Business Series.

Statista. (2018a). *Digital Economy Compass 2018*. Retrieved December 25, 2018 from https://de.statista.com/.

Statista. (2018b). *Number of apps available in leading app stores as of 3rd quarter 2018*. Retrieved December 25, 2018 from https://www.statista.com/statistics/276623/number-of-apps-available-in-leading-app-stores/.

Statista. (2018c, October). *Global digital population as of October 2018 (in millions)*. Retrieved December 25, 2018 from https://www.statista.com/statistics/617136/digital-population-worldwide/.

Hanane Bouzidi is a Senior Expert for Leadership Development Framework and Design at Deutsche Telekom AG. Working in a high technological environment where acceleration and transformation are a constant companion, she develops and implements digital and non-digital programs for top management executives group wide to add on the corporates' digital transformation strategy. Her mission is to inspire and enhance leaders' digital literacy and forward-thinking mindset within the company. She empowers leaders to bring value to their employees, customers and peers to foster a future-oriented culture within the company. Formerly working several years as a Management Consultant, she has performed major projects with diverse topics, such as innovation, human resources, sustainability and smart city always with a clear focus on digital transformation.

Chapter 15
The Role of Business Schools and Their Challenges in Educating Future Leaders: Looking Back to Move Forward

Leonardo Caporarello and Beatrice Manzoni

Abstract In this chapter, we discuss the most debated challenges for Business Schools and management education. If Business Schools want to develop responsible future leaders, they need to rethink their role. Through a structured content analysis of the past 3-years academic research on the role of Business Schools, we describe five major challenges and approaches to them. The first one is innovation (in terms of what and how to teach, and governance). The second one is relevance for practices (in terms of employability and impactful teaching and research). The third one is academic reputation (in terms of accreditation pressure and accessibility). The fourth one is promotion of intercultural differences, while the last one is interdisciplinarity (in terms of contaminating different disciplines).

15.1 Introduction

The world is undergoing a Fourth Industrial Revolution as the World Economic Forum named in 2016 the technological shift that is altering the way we live, work, and relate to one another. Several economic, social, technological and cultural changes are challenging the labor market (e.g. Dyllick 2015) and should consequently have an impact on the institutions that offer management education. The transitions from school to work are not as distinct and linear as they once were; the notion of life-long employability is replacing the notion of lifetime employment within the same organization and Business Schools should support an ongoing process of upskilling and reskilling (Manuti et al. 2015).

Given this context, we could expect Business Schools being a forge of ground-breaking innovation and rethinking. Paradoxically, this is not the reality. Many Business Schools are still in the 20th century for what and how they teach and for the way they are managed, and they engage their faculty and other key stakeholders.

L. Caporarello (✉) · B. Manzoni
SDA Bocconi School of Management, Via Bocconi, 8, 20136 Milan, Italy
e-mail: leonardo.caporarello@unibocconi.it

B. Manzoni
e-mail: beatrice.manzoni@unibocconi.it

© Springer Nature Switzerland AG 2020   209
N. Pfeffermann (ed.), *New Leadership in Strategy and Communication*,
https://doi.org/10.1007/978-3-030-19681-3_15

Yet their ultimate challenge still remains developing responsible leaders who are capable to navigate an increasingly complex—or at least different compared to the past—economy and society. This calls for rethinking what the role of Business Schools is today and will be in the future and what are the trends and the challenges in management education.

If not common practice, this is at least common sense. The need for rethinking the Business Schools' role is widely acknowledged both in research (e.g. Dyllick 2015; Starkey et al. 2004) and practice (Denning 2018; Lorange et al. 2014; Weikle 2018) where Business Schools are periodically under attack. Criticism towards them ranges from being not enough relevant for practice (e.g. Griffiths et al. 2018; Figueiredo et al. 2017) and too academically narrow and specialized in the curricula their offer (e.g. Leahey et al. 2017), to being too market driven or having sold out to the "tyranny of rankings" (e.g. Morphew et al. 2018).

Addressing this criticism in a constructive way calls for a deep understanding of the challenges in management education, how Business Schools should approach them and more in general their (future) role?

While these are frequently debated topics (e.g. Dyllick 2015; Starkey et al. 2004), there are no studies that present them in a very systematic way. For this reason, we used a structured content analysis in order to review the past three years of management literature, to depict what are the most debated challenges for Business Schools and management education in the academic debate. Research emphasizes issues and challenges of innovation, academic reputation, relevance for practice, openness to intercultural differences and interdisciplinarity. In reviewing each challenge, we unfold it and we highlight how it is currently managed by Business Schools.

The rest of the chapter is organized as follows: in Sect. 15.1.1 we explain our methods, the journals that have been included in the sample and how we coded the data. In Sect. 15.1.2 we present our results, namely the challenges Business Schools are facing and how they approach them. In Sect. 15.1.3 we draw some implications and conclusions.

15.1.1   Methods

To reflect on the role of business schools and the future of education we review the existing debate, by analyzing the past three years of academic management research.

We used content analysis (Krippendorff 2013), applying methods adopted in other recent reviews (e.g. Schad et al. 2016; Caligiuri and Thomas 2013). To ensure theoretical transparency, reliability, and validity (Krippendorff 2013), we went through four structured phases of sampling, coding, analysis and interpretation (Duriau et al. 2007) to inductively identify key themes.

First, we rigorously identified 233 academic articles on the role of business school that have been published over a three-year period, from 2016 (January 1) to 2019 (January 31). We thought that examining research published in the last three years allow us to capture latest trends which are currently debated. We focused on contributions in peer-reviewed journals in the domain of management as identified by the Academic Journal Guide 2018, published by the Chartered Association of Business Schools, with a rating of 4* and 4 (those rated 3 were included only when specific for management education).[1] The articles were identified through the Business Source Complete (EBSCO) database, through a search within the Journals previously identified (journal per journal search). We used the following keywords (and their derivates): "business school/s", "management education", "learning", "education" combined with "role", "future", "trend/s". We selected those articles that critically reflect on the trends.

Second, we read, a we coded all of the articles. In doing so, we were guided by the following questions: What are the trends and the challenges in management education and in the future role of Business Schools? How do Business Schools approach these challenges? The coding scheme was developed during the analysis. Instead of using a content analysis software, we coded the materials using self-made thematic tables, in order to keep a perspective on all of the data without losing closeness to its original context. We initially used an open coding using a concept representing the idea below the statement. This open coding resembled the concept of Gioia et al.'s (2013) first-order terms, which are more informant-centric, while the second-order themes are more researcher-centered and suggest concepts able to describe and explain phenomena. We converged on five themes which are the challenges Business Schools confront with: striving for innovation, making an impact and being relevant for practice, continuously reinforcing academic reputation and being strongly research grounded, promoting an intercultural dialogue and being interdisciplinary (see Table 15.1).

Finally, in the writing up, we further interpreted the themes, returning to the literature.

15.1.2   Challenges for Business Schools

15.1.2.1   Striving for Innovation

A constant push towards innovation appears as the major challenge for Business Schools, being addressed by 37% of studies in the sample. Innovation can occur in

---

[1]This is the list of the journals we considered: Academy of Management Learning and Education, British Journal of Industrial Relations, Human Resource Management (USA), Human Resource Management Journal (UK), Industrial Relations: A Journal of Economy and Society, Work, Employment and Society, Academy of Management Journal, Academy of Management Review, Administrative Science Quarterly, Journal of Management, Academy of Management Annals, British Journal of Management, Business Ethics Quarterly, Journal of Management Studies, Organization Science, Human Relations, Leadership Quarterly, Organization Studies, Organizational Research Methods. We added to this list the following journals with a rating of 3 because they belonged to the field of Management Development and Education: British Educational Research Journal, Management Learning, Studies in Higher Education.

Table 15.1 Challenges for Business Schools: themes in the past three years of research

Themes	% Articles discussing the theme	Exemplary articles
Striving for innovation	**37**	
Innovating learning and teaching models and methods	16	Vince et al. (2018), Tomkins and Ulus (2016), Goumaa et al. (2018), Rodgers et al. (2017), O'Neill et al. (2017)
Innovating the curriculum	12	Tan (2017), Skalicky et al. (2018), Bradley and Conway (2016), Fernandez-Sainz et al. (2016)
Innovating the institutional governance	9	Kok and McDonald (2017), Shepherd (2018), Jeanes et al. (2018)
Making an impact and being relevant for practice	**35**	
Ensuring students' employability	18	Griffiths et al. (2018), Figueiredo et al. (2017), Piróg (2016), Nelson and Sandberg (2017), Banks et al. (2016)
Impacting student's present and future behaviors through teaching	9	Vingaard Johansen et al. (2017), Aragon-Correa et al. (2017), Jung and Shin (2018), Weinberg and Flinders (2018), Janmaat (2018)
Impacting the societal and economic contexts through research	8	Moss (2016), Rhodes and Carlsen (2018), Kim et al. (2017), Cunliffe and Scaratti (2017), Lilles and Rõigas (2017)
Continuously reinforcing academic reputation and being strongly research grounded	**14**	
Avoiding becoming an institution for the élite only	6	Feeney et al. (2017), Castilla and Rissing (2018), Iannelli et al. (2016), Bowl and Hughes (2016)
Coping with the rankings and accreditation pressure, as an institution and as faculty members	5	Leckie and Goldstein (2017), Lomer et al. (2018), Noaman et al. (2017), Shukla and Singh (2016)
Being strongly theory grounded in the academic debate	3	Mehrpouya and Willmott (2018), Roohr et al. (2017), Bernstein (2017)

(continued)

Table 15.1 (continued)

Themes	% Articles discussing the theme	Exemplary articles
Promoting an intercultural dialogue	**8**	
Dealing with diverse students on campus	4	Roy et al. (2018), Bordia et al. (2018), Hammersley-Fletcher and Hanley (2016), Finn (2017)
Engaging in a dialogue local stakeholders and communities, minorities in particular	4	Wilkins (2017), Bell et al. (2018), Seeber et al. (2017)
Being interdisciplinary	**6**	
Bringing interdisciplinarity in class	4	Pountney and McPhail (2017), Lindvig et al. (2017), Seibert et al. (2017)
Embedding arts in management	2	Purg and Sutherland (2017), Roberts and Woods (2018), Gallagher et al. (2017)

different realms that relates to "what" we teach, "how" we teach and how we should manage Business Schools as organizations.

With regard to what we teach, scholars call for a curriculum change (12% of the articles address this point). The relevance of the Business School's curriculum has often been questioned in recent years, being accused not to be able to provide students with the skills organizations really need especially from executive profiles. The world is changing, and new sets of skills emerge as relevant, such as leadership and dealing with others (e.g. Skalicky et al. 2018), critical thinking (e.g. Roohr et al. 2017), making decisions and solving problems creatively (e.g. Söderhjelm et al. 2018; Fernandez-Sainz et al. 2016), empathizing with the points of view of others (e.g. Mowles 2017; Millar and Price 2018), and acting in a sustainable way (e.g. De Los Reyes 2017). Other studies highlight the importance of teaching entrepreneurship (e.g. Nabi et al. 2018; Bureau and Komporozos-Athanasiou 2017), as a source of economic renewal and growth that requires from the students' side a proactive approach towards life, learning and work (Nabi et al. 2018).

With regard to how we teach, 16% of the articles in our sample illustrate and analyze learning experiences which have been designed using a variety of learning models and methods that have in common the experiential nature of learning. In particular, scholars discuss the use of groupworks (e.g. Fandos-Herrera et al. 2017; O'Neill et al. 2017), action learning (e.g. Mughal et al. 2018; Vince et al. 2018), case studies (e.g. Skalicky et al. 2018; Bridgman et al. 2016), co-creation and dialogical process activities (Adie et al. 2018; McPhail 2016), peer feedback and role modelling (Parker and Levinson 2018) but also extracurricular activities that bring job-related experience into the curricula (Arranz et al. 2017). Some studies specifically address the role technology plays in supporting the learning in class (Henderson et al. 2017; Goumaa et al. 2018), the teamworking (Thatcher et al. 2016) and the assessment (Timmis et al. 2016).

Finally, the Business Schools' governance should also be improved, and this is a third area of innovation mentioned by 9% of the sample. Some studies discuss the potentially powerful role of the Deans in driving a transformational change (Sutphen et al. 2018; Shepherd 2018). As a result of increased pressure on them, Business Schools are becoming more and more managerialised, but this is not always productive in terms of faculty engagement (Macheridis and Paulsson 2017).

While generally speaking innovation is something good, there is also a risk in pushing too much towards innovation at multiple levels: losing the tradition in building a new identity and dismissing practices that could still work even if they are not new. To take this issue into account, Business Schools mix for example old and new learning methods (Rodgers et al. 2017) and criteria to innovate the curriculum often follow same dynamics (Annala and Mäkinen 2017). At the governance level, models of centralization and autonomy co-exist, and the model of professionalism is suggested, leaving faculty members a relatively wide degree of freedom with regard to teaching, research and some operational activities (Jones and Patton 2018).

15.1.2.2 Making an Impact and Being Relevant for Practice

A second relevant challenge is concerned with making an impact on the business and on the society. 35% of the studies in our sample deals with it, addressing how to make an impact with teaching as well as with research on each student and on the broader society.

Business Schools should guarantee quick employability, thanks to contents' and methods' ongoing innovation. Yet in reality Business Schools sometimes fail in doing so and private schools are the ones that still offer better employability compared to the one of public schools (Green et al. 2017), even if all Business Schools are also measured against their capability of being relevant for the labor market (Noaman et al. 2017). This is a challenge 18% of our sample discuss.

Several studies observed that the science-practice gap makes the transition from school to work harder, because students are not equipped with proper skills (e.g. Moore and Morton 2017; Pastore 2018). The gap is a consequence of not enough training on critical, integrative and interdisciplinary thinking (Piróg 2016); soft skills (social and personal) (e.g. Deaconu and Nistor 2017; Fearon et al. 2018); of a lack of attention to self-knowledge and to the reflective exploration of meaning (e.g. Griggs et al. 2018; Fullana et al. 2016); and of a stronger focus on teaching than on learning (Dyllick 2015). This implies innovating the curriculum as above mentioned, but also asking faculty members to develop more professional business-related expertise as well (Noaman et al. 2017).

Some papers report that frustrated students sometimes leave school demonstrating entrepreneurial behaviors, that business schools had not been able to intercept and develop their inclination (e.g. Fischer et al. 2016; Stenard and Sauermann 2016).

Through what they teach and how they teach, Business Schools have the possibility to really impact on the present and future behaviors of their students, addressing new ways for sustainability and ethics in business. This is a topic that occurs in 9% of the sample. Also, a call for a dialogue with business entities was launched in thinking about impact and society (Kellard and Śliwa 2016). The business school is perceived as the place where to promote these values. Scholars agree that the promotion is vehiculated by both adopting societal values and teaching them to form future managers (e.g. Sales de Aguiar and Paterson 2018; Snelson-Powell et al. 2016). This asks for faculty and staff being role models in terms of ethics (e.g. Janmaat 2018; Gupta and Sayeed 2016) and guiding their thinking and future actions (Anderson et al. 2017). Research reports that the new generation of leaders already shows some behavioral change in this sense (Jung and Shin 2018).

The research carried on by the Business Schools should also be more relevant for practice (8% of the sample deal with this issue). In this sense, some articles call for a shift in the role of Business Schools which should move from being a place where knowledge is generated to be a place where knowledge is disseminated to make an impact on the society as a whole (Williams 2016; George 2016).

Business Schools that don't address these challenges risk remaining auto-referential and engulfed in research that is detached from practice and losing the privilege to educate future managers (Nelson and Sandberg 2017). Moreover, not following sustainable and ethical practices can concur to generate new managers with lack of social concern on the market, increasing negative behaviors for the society (Snelson-Powell et al. 2016).

Many studies in this subsample offer best practices to deal with this challenge, that specifically relates to the use of learning methods that facilitate theory/practice integration. Scholars discuss the use of business games and case studies (e.g. Schonell and Macklin 2018; Grant and Baden-Fuller 2018), guest speakers from the business world (e.g. Maguire et al. 2018; Finch et al. 2017) but also more support in terms of career counseling service (Clements and Kamau 2018; Donald et al. 2017). A curriculum change is also proposed to include more interpersonal communication skills (Deaconu and Nistor 2017) and tools for critical reflection (Harker et al. 2016). To specifically make an impact on sustainability, scholar recommend the use of empathy tools (Michaelson 2016) and the development of fresh problem-solving techniques to face complex social problems (Sales de Aguiar and Paterson 2018).

15.1.2.3   Continuously Reinforcing Academic Reputation and Being Strongly Research Grounded

Building and consolidating a strong academic reputation, leveraging on a solid theoretical ground, is a third crucial concern for business schools and 14% of studies are focused on this topic. This challenge unfolds in remaining an accessible institution even when the prestige grows; facing the pressures that come from the rankings and the accreditation processes; being strongly present and grounded in the academic research debate.

Some studies (6% of our sample) critically question whether the process of building reputation and climbing the rankings confines Business Schools within their walls encouraging elite formation and auto-referentiality. Prospective students from wealthy families apply for A-level schools to maintain their status (Iannelli et al. 2016) and to acquire future elite positions in the business world (e.g. Nixon et al. 2018; Feeney et al. 2017). With regard to this issue, Business Schools face the challenge of making university education accessible. Limitation of accessibility to universities are connected to other elements such as tuitions, asymmetry of information and selective orientation of students for higher education so for instance, channel of communication and promotion of certain business schools are limited to some elite colleges (Dilnot 2016). The risk is that Business Schools become a place for rich and isolated students, disconnected from the rest of the world (Currie et al. 2016).

The second topic (5% of the articles) deals with the pressure coming from the rankings and the international accreditations, even if Business Schools do not all feel the pressure in the same way everywhere. Where the pressure is strong, Business Schools feel their reputation is at question every year and institutions such as the Financial Times, Forbes, Bloomberg Businessweek and The Economist have the last say. They make a ranking which is based on certain criteria, such as for instance the number of top-tier publications (e.g. Morphew et al. 2018; Mehrpouya and Willmott 2018).

This connects to the third topic of being actively present in the academic research debate, asking the faculty to publish more and more on top tier journal, without giving into anti-intellectual behaviors and academic misconduct, especially in research where they "play" with data, making minor changes in data and sources presenting them as new (Butler et al. 2017).

Other accreditation stress come from the increasing managerialism that introduce private corporations' rules and practices in the university, with performance management processes and formal managers in charge of managing people and resources (Hamlin and Patel 2017). This is perceived as critical by 3% in the sample which note that the faculty is forced to adhere to some form of goal setting and quantitative targets that are in conflict with the traditional values of freedom and sense of community (Kallio et al. 2016).

Literature is still fragmented on how to deal with this challenge. Some scholars recommend engaging with several forms of knowledge dissemination in open journals, conferences, workshops, and critically debate research problems with the public (Mehrpouya and Willmott 2018). Some other scholars discuss possible forms of decentralization in the governance, that allow the faculty to be more engaged in decision making processes instead of feeling controlled (e.g. Macheridis and Paulsson 2017).

15.1.2.4 Promoting an Intercultural Dialogue

Another challenge for Business Schools deals with their role in promoting an intercultural dialogue and creating an environment that is open to cultural differences, as a consequence of an increasingly globalized world. 8% of the studies in our sample deal with this topic.

On the one hand, intercultural diversity exists on campus, being the number of students spending a period abroad steadily increasing (Roy et al. 2018). The presence of international students on campus is perceived as an occasion of inclusion, as an opportunity to develop stronger professional and managerial skills (Finn 2017) and as a chance to promote global citizenship (Lehtomäki et al. 2016). With foreign students incoming, the campus becomes a place representing our multicultural society and international students. An issue is that international students are often spending a limited amount of time in a foreign country and this can produce a low cross-cultural communication and only a few interactions with domestic students (Finn 2017). Given cultural differences, these students also require a different assistance when making education-related and career-related choices (e.g. Nada and Araújo 2018; Hammersley-Fletcher and Hanley 2016; Bordia et al. 2018).

On the other hand, intercultural diversity is evident in the interactions with local community and society. Business Schools play a role within their territory encouraging inclusion of diversity and an intercultural dialogue. The idea is that the university becomes a neutral place, that promotes values of tolerance and inclusion (e.g. Beamish 2017; Seeber et al. 2017). Particularly sensitive cases are present in war zones, as for instance in Israeli where the Arab population is a minority represented in universities (Yirmiyahu et al. 2017) or in Australia where indigenous communities should be better included in governance-related decision-making processes (Wise et al. 2018).

Moreover, business schools are moving at an international level, opening new branches or growing in less developed countries. This new wave of globalization generated a series of problems dealing with the local culture. Problems are concerned with the respect of the local tradition when transferring means and values from the Western world (e.g. Beamish 2017; Bell et al. 2018). With regard to this, some scholars point out that keeping the balance between international concerns and demands and local traditions is among the hardest challenges (e.g. Finn 2017; Kothiyal et al. 2018).

Among the management approaches to deal with these challenges, existing studies recommend the use of better tutorship services for international students, from everyday support to career orientation (Nada and Araújo 2018); while the promotion of doctoral studies on the identity, multilingual scholarships (Kothiyal et al. 2018) and an integration of teaching approaches and values from different cultures can help in making Business Schools truly more international, without losing its traditional identity (e.g. Kothiyal et al. 2018; Hammersley-Fletcher and Hanley 2016).

15.1.2.5   Being Interdisciplinary

The last challenge for Business Schools which emerge in 6% of our articles asks Business Schools to be more interdisciplinary in what and how they teach. Being interdisciplinary implies bringing together two or more fields and disciplines in order to enrich learning and knowledge generation.

More than a study claims that the integration of other fields different from management can improve management research (e.g. Seibert et al. 2017; Schad et al. 2016) and contribute to better develop the skills the job market asks for (Power and Handley 2017). Psychology, philosophy and neuroscience can highlight new paradoxes in management research and they already demonstrated great application respectively in marketing and economics theory (Schad et al. 2016; Lindebaum 2016).

The challenge for the Business Schools relates to how to make this integration among disciplines possible and real. Two options exist. The first option is that faculty members can invest on becoming themselves more interdisciplinary, working by their own in different fields (Safavi and Håkanson 2018), which however does not prove beneficial for publishing, because a highly specialized expertise and a clear positioning in a very specific and narrow domain make publishing and belonging to a specific research community easier (Mehrpouya and Willmott 2018). The second option is that Business Schools can sustain e promote the collaboration between faculty members with a very diverse expertise (Pountney and McPhail 2017).

Interdisciplinarity is challenging for the faculty, but also for the students because it asks for greater learning agility (Power and Handley 2017; Knewstubb 2016).

Some studies (2% of our sample) specifically discuss the integration of arts into management curricula with the inclusion of arts courses, artistic practices (e.g. collage) and arts-based workshops. This is a promising and growing area in the literature (Purg and Sutherland 2017). Art has the power to challenge management practices stimulating deeper reflection and critical understanding of certain phenomena within the business world (Purg and Sutherland 2017). For instance, art best actives a reflection on leadership (Roberts and Woods 2018) and self-reflection skills (Purg and Sutherland 2017).

In the articles we reviewed, scholars do not see risks in being more interdisciplinary or at least the benefits strongly overcome potential drawbacks in terms of losing a specialized and focused expertise.

Among the approaches proposed to deal with this challenge, scholars discuss best practices of cooperation and collaboration across faculty members that engage in cross-disciplinary projects (e.g. Rienties and Héliot 2018; Power and Handley 2017), that implies for example co-authoring research papers (Seibert et al. 2017), or having interfaculty group working (Rienties and Héliot 2018) or workshops (Gallagher et al. 2017). Scholars also recommend developing a climate for integration of diverse fields and to establish rewards schemes that encourage cooperation and interdisciplinarity (Power and Handley 2017). Moreover, they suggest adopting special dialogical approach with students making practical examples (Pountney and McPhail 2017) or using hermeneutics (Knewstubb 2016).

15.1.3   Conclusions

Business Schools stand at a pivotal point in their histories. Given the ongoing eco-nomic, social, technological and cultural changes, they can be among the key insti-tutions contributing to and fostering the ongoing transformation by preparing new generations of leaders. This asks for renovating their roles and effectively managing the challenges they face.

While the challenges per se may not new in the international debate, given the continuous evolution of the socio-economic context it becomes relevant for Business Schools to first self-evaluate their capability to face such challenges, and to be able, more than the others, to continuously evolve and adapt to the future state. The digital and technological evolution, the generation effect, the open online course market are just a few examples of driving factors that call for a renovation of Business Schools' model. And this renovation is already ongoing now. Consequently, the question is: what are the major challenges that Business schools must face today? As there is a wide debate in literature, in this chapter we outlined the future role of Business Schools, by shedding light on the challenges they face according to the most recent academic publications.

According to research, Business Schools face multiple and often competing demands: striving for continuous innovation which is inevitably onerous in terms of any type of resources; making an impact and being relevant for practice while reinforcing academic reputation and being strongly research grounded; promoting an intercultural dialogue while respecting local and long-standing traditions; being interdisciplinary while also highly specialized. Coping with them and managing them is often far from being easy and it often translates into making a choice between dif-ferent priorities which are perceived as inevitably alternative one to each other.

Instead, we recommend a shift in the way these tensions are experienced and approached. In the immediate future the role of Business Schools will be, even more than in the past, integrating these competing demands managing them all, by leverag-ing the opposites, stressing interdependence and making synergies. The challenges we outlined in this chapter are interconnected and mutually reinforcing. Creating good research-grounded knowledge inspires the designing and the implementation of effective learning experiences that informs business leaders' action and ultimately improves the society. Moreover, being interdisciplinary and sensitive to intercultural differences can only improve the learning experience Business School offers.

These is in fact a common denominator in the challenges we examined, which is the challenge of creating valuable learning experiences for future leaders. This contributes furthermore to the debate, and it also becomes interesting exploring how these learning experiences are designed, delivered and perceived by learners, and what are the learning methods most appropriate with the intended learning goals to achieve. These aspects can be part of further research and studies. Ultimately speaking, this is also part of the role of Business Schools.

Bibliography

Adie, L., van der Kleij, F., & Cumming, J. (2018). The development and application of coding frame-works to explore dialogic feedback interactions and self-regulated learning. *British Educational Research Journal, 44*(4), 704–723.

Anderson, L., Ellwood, P., & Coleman, C. (2017). The impactful academic: Relational management education as an intervention for impact. *British Journal of Management, 28*(1), 14–28.

Annala, J., & Mäkinen, M. (2017). Communities of practice in higher education: Contradictory nar-ratives of a university-wide curriculum reform. *Studies in Higher Education, 42*(11), 1941–1957.

Aragon-Correa, J. A., Marcus, A. A., Rivera, J. E., & Kenworthy, A. L. (2017). Sustainability management teaching resources and the challenge of balancing planet, people, and profits.

Arranz, N., Ubierna, F., Arroyabe, M. F., Perez, C., & de Arroyabe, J. C. F. (2017). The effect of curricular and extracurricular activities on university students' entrepreneurial intention and competences. *Studies in Higher Education, 42*(11), 1979–2008.

Banks, G. C., Pollack, J. M., Bochantin, J. E., Kirkman, B. L., Whelpley, C. E., & O'Boyle, E. H. (2016). Management's science–practice gap: A grand challenge for all stakeholders. *Academy of Management Journal, 59*(6), 2205–2231.

Beamish, P. W. (2017). The transferability of western business education to the east. *Journal of Management Studies.*

Bell, R. G., Filatotchev, I., Krause, R., & Hitt, M. (2018). From the guest editors: Opportunities and challenges for advancing strategic management education. *Academy of Management Learning and Education, 17*(3), 233–240.

Bernstein, E. S. (2017). Making transparency transparent: The evolution of observation in manage-ment theory. *Academy of Management Annals, 11*(1), 217–266.

Bordia, S., Bordia, P., Milkovitz, M., Shen, Y., & Restubog, S. L. D. (2018). What do international students really want? An exploration of the content of international students' psychological contract in business education. *Studies in Higher Education*, 1–15.

Bowl, M., & Hughes, J. (2016). Fair access and fee setting in English universities: What do institu-tional statements suggest about university strategies in a stratified quasi-market? *Studies in Higher Education, 41*(2), 269–287.

Bradley, J. L., & Conway, P. F. (2016). A dual step transfer model: Sport and non-sport extracurricular activities and the enhancement of academic achievement. *British Educational Research Journal, 42*(4), 703–728.

Bridgman, T., Cummings, S., & McLaughlin, C. (2016). Restating the case: How revisiting the development of the case method can help us think differently about the future of the business school. *Academy of Management Learning & Education, 15*(4), 724–741.

Bureau, S. P., & Komporozos-Athanasiou, A. (2017). Learning subversion in the business school: An 'improbable' encounter. *Management Learning, 48*(1), 39–56.

Butler, N., Delaney, H., & Spoelstra, S. (2017). The gray zone: Questionable research practices in the business school. *Academy of Management Learning & Education, 16*(1), 94–109.

Caligiuri, P., & Thomas, D. C. (2013). From the editors: How to write a high-quality review. *Journal of International Business Studies, 44*(6), 547–553.

Castilla, E. J., & Rissing, B. A. (2018). Best in Class: The returns on application endorsements in higher education. *Administrative Science Quarterly,* 0001839218759965.

Clements, A. J., & Kamau, C. (2018). Understanding students' motivation towards proactive career behaviours through goal-setting theory and the job demands–resources model. *Studies in Higher Education, 43*(12), 2279–2293.

Cunliffe, A. L., & Scaratti, G. (2017). Embedding impact in engaged research: Developing socially useful knowledge through dialogical sensemaking. *British Journal of Management, 28*(1), 29–44.

Currie, G., Davies, J., & Ferlie, E. (2016). A call for university-based business schools to "lower their walls:" Collaborating with other academic departments in pursuit of social value. *Academy of Management Learning & Education, 15*(4), 742–755.

De Los Reyes, G., Jr., Kim, T. W., & Weaver, G. R. (2017). Teaching ethics in business schools: A conversation on disciplinary differences, academic provincialism, and the case for integrated pedagogy. *Academy of Management Learning & Education, 16*(2), 314–336.

Deaconu, A., & Nistor, C. S. (2017). Competences in Romanian higher education–an empirical investigation for the business sector. *Studies in Higher Education, 42*(11), 1917–1940.

Denning, S. (2018). Why today's Business Schools teach yesterday's expertise, Forbes. https://www.forbes.com/sites/stevedenning/2018/05/27/why-todays-business-schools-teach-yesterdays-expertise.

Dilnot, C. (2016). How does the choice of A-level subjects vary with students' socio-economic status in English state schools? *British Educational Research Journal, 42*(6), 1081–1106.

Donald, W. E., Baruch, Y., & Ashleigh, M. (2017). The undergraduate self-perception of employability: Human capital, careers advice, and career ownership. *Studies in Higher Education,* 1–16.

Duriau, V. J., Reger, R. K., & Pfarrer, M. D. (2007). A content analysis of the content analysis literature in organization studies—Research themes, data sources, and methodological refinements. *Organizational Research Methods, 10*(1), 5–34.

Dyllick, T. (2015). Responsible management education for a sustainable world: The challenges for business schools. *Journal of Management Development, 34*(1), 16–33.

Fandos-Herrera, C., Jiménez-Martínez, J., Orús, C., & Pina, J. M. (2017). Introducing the discussant role to stimulate debate in the classroom: Effects on interactivity, learning outcomes, satisfaction and attitudes. *Studies in Higher Education,* 1–17.

Fearon, C., Nachmias, S., McLaughlin, H., & Jackson, S. (2018). Personal values, social capital, and higher education student career decidedness: A new 'protean'-informed model. *Studies in Higher Education, 43*(2), 269–291.

Feeney, S., Hogan, J., & O'Rourke, B. K. (2017). Elite formation in the higher education systems of Ireland and the UK: Measuring, comparing and decomposing longitudinal patterns of cabinet members. *British Educational Research Journal, 43*(4), 720–742.

Fernandez-Sainz, A., García-Merino, J. D., & Urionabarrenetxea, S. (2016). Has the Bologna process been worthwhile? An analysis of the learning society-adapted outcome index through quantile regression. *Studies in Higher Education, 41*(9), 1579–1594.

Figueiredo, H., Biscaia, R., Rocha, V., & Teixeira, P. (2017). Should we start worrying? Mass higher education, skill demand and the increasingly complex landscape of young graduates' employment. *Studies in Higher Education, 42*(8), 1401–1420.

Finch, D., Deephouse, D. L., O'Reilly, N., Foster, W. M., Falkenberg, L., & Strong, M. (2017). Institutional biography and knowledge dissemination: An analysis of Canadian business school faculty. *Academy of Management Learning & Education, 16*(2), 237–256.

Finn, K. (2017). Multiple, relational and emotional mobilities: Understanding student mobilities in higher education as more than 'staying local' and 'going away'. *British Educational Research Journal, 43*(4), 743–758.

Fischer, M. D., Dopson, S., Fitzgerald, L., Bennett, C., Ferlie, E., Ledger, J., et al. (2016). Knowledge leadership: Mobilizing management research by becoming the knowledge object. *Human Relations, 69*(7), 1563–1585.

Fullana, J., Pallisera, M., Colomer, J., Fernández Peña, R., & Pérez-Burriel, M. (2016). Reflective learning in higher education: A qualitative study on students' perceptions. *Studies in Higher Education, 41*(6), 1008–1022.

Gallagher, M., Prior, J., Needham, M., & Holmes, R. (2017). Listening differently: A pedagogy for expanded listening. *British Educational Research Journal, 43*(6), 1246–1265.

George, G. (2016). Management research in AMJ: Celebrating impact while striving for more. *Academy of Management Journal, 59*(6), 1869–1877.

Gioia, D. A., Corley, K. G., & Hamilton, A. L. (2013). Seeking qualitative rigor in inductive research notes on the Gioia methodology. *Organizational Research Methods, 16*(1), 15–31.

Goumaa, R., Anderson, L., & Zundel, M. (2018). What can managers learn online? Investigating possibilities for active understanding in the online MBA classroom. *Management Learning,* 1350507618800602.

Grant, R. M., & Baden-Fuller, C. (2018). How to develop strategic management competency: Reconsidering the learning goals and knowledge requirements of the core strategy course. *Academy of Management Learning & Education*.

Green, F., Henseke, G., & Vignoles, A. (2017). Private schooling and labour market outcomes. *British Educational Research Journal, 43*(1), 7–28.

Griffiths, D. A., Inman, M., Rojas, H., & Williams, K. (2018). Transitioning student identity and sense of place: Future possibilities for assessment and development of student employability skills. *Studies in Higher Education, 43*(5), 891–913.

Griggs, V., Holden, R., Lawless, A., & Rae, J. (2018). From reflective learning to reflective practice: Assessing transfer. *Studies in Higher Education, 43*(7), 1172–1183.

Gupta, M., & Sayeed, O. (2016). Social responsibility and commitment in management institutes: Mediation by engagement. *Business: Theory and Practice, 17*, 280.

Hamlin, R. G., & Patel, T. (2017). Perceived managerial and leadership effectiveness within higher education in France. *Studies in Higher Education, 42*(2), 292–314.

Hammersley-Fletcher, L., & Hanley, C. (2016). The use of critical thinking in higher education in relation to the international student: Shifting policy and practice. *British Educational Research Journal, 42*(6), 978–992.

Harker, M. J., Caemmerer, B., & Hynes, N. (2016). Management education by the French Grandes Ecoles de Commerce: Past, present, and an uncertain future. *Academy of Management Learning & Education, 15*(3), 549–568.

Henderson, M., Selwyn, N., & Aston, R. (2017). What works and why? Student perceptions of 'useful' digital technology in university teaching and learning. *Studies in Higher Education, 42*(8), 1567–1579.

Iannelli, C., Smyth, E., & Klein, M. (2016). Curriculum differentiation and social inequality in higher education entry in Scotland and Ireland. *British Educational Research Journal, 42*(4), 561–581.

Janmaat, J. G. (2018). Educational influences on young people's support for fundamental British values. *British Educational Research Journal, 44*(2), 251–273.

Jeanes, E., Loacker, B., & Śliwa, M. (2018). Complexities, challenges and implications of collaborative work within a regime of performance measurement: The case of management and organisation studies. *Studies in Higher Education*, 1–15.

Jones, D. R., & Patton, D. (2018). An academic challenge to the entrepreneurial university: The spatial power of the 'Slow Swimming Club'. *Studies in Higher Education*, 1–15.

Jung, J., & Shin, T. (2018). Learning not to diversify: The transformation of graduate business education and the decline of diversifying acquisitions. *Administrative Science Quarterly*, 0001839218768520.

Kallio, K. M., Kallio, T. J., Tienari, J., & Hyvönen, T. (2016). Ethos at stake: Performance management and academic work in universities. *Human Relations, 69*(3), 685–709.

Kellard, N. M., & Śliwa, M. (2016). Business and management impact assessment in research excellence framework 2014: Analysis and reflection. *British Journal of Management, 27*(4), 693–711.

Kim, Y., Horta, H., & Jung, J. (2017). Higher education research in Hong Kong, Japan, China, and Malaysia: Exploring research community cohesion and the integration of thematic approaches. *Studies in Higher Education, 42*(1), 149–168.

Knewstubb, B. (2016). The learning–teaching nexus: Modelling the learning–teaching relationship in higher education. *Studies in Higher Education, 41*(3), 525–540.

Kok, S. K., & McDonald, C. (2017). Underpinning excellence in higher education–an investigation into the leadership, governance and management behaviours of high-performing academic departments. *Studies in Higher Education, 42*(2), 210–231.

Kothiyal, N., Bell, E., & Clarke, C. (2018). Moving beyond mimicry: Developing hybrid spaces in Indian business schools. *Academy of Management Learning & Education, 17*(2), 137–154.

Krippendorff, K. (2013). *Content analysis: An introduction to its methodology* (3rd ed.). Los Angeles, CA: SAGE.

Leahey, E., Beckman, C. M., & Stanko, T. L. (2017). Prominent but less productive: The impact of interdisciplinarity on scientists' research. *Administrative Science Quarterly, 62*(1), 105–139.

Leckie, G., & Goldstein, H. (2017). The evolution of school league tables in England 1992–2016: 'Contextual value-added', 'expected progress' and 'progress 8'. *British Educational Research Journal, 43*(2), 193–212.

Lehtomäki, E., Moate, J., & Posti-Ahokas, H. (2016). Global connectedness in higher education: Student voices on the value of cross-cultural learning dialogue. *Studies in Higher Education, 41*(11), 2011–2027.

Lilles, A., & Rõigas, K. (2017). How higher education institutions contribute to the growth in regions of Europe? *Studies in Higher Education, 42*(1), 65–78.

Lindebaum, D. (2016). Critical essay: Building new management theories on sound data? The case of neuroscience. *Human Relations, 69*(3), 537–550.

Lindvig, K., Lyall, C., & Meagher, L. R. (2017). Creating interdisciplinary education within monodisciplinary structures: The art of managing interstitiality. *Studies in Higher Education,* 1–14.

Lomer, S., Papatsiba, V., & Naidoo, R. (2018). Constructing a national higher education brand for the UK: Positional competition and promised capitals. *Studies in Higher Education, 43*(1), 134–153.

Lorange, P., Sheth, J. N., & Thomas, H. (2014). New models and the changing contexts of business schools. *The European Business Review.* http://www.europeanbusinessreview.com/new-business-models-and-the-changing-contexts-of-business-school-2/.

Macheridis, N., & Paulsson, A. (2017). Professionalism between profession and governance: How university teachers' professionalism shapes coordination. *Studies in Higher Education,* 1–16.

Maguire, K., Prodi, E., & Gibbs, P. (2018). Minding the gap in doctoral supervision for a contemporary world: A case from Italy. *Studies in Higher Education, 43*(5), 867–877.

Manuti, A., Pastore, S., Scardigno, A. F., Giancaspro, M. L., & Morciano, D. (2015). Formal and informal learning in the workplace: A research review. *International Journal of Training and Development, 19*(1), 1–17.

McPhail, G. (2016). The fault lines of recontextualisation: The limits of constructivism in education. *British Educational Research Journal, 42*(2), 294–313.

Mehrpouya, A., & Willmott, H. (2018). Making a niche: The marketization of management research and the rise of 'knowledge branding'. *Journal of Management Studies, 55*(4), 728–734.

Michaelson, C. (2016). A novel approach to business ethics education: Exploring how to live and work in the 21st century. *Academy of Management Learning & Education, 15*(3), 588–606.

Millar, J., & Price, M. (2018). Imagining management education: A critique of the contribution of the United Nations PRME to critical reflexivity and rethinking management education. *Management Learning,* 1350507618759828.

Moore, T., & Morton, J. (2017). The myth of job readiness? Written communication, employability, and the 'skills gap' in higher education. *Studies in Higher Education, 42*(3), 591–609.

Morphew, C. C., Fumasoli, T., & Stensaker, B. (2018). Changing missions? How the strategic plans of research-intensive universities in Northern Europe and North America balance competing identities. *Studies in Higher Education, 43*(6), 1074–1088.

Moss, G. (2016). Knowledge, education and research: Making common cause across communities of practice. *British Educational Research Journal, 42*(6), 927–944.

Mowles, C. (2017). Experiencing uncertainty: On the potential of groups and a group analytic approach for making management education more critical. *Management Learning, 48*(5), 505–519.

Mughal, F., Gatrell, C., & Stead, V. (2018). Cultural politics and the role of the action learning facilitator: Analysing the negotiation of critical action learning in the Pakistani MBA through a Bourdieusian lens. *Management Learning, 49*(1), 69–85.

Nabi, G., Walmsley, A., Liñán, F., Akhtar, I., & Neame, C. (2018). Does entrepreneurship education in the first year of higher education develop entrepreneurial intentions? The role of learning and inspiration. *Studies in Higher Education, 43*(3), 452–467.

Nada, C. I., & Araújo, H. C. (2018). 'When you welcome students without borders, you need a mentality without borders' internationalisation of higher education: Evidence from Portugal. *Studies in Higher Education,* 1–14.

Nelson, A., & Sandberg, M. (2017). Labour-market orientation and approaches to studying—A study of the first *Bologna Students* at a Swedish Regional University. *Studies in Higher Education, 42*(8), 1545–1566.

Nixon, E., Scullion, R., & Hearn, R. (2018). Her majesty the student: Marketised higher education and the narcissistic (dis) satisfactions of the student-consumer. *Studies in Higher Education, 43*(6), 927–943.

Noaman, A. Y., Ragab, A. H. M., Madbouly, A. I., Khedra, A. M., & Fayoumi, A. G. (2017). Higher education quality assessment model: Towards achieving educational quality standard. *Studies in Higher Education, 42*(1), 23–46.

O'Neill, T. A., Hoffart, G. C., McLarnon, M. M., Woodley, H. J., Eggermont, M., Rosehart, W., et al. (2017). Constructive controversy and reflexivity training promotes effective conflict profiles and team functioning in student learning teams. *Academy of Management Learning & Education, 16*(2), 257–276.

Parker, R., & Levinson, M. P. (2018). Student behaviour, motivation and the potential of attachment-aware schools to redefine the landscape. *British Educational Research Journal, 44*(5), 875–896.

Pastore, F. (2018). Why so slow? The school-to-work transition in Italy. *Studies in Higher Education,* 1–14.

Piróg, D. (2016). The impact of degree programme educational capital on the transition of graduates to the labour market. *Studies in Higher Education, 41*(1), 95–109.

Pountney, R., & McPhail, G. (2017). Researching the interdisciplinary curriculum: The need for 'translation devices'. *British Educational Research Journal, 43*(6), 1068–1082.

Power, E. J., & Handley, J. (2017). A best-practice model for integrating interdisciplinarity into the higher education student experience. *Studies in Higher Education,* 1–17.

Purg, D., & Sutherland, I. (2017). Why art in management education? Questioning meaning.

Rhodes, C., & Carlsen, A. (2018). The teaching of the other: Ethical vulnerability and generous reciprocity in the research process. *Human Relations, 71*(10), 1295–1318.

Rienties, B., & Héliot, Y. (2018). Enhancing (in) formal learning ties in interdisciplinary management courses: A quasi-experimental social network study. *Studies in Higher Education, 43*(3), 437–451.

Roberts, A., & Woods, P. A. (2018). Theorising the value of collage in exploring educational leadership. *British Educational Research Journal, 44*(4), 626–642.

Rodgers, W., Simon, J., & Gabrielsson, J. (2017). Combining experiential and conceptual learning in accounting education: A review with implications. *Management Learning, 48*(2), 187–205.

Roohr, K. C., Liu, H., & Liu, O. L. (2017). Investigating student learning gains in college: A longitudinal study. *Studies in Higher Education, 42*(12), 2284–2300.

Roy, A., Newman, A., Ellenberger, T., & Pyman, A. (2018). Outcomes of international student mobility programs: A systematic review and agenda for future research. *Studies in Higher Education,* 1–15.

Safavi, M., & Håkanson, L. (2018). Advancing theory on knowledge governance in universities: A case study of a higher education merger. *Studies in Higher Education, 43*(3), 500–523.

Sales de Aguiar, T. R., & Paterson, A. S. (2018). Sustainability on campus: Knowledge creation through social and environmental reporting. *Studies in Higher Education, 43*(11), 1882–1894.

Schad, J., Lewis, M. W., Raisch, S., & Smith, W. K. (2016). Paradox research in management science: Looking back to move forward. *The Academy of Management Annals, 10*(1), 5–64.

Schonell, S., & Macklin, R. (2018). Work integrated learning initiatives: Live case studies as a mainstream WIL assessment. *Studies in Higher Education,* 1–12.

Seeber, M., Barberio, V., Huisman, J., & Mampaey, J. (2017). Factors affecting the content of universities' mission statements: an analysis of the United Kingdom higher education system. *Studies in Higher Education,* 1–15.

Seibert, S. E., Kacmar, K. M., Kraimer, M. L., Downes, P. E., & Noble, D. (2017). The role of research strategies and professional networks in management scholars' productivity. *Journal of Management, 43*(4), 1103–1130.

Shepherd, S. (2018). Managerialism: An ideal type. *Studies in Higher Education, 43*(9), 1668–1678.

Shukla, A., & Singh, S. (2016). Facets of academic excellence in management education: Conceptualization and instrument development in India. *Studies in Higher Education, 41*(11), 1883–1899.

Skalicky, J., Warr Pedersen, K., van der Meer, J., Fuglsang, S., Dawson, P., & Stewart, S. (2018). A framework for developing and supporting student leadership in higher education. *Studies in Higher Education,* 1–17.

Snelson-Powell, A., Grosvold, J., & Millington, A. (2016). Business school legitimacy and the challenge of sustainability: A fuzzy set analysis of institutional decoupling. *Academy of Management Learning & Education, 15*(4), 703–723.

Söderhjelm, T., Björklund, C., Sandahl, C., & Bolander-Laksov, K. (2018). Academic leadership: management of groups or leadership of teams? A multiple-case study on designing and implementing a team-based development programme for academic leadership. *Studies in Higher Education, 43*(2), 201–216.

Starkey, K., Hatchuel, A., & Tempest, S. (2004). Rethinking the business school. *Journal of Management Studies, 41*(8), 1521–1531.

Stenard, B. S., & Sauermann, H. (2016). Educational mismatch, work outcomes, and entry into entrepreneurship. *Organization Science, 27*(4), 801–824.

Sutphen, M., Solbrekke, T. D., & Sugrue, C. (2018). Toward articulating an academic praxis by interrogating university strategic plans. *Studies in Higher Education,* 1–13.

Tan, C. (2017). Teaching critical thinking: Cultural challenges and strategies in Singapore. *British Educational Research Journal, 43*(5), 988–1002.

Thatcher, J., Alao, H., Brown, C. J., & Choudhary, S. (2016). Enriching the values of micro and small business research projects: Co-creation service provision as perceived by academic, business and student. *Studies in Higher Education, 41*(3), 560–581.

Timmis, S., Broadfoot, P., Sutherland, R., & Oldfield, A. (2016). Rethinking assessment in a digital age: Opportunities, challenges and risks. *British Educational Research Journal, 42*(3), 454–476.

Tomkins, L., & Ulus, E. (2016). 'Oh, was that "experiential learning"?!' Spaces, synergies and surprises with Kolb's learning cycle. *Management Learning, 47*(2), 158–178.

Vince, R., Abbey, G., Langenhan, M., & Bell, D. (2018). Finding critical action learning through paradox: The role of action learning in the suppression and stimulation of critical reflection. *Management Learning, 49*(1), 86–106.

Vingaard Johansen, U., Knudsen, F. B., Engelbrecht Kristoffersen, C., Stellfeld Rasmussen, J., Saaby Steffen, E., & Sund, K. J. (2017). Political discourse on higher education in Denmark: From enlightened citizen to homo economicus. *Studies in Higher Education, 42*(2), 264–277.

Weikle, B. (2018). How business schools are adapting to the changing world of work, *CBC News.* https://www.cbc.ca/news/business/business-schools-adapting-1.4955802.

Weinberg, J., & Flinders, M. (2018). Learning for democracy: The politics and practice of citizenship education. *British Educational Research Journal, 44*(4), 573–592.

Wilkins, S. (2017). Ethical issues in transnational higher education: the case of international branch campuses. *Studies in Higher Education, 42*(8), 1385–1400.

Williams, J. (2016). A critical exploration of changing definitions of public good in relation to higher education. *Studies in Higher Education, 41*(4), 619–630.

Wise, G., Dickinson, C., Katan, T., & Gallegos, M. C. (2018). Inclusive higher education governance: Managing stakeholders, strategy, structure and function. *Studies in Higher Education.*

Yirmiyahu, A., Rubin, O. D., & Malul, M. (2017). Does greater accessibility to higher education reduce wage inequality? The case of the Arab minority in Israel. *Studies in Higher Education, 42*(6), 1071–1090.

Leonardo Caporarello, Ph.D. is Professor of Practice of Leadership, Organization and HR at SDA Bocconi School of Management (Milano, Italy). He's the Delegate Rector for elearning at Bocconi University (Italy), where is the Director of BUILT (Bocconi University Innovations in Learning and Teaching) and the Director of SDA Bocconi Learning Lab. Leonardo is the Program Director of the Organizational Design, and Negotiation Executive Open Market programs. His main research activities focus on organizational behavior area. Currently, he is focusing on four main topics: leading teams in a co-located and virtual setting; analyzing and redesigning organizational structures and processes; designing effective online and blended learning experiences; managing organizational change initiatives (macro and micro levels). Leonardo has published in journals and books, at both national and international levels.

Dr. Beatrice Manzoni is Associate Professor of Practice of Leadership, Organization and HR at SDA Bocconi School of Management in Italy. She is Program Director of the HR Management and People Management Executive Open Market programs. Her research focuses on the following areas: designing and implementing effective learning experiences, leveraging on innovative learning models and methods; understanding and managing generational differences (values, working styles, learning styles, relationship with technology) in the workplace; fostering creativity and innovation at the individual, team and organizational level. Her works have been published in international journals and in edited books published by Routledge, Springer, Wiley and Egea.

Chapter 16
Tackling Executive Challenges Through Graduate Crowdsourcing

Henrik Totterman

Abstract This chapter describes how executives benefit from exploration when solving business challenges. Similar to the notion of crowdsourcing, exploratory learning is a process, which progresses from problem scoping to executable outcomes. Graduate crowdsourcing can drive innovation and business transformation, and opens opportunities for business, science, and social collaboration. Graduates typically represent a range of competences and cultures, and are typically equipped with a curiosity and eagerness to solve real business challenges. The ultimate challenge for most executives and organizations is to transform or create new business. A business school environment is great for such challenges, as learning takes often place through real-world problems solving with practical implementation in mind.

16.1 Introduction

This chapter describes how executives can benefit form applying an exploratory graduate learning approach when solving their most pressing business challenges. Similar to the notion of crowdsourcing, exploratory learning builds on distinct process steps, which progress from initial problem scoping to presenting concrete and preferably executable outcomes. Graduate crowdsourcing can at best drive innovation and business transformation, and opens opportunities for collaboration in the areas of business, science, and the society at large. Crowdsourcing is defined broadly as an organizational business process, which, for instance, Howe (2006) specifies as "the act of a company or institution taking a function once performed by employees and outsourcing it to an undefined network of people in the form of an open call." Consequently, crowdsourcing is here defined as an organizational process, where executive clients match their challenges or needs of their organizations with the talent and knowledge of business school graduates. Companies get access to creative and knowledge resources by reaching out to the crowd. As proposed by Jain (2010), the resource of a crowd is the 'crowd wisdom', and the essence of crowdsourcing

H. Totterman (✉)
One Education Street, Cambridge, MA 02141, USA
e-mail: henrik.totterman@hult.edu

© Springer Nature Switzerland AG 2020
N. Pfeffermann (ed.), *New Leadership in Strategy and Communication*,
https://doi.org/10.1007/978-3-030-19681-3_16

is to make use of ideas, resources, and competencies of people interested to solve problems and create new solutions. At best, crowdsourcing is connected with organizational learning, as suggested by Schlagwein and Bjorn-Andersen (2014).

Below, I will shed light on three main categories of graduate crowdsourcing, ranging from solving operational challenges e.g. business process improvement, tactical challenges e.g. creating open innovations, or strategic challenges e.g. building competitive advantage. In terms of operational challenges, a crowdsourcing approach is particularly valuable when graduates can associate with the target audience or the situation at hand. At best they can bring new insights to organizational business innovation and systems (re-)design, through optimization, scenarios, or creating alternative solutions through a creative processes. For instance, the work we did together with Hanken School of Economic's (Hanken[1]) EMBA students for the President and CEO of The Switch,[2] specializing in electrical drive train technology for renewable and industrial applications. One of the key questions we addressed, was: how should highly skilled technical employees be supported to adapt the "Switch Way" of working closely with customers on solving their problems? Insights from the class were highly regarded and helped to shape employee development and client project performance metrics.

When thinking of tactical challenges, it's particularly important for graduates to make a deep dive into the particular industry and business vertical. For instance, some years ago a Senior Strategy Manager working for IBM's US State and Local Government Business[3] faced challenges in producing revenue growth in a marketplace that was flat with declining customer IT spend. In order to respond to this challenge, Hult International Business School (Hult[4]) Boston graduate students had to dig deep and learn new insights to figure out key sales and marketing strategies for the public sector. They conducted a critical analysis of the strategy and current performance by focusing on development areas, critical factors, unexplored opportunities, fields of play, and suggested measures of success. Following this, they were able to propose several tactical approaches that emphasized cost savings, flexible subscription solutions, along with considering the emergence of new marketing and communication means that would improve sales and marketing execution in this particular vertical.

Finally, strategic challenges are typically more extensive in terms of complexity in understanding the landscape thoroughly to build and test comprehensive solutions. Hult is famous for its consulting driven practical curriculum, where graduate teams compete while solving executive client challenges for large global clients (e.g. Verizon, Philips, Domino's etc.). Most recently, I designed a Capstone Program in Management for the Harvard University Extension School[5] masters degree, where students advice business clients. Another great example is SHH Academic Business

---

[1] www.hanken.fi.

[2] www.theswitch.com.

[3] www.ibm.com/industries/government.

[4] www.hult.edu.

[5] www.extension.harvard.edu.

Consulting (ABC[6]) Ltd. at Hanken, which I oversaw for several years as faculty member before moving to Boston. ABC is a company solely managed by graduate students, applying latest academic knowledge, with the support of professors to solve current business problems that organizations are facing. Impressively each team acquires their own paying clients, scopes the project, and engages in developing the ABC company. Executives benefit from the project outcomes, and they get introduced to young professionals in action by utilizing their services at reasonable prices.

In this chapter, the underlying assumption is that international graduate students typically represent a diverse mix of competences, experiences, and cultures, and that they are equipped with a strong curiosity and eagerness to learn from solving problems, real-world business challenges in particular. At best, solving such challenges allow business school graduates to nurture and eventually master skills that make them more employable and well-rounded in their careers. In addition, a business school environment is particularly stimulating for solving real-world challenges, as problem scoping and action learning are often at the core of the educational culture. Faculty are also typically seasoned in teaching business cases and working to at least some extent with executives.

Consequently, progressive educators are likely to see crowdsourcing of executive challenges as a great opportunity to enhance practical learning in and outside the classroom environment. Some of the more conservative faculty may feel troubled of loosing immediate control of what is being taught and how learning is exactly captured in the real-world scenario. That said, the role of faculty continues to transform from being the force of supervision and control, to becoming increasingly the facilitator calling together stakeholders for meaningful learning engagements. As stated by Beard and Wilson (2013), this implies high quality coaching and facilitation, which builds on good practice and emphasizes the importance of ethics. In that sense, weight has shifted from transmitting knowledge and experience, to stimulating creativity and problem solving skills, while guiding the learning experience with educational insights and implementation in mind. Regardless of the scoping, my goal is for management students to gain an in depth understanding of what it takes to build and lead modern organizations, or to work in engaging enterprises after graduation.

My chapter is organized as follows: First, I introduce underlying objectives and present the explorative approach. Next, I demonstrate how the distinct process steps play out in graduate crowdsourcing of executive challenges. Then, I discuss how this combination of executive crowdsourcing and graduate explorative learning results in distinctive outcomes for executives and learners alike. Finally, I discuss the emergence of the field of crowdsourcing research, and why the proposed approach is particularly valuable in bringing new insights to organizational business innovation and systems (re-)design.

---

[6]www.academicbusinessconsulting.fi.

16.2   Underlying Corporate Objectives

The ultimate challenge for most executives and organizations is to identify new business in order to survive or support sustainable growth. Quoting my great grandfather H. J. Helkama, the founder of the Finnish 115-year old Helkama[7] family of companies in 1905: "A business leader or owner is a servant of the community who earns a pay for being able to create something new, respectable and durable which everyone needs." The prospect of entering a new market continues to offer appealing business opportunities, but as pointed out by Marquardt (2011), rapid globalization and fierce competition in the marketplace causes increased complexity of organizational problems. Leaders need new attributes and capabilities to cope with these challenges, such as transformational abilities, strong learning skills, emotional intelligence, ethical standards, problem-solving and project management strengths, self-awareness, and building capabilities of being humble yet confident. Successful business leaders blend increasingly strategic thinking with agile and dynamic approaches to adopt, transform and disrupt, while addressing associated risks and uncertainties to the best of their capabilities. However, business transformation, market entry and introducing new business models can be a daunting experience, which is why executives can benefit from crowdsourcing ideas from graduates and other outsiders whose minds are not primarily driven by industry norms and traditions. In addition, business executives may actually learn new approaches and useful tools, while collaborating with graduates and their faculty in an exploratory fashion.

As proposed by Soll et al. (2015), sometimes its gainful to get a fresh view to explore and experiment in order to support transformation and alignment with emerging challenges. Working with graduate students provides also a relatively safe and effective testbed to enhance next generation leadership attributes and capabilities, along with providing a low cost—high transparency access to an interesting talent pool. What is more, multiple teams of students are likely to possess valuable knowledge and collective wisdom to understand and propose services and products by external customers. The challenge remains of course to nurture, capture and transfer valued knowledge, especially as most in-house development projects in themselves are either dysfunctional or delayed in developing and delivering within the limited times imposed by the marketplace (cf. Marquardt 2011). That said, the majority of business leaders today are impacted at least indirectly by globalization, unpredictable business cycles, and increasingly rapid technology adoption rates.

As a crowd, graduate students can assist business executives in thinking way beyond their own frameworks. Identifying emerging megatrends and nurturing strategic options early, even in a highly exploratory fashion makes sense. It's a resource effective way to seek alternative truths, without committing the organization and its stakeholders until initial screening is completed. At best, this enables organizations and their people to maintain performance levels during times of inevitable change and associated transformational complexity. That said, the challenge remains to adapt a successful approach and pace of transformation, while keeping people engaged and

---

[7]www.helkama.com.

maintaining a business as usual mentality. This is particularly hard when organizations are surrounded by success under the watchful eye of various inpatient stakeholders.

Ever since knowledge management emerged as a scientific field in the late 1980s (e.g. Sveiby and Lloyd 1987; Nonaka 1991 in 2007 etc.) scholars and practitioners alike have paid particular attention to knowledge as a core competitive asset of modern organizations. Today more than ever, organizations must learn to transmit and apply knowledge from internal and external sources. In that sense, the question is less about why we as executives should engage in explorative learning by engaging smart and driven graduates, but more why international graduate students would want to volunteer their time to work with outside organizations? Especially good question, when typically graduates are not compensated for such engagements. In my view, graduates are generally equipped with a creative and curious mindset, and eager to act to enhance learning and post-graduate employability.

16.3 The Foundation of Explorative Learning

During my 16 years of teaching highly international undergraduates, graduates, and executive students, I recall very few instances were students would not have been eager and excited of the prospects to work on real-world business challenges. In fact, one of the few instances that come to mind is a seasoned medical doctor doing his MBA at Hult in Boston many years ago. This particular student had already a successful career in medicine and was indeed eager to work with a particular business challenge, posed at the time by the international pharmaceutical company Sanofi.[8] However, in line with the MBA program policy, our academic team had randomly chosen that his team would work on a challenge posed by Harsco Steel.[9] The student was at first extremely upset, since working for a large Pharma company would have aligned with his extensive past experience, and potentially lead to an interesting job opportunity. Even worse, solving a strategic challenge for a large steel company felt far away from the students comfort zone, and definitely within an industry that he had no incentive to learn more about.

When our program management refused to adjust, open frustration transformed first into blunt silence, until gradually the intellectual curiosity awakened. It was inspiring to see a strong skepticism turn into the joy of learning something new while providing strategic advice to a major client company. As his Dean at the time, my persistence was awarded a few weeks after graduation, when the newly minted MBA-MD called to thank us for opening his eyes. He was also asking for a reference to land a manager role with the medical instrumentation company Olympus.[10] The hiring manager was excited by his prospects, but ironically concerned with the lack

---

[8] www.sanofi.com.

[9] www.harsco.com.

[10] www.medical.olympusamerica.com.

of broader industry experience. As a consequence, we agreed that I was allowed to reveal the business challenge story, and tell how it commenced from pure frustration and turned into new insights and very positive client feedback. Olympus eventually made the hire, and since then the graduate has been promoted several times, working today as Executive Medical Director at Olympus.

It's part of the business school DNA and at the core of most business degrees to understand how businesses function in general, in and across a range of industries internationally. As suggested by Marquardt (2011) action oriented learning incorporates adult learning principles. Similar to having an international mindset, a broad experience base is particularly helpful to make association to take place between seemingly disparate fields or areas of expertise (cf. Dyer et al. 2009). Typically, a broad ranging basic understanding of phenomena stimulates innovation and enables implementation of systematic change in practice. This is why learning to understand current affairs of a real-company, and engaging with their executives to solve an associated challenge can be such an enriching career eye opener. I would argue that a descent real-life executive challenge is better, or at least comparable to the power of learning from a particularly well written business case.

I agree with Mellander (1993:11) that a powerful learning process derives from "Attention, making us receptive to information, which we process together with prior knowledge, until we arrive at conclusions and understanding, which we then apply and test for confirmation." Another similar classic framework is Kolb's Learning Cycle, which is cited in many different works (Kolb 1984). When implementing an explorative learning process, I typically derive from a design thinking approach, where the first step is to identify a challenge or problem area, which can be posed by an executive client, entrepreneur, or me as the faculty. This initial phase is followed by creative and explorative work to broaden and then narrow focus to propose solutions. At best, sustainable interest among graduates is awakened that carries from the beginning and way past the actual learning exercise. The challenge is that most of us are not spontaneous learners in any true pedagogical sense, which is why we benefit from a triggering event or activity that is worthy of our attention and stimulates our curiosity to learn more about the underlying phenomena or challenge in a facilitated learning process format. For most business graduate students in this modern day and age, solving business challenges hits this sweet spot, in contrast with learning from reading, lectures, or even solving written business cases.

The notion of a physical classroom is increasingly blurred, as exploratory learning takes place online, on-site, and offsite. So in addition to determining what learners need to know in advance and how to boost emotional engagement, the facilitator must consider carefully where learning will take place (cf. Beard and Wilson 2013). For instance, in my New Business Venturing course at Harvard University Extension School, sessions are mostly online with the exception of a long campus weekend in Cambridge, MA. This weekend takes place at the outset of the course to build trust and trigger curiosity among students to drive learning on distance, supported by extensive field studies conducted by students in their home regions. My other courses gather primarily physically in flexible classrooms, where learning is actively facilitated by me and primarily executed in teams and as a cohort. I am a big believer in the importance of the environment as a stimulant for learning, and yes, it's often about the space and place. In particular, the social aspects of learning in the right space with the right people at the right time. Learners must be prepared for the task or activity at hand, which may happen in a preparatory fashion or duing a particular session. In practice, I urge and require that students get out from the building to explore the surroundings through different assigned market safaris.

16.4   Planning and Scoping (Understand)

As emphasized by Biggs (2003), practical problem based learning is likely to max-imize engagement across involved stakeholders, who are typically the graduates, executives, and faculty as facilitators. Learning to frame problems and learning to solve them in a real life context is seldom as clear cut and analytically approachable as a written case study leads to believe. This can be a problem at times, since our objective as practice oriented educators should foremost be to teach highly applica-ble skills that last over time. This way graduates gain practical experience in using simple yet powerful tools, frameworks and concepts on a real problem. Rather than just debating about them conceptually, or based on a simplified world view in a case classroom setting. A real-world problem setting will always remain fuzzy to begin with. This is why I emphasize so much the ability to even identify there is a problem, and then request students to define what it could be. In my view, we often take the most crucial part of an executives task away by stating clear and very specific problem statements, which we then expect students to analyze and then assemble resources to address. Rather than teaching to solely apply analytical tools, my ambition as a teacher is to teach problem scoping, reasoning, and decision-making along with other useful skills, while being surrounded by ambiguity, risk, and uncertainty. On this note, for readers interested in pedagogy, a great read on choosing experimental methods for management education is offered by Herbert and Stenfors (2007).

Marquardt (2011: 27–44) provides an excellent approach to problem scoping, by associating it to explorative action learning. As in design thinking (e.g. Banfield et al. 2016), the problem should be important, urgent, and significant, and moreover, barely within the ability of the team to solve to provide opportunities for everyone to learn. Regardless of the scoping, typically the starting point should be in identifying a problem worth solving. I recall, for instance, a scoping call with the Vice President of Business Integration at Fidelity Investments,[11] who proposed 3 different challenges that needed addressing. Each one of them was intriguing and complex enough for a management consulting firm to be truly excited. That said, only one of them resonated with the course objectives and qualifications of the graduate students in that particular program.

As a consequence, we worked closely together on ensuring that the challenge was in balance with emphasis on the student learning experience, without forgetting the

---

[11]www.fidelity.com.

needs of Fidelity. Another example was working with the former location and mapping division of Nokia called Here.[12] Basically they wanted Hult students to figure out how Here could launch their consumer mapping services among international graduate students in the Boston region. In other words, a perfect challenge for our international student body. No wonder one of the student teams was afterwards asked to present their proposal to the Here executives and staff. So never underestimate the crowd in solving burning corporate challenges.

Problems may focus on almost anything of relevance for the executive client, as long as solving it aligns with the underlying learning objectives. Challenges that I facilitate typically address, in line with e.g. Keeley et al. (2013), the business core, aiming to reconfigure the profit model, network, organizational structure, or business related models and processes. I also have students frequently work with executives on the core offering or collection of products, services or solutions to explore ways to introduce new offering or enhance product performance or the product system. I often organize business challenges with multiple verticals, allowing student teams to select a particular approach and scoping to solve a larger strategic problem. This serves multiple purposes, one being the benefit of variety when learning from each other, another being the opportunity for learners to choose an approach that feeds their curiosity and resonates with they previous knowledge and experience.

What is more, I have in recent years started to foster the class to regard themselves as a professional consultancy or department of the assigned client company. This learning process approach is in line with Marquardt (2011), who emphasizes single and multiple problem action learning. Instead of everyone solving the same problem, this type of setup lends itself perfectly to establishing a facilitated action learning group, where people work on their own assignments individually and in teams, and also engages in advising others to progress ideas and learn from others. For instance, I worked in this fashion as we developed insightful new mobile enterprise solutions for one of the worlds largest electronics manufacturers (undefined), or when proposing new business solutions for Verizon's[13] 5G mobile consumer business. Working as a company or department serves multiple purposes, one is to simulate the benefits of working simultaneously in lead teams on multiple project with positive cross-team spillovers (as is often the case in real companies), another one is to enhance learning of everyone, by being transparent of obstacles and solutions being generated across everyone in the course.

In this model, teams are equally responsible for their own success and for making the entire 'company' shine in front of the real-world client. The process is facilitated in a way where individuals become familiar with gained insights, verified assumptions, and also problems that other teams are facing and they are asked to help address these as peer support. This shared responsibility will feed energy and knowledge to independent teamwork, but also raises significantly the quality of all team assignments on the aggregate, mostly by helping to balance chaos and by creating order. In practice, the facilitated support follows something similar to the Shewhart/Deming

---

[12] www.here.com.

[13] https://www.verizonwireless.com.

cycle of continuous improvement, where individuals and teams help themselves and other teams on the aggregate, by driving increased understanding through a continuous cycle of planning, doing, checking, and acting (cf. Beard and Wilson 2013). Similarly, as showcased already in Totterman (2018a, b, c), management, innovation, design, and entrepreneurship literature are crowded with process models tailored for particular purposes and defined to make order out of initial chaos, by dealing with e.g. ambiguity, risk, and uncertainty.

In order for the executive-graduate interaction to succeed, I typically prepare students in advance, so they are aware of the industry basics and feel comfortable from the start to break boundaries. In addition, one powerful technique is to study something seemingly unrelated and then force gained learning into another context (cf. Dyer et al. 2009). For instance, in Totterman (2018a), I explain how students explored the unique values of IKEA[14] and then considered what would be fundamentally different if our client company Verizon was founded by Ingvar Kamprad and Verizon was operated as his company IKEA. By comparing value attributes across the two companies, students learned something new while stretching their imagination well beyond the evident (Totterman 2018b). The underlying reasoning is that in order to assist executives, students must think beyond immediate products, services, or organizational structures. Freeing their minds to think of the unimaginable will ease the process moving forward, and at best seed some initial problems-solution combinations for further investigation along the process. Breakthrough solutions may well emerge from combining products or ideas with alternative business models, and sometimes they emerge from the seemingly strange combinations at the outset.

16.5   Building Assumptions (Ideate)

Design thinking (e.g. Kelley and Littman 2001, 2005, or Banfield et al. 2016) combined with the lean startup methodology (Ries 2011), offer a powerful tool to drive creativity and engage organizations in problem solving activities. The challenge in the corporate setting is that innovators, change agents of different kinds, or so called intrapreneurs are often seen as trouble makers as they challenge the existing. Goes without saying that openness and trust are needed for ideation to blossom, which can be hard to achieve in a competitive organizational setting. Careers and busy schedules may steer priorities more than collegiality, and shared learning, or drive for change.

Similarly in education, far too often consensus driven learning is prioritized, when debate, disruption, and creativity may yield engaging and insightful opportunities to progress something that matters in real life and learn while making a difference. Executives typically know that the unpredictability of innovation and (re)design activities call for experimentation and ecosystem level engagement, so why not develop these insights in a safe learning environment by crowdsourcing to graduates? This can be particularly effective, as long as graduates are guided by simple ground rules to stay

---

[14]www.ikea.com.

focused on the executive challenge throughout the process (cf. Kupp et al. 2017). As pointed out by Ries (2017), there is no assurance where or by whom the next great venture will emerge. This is also evidenced by my experience from working with executives, entrepreneurs and venture capitalists, who are generally eager to join us on campus to listen and give feedback on student work.

In the ideate phase, through a range of exercises, the entire class learn different ways to creatively address the challenge, and they become familiar with the clients objectives and goals, development and other priorities, and learn how the client defines performance and successful outcomes. This conversation is as valuable for the teams as it is for me as the faculty member. Typically, the clients enjoy these early brainstorming sessions a lot as well, as ideation requires alternative thinking and graduates stimulate thinking and provides first insights in a timely and effective manner. Some of these exercises can seem a bit disconnected at the outset, but they are central in teaching the important lesson of freeing creativity, but also to learn from failure, and foremost, how to turn failure into success. In addition, individuals will step out from their comfort zones, since assignments seem fun and involves limited risks of loosing face. They teach the importance of maximizing creativity, combined with realism, and all of this under a tight timeline using different scarce resources. Focus is from the outset on the underlying customer problem, initially to understand and grasp it, followed by identifying related themes and grouping these together to discuss early findings and refined views of the problem.

Teams work in the beginning on loads of ideas, rather than digging into too much detail. By the time they narrow down, they will have a detailed collective understanding of the core problem, who has it, how many, is it worth solving, and how big is the market? If previous ok, then initiate more robust investigation of business potential. Search and benchmark, if not passing then back to understanding and solving it in another way. Gradually students will come up with their distinct view of a specific problem at hand, consider if it is worth solving or not. The main challenge is to become truly creative in order to visualize and seek inspiration from unimaginable places, by stepping truly out from comfort zones. For each idea, they identify what is known currently, what is still unknown, and importantly what assumptions are made to avoid confirmation bias and knowledge gaps.

Then graduates define in a creative fashion how their client can overcome any identified issues by standing in the stakeholder and/or customers shoes. The internal stakeholder approach is particularly important when considering larger organizations with many layers of influencers. Graduates wont know all the complexities and personalities that comes to play, but I've seen them resonate surprisingly well with addressing immediate concerns of our executive clients. For instance, a team I worked with recently proposed to Verizon, how they could enter the 5G mobile gaming scene. The team first presented a compelling solution and then how our executive client, Verizon's Head of Consumer Products Innovation and Insights, should convince her colleagues internally to proceed with their proposal. As stated by the executive client, "You guys really thought about everything" (Totterman 2018b), and as a follow up, the team was invited to present their proposal to a larger group of Verizon's senior executives.

16.6 Field Studies (Refine)

Teams are encouraged to find inspiration from the market place, by building, combining, pivoting, simplifying, and innovating based on findings and creative input from team reflection. The challenge with explorative learning is to ensure students are not overwhelmed and stay on track with course objectives in mind. This is why the assignments along the journey play a key role, as they steer students to actively engage with stakeholders in their communities, who can verify assumptions and steer participants in the right direction to truly understand the problem they are trying to address. Sometime less scoping material is actually favorable, as feeding a lot of corporate data and insights may overwhelm and handicap the graduates creative urge to seek for alternative truths by conducting extensive field studies. Even worse, the executive client will basically receive refined versions of facts they were already well aware of.

Most real-world business problems are not neatly stacked and identifiable. Rather the opposite, which is why it's elementary that students are forced to deal with high levels of ambiguity, while using their imagination and creativity in sourcing and vetting the quality of suitable information sources. Students are prone to enjoy the comfort of their lounge chairs or the campus environment, but problem solving requires observation and interviewing people who are facing the problem in the first place. Creating ideas and defining assumptions is key to the process, followed by running experiments to test and nurture these assumptions based on retrieved insights towards final verifications. I see it as essential to stimulate the creative process by applying a sense of urgency and rapid iterations, by applying suitable tools along the highly explorative journey. This implies that assumptions are being defined, tested, and validated while failure is tolerated and learning is documented towards building a solution that aligns with emerging questions and continuous reflections.

I recall particularly well working with a six week long client assignment at Hult, which kicked-off with a weekend long design sprint based on the client posing the problem and me describing the process to a large crowd of enthusiastic volunteering graduates. Our client was the Vice President of Strategy and Corporate Development at Zipcar,[15] the pioneering car sharing company. Following this, the 15 teams were supposed to work on campus to ideate around the challenge scoping, but after the client briefing I could only locate 14 teams. It turned out that one team took my strong urge to "get out from the building to find answers" literarily. Instead of discussing the problem, they immediately signed up for Zipcar membership and rented a vehicle for that afternoon to understand better the underlying problem environment. Goes without saying that their proactivity and creativity paid off significantly, as they went all the way to understand the problem, rather than first falling in love with their early ideas. Similarly, another team I worked with recently at Harvard University Extension School, wanted to reimagine the barber and hair saloon experience, and in order to do they spent significant amounts of time exploring and visiting stores, observing, and talking to employees as well as clients. Interestingly, they eventually

---

[15]www.zipcar.com.

realized that their intended solution was not the right answer to the problem, and instead proposed a completely different and much more unique solution. This pivot was only possible, because they truly understood the problem area and were willing to desert their original solution.

16.7 Outcomes for Action (Launch)

Several years ago, I worked with an EMBA cohort at Hanken School of Economics, who I tasked to solve both performance and succession related issues for a Nordic family business called Norex International,[16] which is active in the food and beverage industry. Teams were divided to prepare certain areas of concern prior to the client meeting, and when we got together we worked with the client for a day on creating 5 strategy maps. These emphasized: (1) the creation of value for customers, (2) organization, development and utilization of internal and external resources, (3) future revenue growth and productivity, and (4) differentiation of products and services, and (5) concluding map bringing together the overall vision and strategy. The day with the client was highly memorable for all participants, and what is most important, the client developed their long term strategy based on gathered insights. The implementation resulted in a relatively painless and well prepared leadership succession, and a refined strategy to support strong continued growth.

Over the years, I've experienced several great ways that executive clients and their organizations capture insights through graduate crowdsourcing and I have seen first hand how they implement these in their operations. That said, the weak point is often figuring out the alignment between the proposed graduate solutions, and the strategic and organizational fit. At worst, I've seen great solutions to well understood problems fall on the floor, primarily because the client executive has not been able to find support in house, or the offered solution does not quite align without significant process tailoring. That said, it's not uncommon that executive clients are not too concerned of implementation, as they might primarily be exploring alternative insights, rather than considering successful execution. For them learning how young professionals demonstrate feasibility and sustainability related to the original problem scoping, provide alternative ways to solve a particular customer pain. Regardless of the ultimate expectations, an effective way to transfer insights is to meet regularly with the student teams as they progress executive challenges. Preferably, the client is able to engage representatives from different levels of leadership and functional areas to ensure shared excitement and commitment. Influential agents and sponsors of change are particularly important, along with people who can capture essential insights and recreate them in the organizational context. Perhaps human resources can also identify some of the strongest students, to continue working on the problem scoping as an internship or post graduation.

---

[16]www.norex.fi.

Another great way for companies to ensure they get the essence from graduate crowdsourcing, is to establish internal team(s) who work simultaneously with associated challenges. For instance, the Hult Boston campus was approached by Janssen Labs[17] (Johnson & Johnson company) some years ago. Janssen is among other things accelerating business related to decease interception.[18] I will never forget the client scoping meeting together with two of their Vice Presidents. Despite my initial concerns, they saw great potential in international business students assisting them in developing innovative business models, which aimed to promote proactive prevention of a range of exotic deceases. The challenge was then broken into 3 phases: identification of suitable deceases, learning about the chosen decease, and then building a scalable and innovative business model to address a particular decease. Still today, I remain amazed of the outcomes and the dedication of the Janssen senior executives, who spent two weekends on campus working together with the students. The alignment with their internal projects and needs were evident. This way, they were able to get young professionals to assist in framing complex medical dilemmas, so they would be understood by investors and business people in general. A great client to work for and the students of course felt particularly privileged of the close executive contact, combined with the immediate connection between their solutions and the clients accelerator program.

16.8   Crowdsourcing as an Emerging Field

In the modern age of digitalization and social networks, crowdsourcing is seen, for instance, by Lenart-Gansiniec (2019), as one of the key mega trends, which drive innovation, and steers collaboration in the areas of scientific studies, business, and society. Similarly, other authors underline the emergence of crowdsourcing due to the rapid expansion of virtual environments (cf. Schemmann et al. 2016). No wonder that Leimeister et al. (2009) stress that more and more organizations aim for crowdsourcing, which is also something I have noticed when advising international business executives. This is much in line with findings from Yang et al. (2008) who discuss the benefits from connecting companies to virtual communities. In line with this, I recently finalized a business case describing a dramatic turnaround of a public, international engineering and management consulting firm called Pöyry.[19] After the transformation, their 5500 employees are no acting and behaving as 'intrapreneurs', the company has a virtual innovation platform, and it operates open innovation hubs in several of the 45 countries with regional offices. Most importantly, their senior level advisory board is highly supportive of in-house and extended ecosystem driven innovation, with direct insight from the President CEO.

---

[17]www.janssen.com.

[18]www.janssen.com/disease-interception-accelerator.

[19]www.poyry.com.

At the core, crowdsourcing involves transactions that expand the boundaries of the organization, by connecting outside and often previously untapped resources. Crowdsourcing can be done through experimentation, typically in a frugal and quicker fashion than in house. In addition, as indicated previously, crowdsourcing offers general access to experience, innovativeness, information, crowd skills and work, which are located outside the immediate organization (Aitamurto et al. 2011). As emphasized in this chapter, and defined by Marjanovic et al. (2012) in their influential framework, at the essence crowdsourcing is about input, process, output, and outcomes (cf. Muhdi et al. 2011). The process starts from defining a problem, identifying suppliers of ideas/knowledge, and stating a call for action. This is followed by organizing, managing, coordinating, and monitoring actions between the company and the crowd, with focus on building community that encourages knowledge and skills transfer. The assumption is that the receiver is ready to use solutions to problem, and that there are bilateral benefits generated (cf. Saxton and Kishore 2013).

In my experience, crowdsourcing is particularly useful when organizations are looking for a fresh view on their challenges, and as argued by Afuah and Tucci (2012), to solve problems in a particularly innovative fashion. However, perhaps somewhat naively we often assume that the crowd is always wise, rational, kind, investing in social capital, useful, and ready to solve problems (Wexler 2011). As a fact, we know that crowds can also be highly irrational, and at worst cause more damage than good. What is more, even if a crowd is generally willing to assist, it remains important to carefully align the scoping with their abilities to solve given problems and ensure proper organization is in place to undertake given activity. A critical mind may also argue that crowdsourcing is not all that different from ordinary business or talent related exchange (cf. Afuah and Tucci 2012). Then there is also the question of appropriate level of analysis, should we always consider the organization, or does the intermediary, user, system or application also qualify. On that note, in this chapter, focus is primarily on the executive client and their interaction with the graduate crowd, as part of the larger context of the organization and business school environment.

However, as a research field crowdsourcing is still emerging in its early phases, with inputs both from practice and diverse scientific fields like medicine, technical, and economic sciences with growing attention since 2010 (cf. Lenart-Gansiniec 2019). Despite the growing interest, the field remains still highly conceptual, with empirical evidence being shaped around narratives, rather than systemic and rigorous data collection that would measure the true value and effectiveness of a crowdsources approach to, for instance, innovation, problem solving, or optimizing organizational processes. In addition, as any emerging research venue, it's still defined rather broadly, and this chapter adds to the mix intentionally, rather than trying to drive a more unified definition. Others need to follow, who will continue to conceptually frame the field, and back these assumptions with empirical evidence. There are currently two emerging streams of crowdsourcing related research, one focuses primarily on the application of digital technologies, including online learning. The other one is in line with this chapter and most other literature sources on crowdsourcing, with emphasis on innovation, problem solving, cost optimizing, organizational activity, and business process improvement. As in entrepreneurship research, the field of crowdsourcing will in my view benefit from an open and welcoming approach, and per definition allow the crowd to do what it does best when bringing together the 'wisdom of the crowd'. Eventually, contradictions and differing views will be organized into a conceptual framework, where different disciplinary inputs play a major role. Most urgently, as in entrepreneurship research, we should avoid trying to conceptually define who the executive and crowd is and instead gain empirical evidence of the role they play and what they actually do. This is similar to entrepreneurs, as wisely stated by Gartner (1989), when considering the emergence of entrepreneurship as a research field.

16.9 Concluding Remarks

This chapter presented the practice of introducing, implementing, and sustaining explorative learning, as an intermediate which combines graduate students with executive challenges through crowdsourcing. I strongly believe that students become better leaders, consultants, and team players through experimental learning, and I am not alone on this one (e.g. Neumann 2007). For executives and their organizations, I recommend that you start actively considering ways to break institutional barriers that hinders crowdsourced experimental learning from flourishing. For managers, employees, and recent graduates, I recommend that you become a bit naughty and despite the existing paradigm cautiously infuse elements of explorative learning and crowdsourcing. For researchers, I recommend that you engage in this transformative and emerging field of research, which is exciting and yet desperately in need of more conceptual and especially empirical knowledge.

Working with graduate students provides more opportunities for leadership to solve heterogeneous problems with speed and greater agility, than working on them solely with internal experts who may prioritize risk mitigation, compliance, and short term gains through economies of scale. Graduate student involvement is a frugal approach per definition, which lends itself to adaptation and agility by reducing waste of time/energy with natural ways to boost or snooze projects. Ensure outcomes are aligned with organizations system to innovate at scale, this requires clear response areas and accountability as foundation of management, with well working tools and tactics, and support from motivated people who value the cultural norms and beliefs of the organization. Businesses do not operate in isolation, as they are integrated parts of supply-chains, industry, societies regionally and globally. As I tend to say, TOGETHER WE CAN!

Bibliography

Afuah, A., & Tucci, C. L. (2012). Crowdsourcing as a solution to distant search. *Academy of Management Review, 37*(3), 355–375.

Aitamurto, T., Leiponen, A., & Tee, R. (2011). *The promise of idea crowdsourcing—Benefits, contexts, limitations.* Whitepaper for Nokia. Publications by Nokia Ideas Project. http://www.tanjaaitamurto.com/Publications.html. Accessed 031019.

Banfield, R., Lombardo, C. T., & Wax, T. (2016). *Design sprint: A practical guidebook for building great digital products.* O'Reilly.

Beard, C., & Wilson, J. P. (2013). *Experiential learning: A handbook for education.* Kogan Page: Training and Coaching.

Biggs, J. (2003). *Teaching for quality learning at university* (2nd Ed.). Open University Press—McGraw-Hill Education.

Dyer, J. H., Gregersen, H., & Christensen, C. M. (2009, December). The innovator's DNA. *Harvard Business Review,* 61–68.

Gartner, W. B. (1989, July). Who is an entrepreneur? Is the wrong question. *Entrepreneurship Theory and Practice.*

Herbert, A., & Stenfors, S. (2007). Choosing experimental methods for management education: The fit of action learning and problem-based learning. In *The handbook of experiential learning & management education.* Oxford University Press.

Howe, J. (2006). The rise of crowdsourcing. *Wired Magazine, 14*(6).

Jain, R. (2010). Investigation of governance mechanisms for crowdsourcing initiatives. *AMCIS Proceedings (Americas Conference on Information Systems).* https://pdfs.semanticscholar.org/7aaf/44d29d314b70cf352f141c22a84d64229431.pdf?_ga=2.186378859.316951234.1561726584117005277.1549315547.

Keeley, L., Pikkel R., Quinn, B., & Walters, H. (2013). *Ten types of innovation: The discipline of building breakthroughs.* Wiley.

Kelley, T., & Littman, J. (2005). *The ten faces of innovation.* Doubleday.

Kelley, T., & Littman, J. (2001). *The art of innovation.* Doubleday.

Kolb, D. A. (1984). *Experiential learning: Experience as the source of learning and development.* Prentice-Hall.

Kupp, M., Anderson, J., & Reckhenrich, J. (2017, Fall). Why design thinking in business needs a rethink. *MIT Sloan Management Review.* https://sloanreview.mit.edu/article/why-design-thinking-in-business-needs-a-rethink/. Accessed March 10 2019.

Leimeister, J. M., Huber, M., Bretschneider, U., & Krcmar, H. (2009, January). Leveraging crowd-sourcing -activation-supporting components for IT-based idea competitions. *Journal of Management Information Systems.*

Lenart-Gansiniec, R. (2019). *Crowdsourcing and knowledge management in contemporary business environments.* A volume in the Advances in Logistics, Operations, and Management Sciences (ALOMS) Book Series, IGI Global.

Marjanovic, S., Fry. C., & Chataway, J. (2012). Crowdsourcing based business models: IN search of evidence for innovation 2.0. *Science & Public Policy, 49*(2), 193–213.

Marquardt, M. J. (2011). *Optimizing the power of action learning* (2nd Ed.). Nicholas Brealey Publishing.

Mellander, K. (1993). *The power of learning: Fostering employee growth.* McGraw-Hill.

Muhdi, L., Daiber, M., Friesike, S., & Boutellier, R. (2011). The crowdsourcing process: An interme-diary mediated idea generation approach in the early phase of innovation. *International Journal of Entrepreneurship and Innovation Management, 14*(4), 315–332.

Neumann, J. E. (2007). Becoming better consultants through varieties of experiential learning. In *The handbook of experiential learning & management education.* Oxford University Press.

Nonaka, I. (2007, June–August). The knowledge-creating company. *Harvard Business Review.* https://hbr.org/2007/07/the-knowledge-creating-company. Accessed 031019.

Ries, E. (2017). *The startup way: How modern companies use entrepreneurial management to transform culture and drive long-term growth*. Currency.

Ries, E. (2011). *The lean startup*. Crown Publishing Group.

Saxton & Kishore (2013).

Saxton, G. D., Oh, O., & Kishore, R. (2013). Rules of crowdsourcing: Models, issues, and systems of control. *Information Systems Management, 30*(1), 2–20.

Schlagwein, D., & Bjorn-Andersen, N. (2014). Organizational learning with crowdsourcing: The revelatory case of LEGO. *Journal of the Association for Information Systems, 15*(11).

Schemmann, B., Herrmann, A. M., Chappin, M. M. H., & Heimeriks, G. J. (2016, July). Crowdsourcing ideas: Involving ordinary users in the ideation phase of new product development. *Research Policy*.

Soll, J., Milkman, K. L., & Payne, J. (2015, May). Outsmart your own biases. *Harvard Business Review*. https://hbr.org/2015/05/outsmart-your-own-biases. Accessed 030919.

Sveiby, K. E., & Lloyd, T. (1987). *Knowhow: Add value by valuing creativity*. Bloomsbury. https://www.amazon.com/Managing-Knowhow-Value-ValuingCreativity/dp/0747500738

Totterman, H. (2018a, September). What would higher education look like if run by IKEA? *New England Journal of Higher Education*. https://nebhe.org/journal/what-would-higher-education-look-like-if-run-by-ikea/. Accessed 030919.

Totterman, H. (2018b). https://www.hult.edu/blog/verizon-challenges-hult-graduate-students/. Accessed 030919.

Totterman, H. (2008c). *From creative ideas to new emerging ventures: Entrepreneurial processes among finnish design entrepreneurs*. Hanken School of Economics.

Yang, J., Adamic, L. A., & Ackerman, M. S. (2008). Crowdsourcing and knowledge sharing: Strategic user behavior on taskcn. In *Proceedings of ACM Electronic Commerce'08* (pp. 246–255). http://www-personal.umich.edu/~ladamic/papers/taskcn/EC2008Witkey.pdf Accessed 030919.

Wexler, M. N. (2011). Reconfiguring the sociology of the crowd: Exploring crowdsourcing. *The International Journal of Sociology and Social Policy, 31*(1/2), 6–20.

Henrik Totterman is a Professor of Practice in entrepreneurship and management at Hult International Business School and a teaching faculty member at Harvard University Division of Continued Education. Henrik is an international entrepreneurial leader and global business school executive, with over 20 years of experience developing his own ventures and advising others. His background enables him to combine academic concepts with the real world of business, exemplified by the work his company LEADX3M LLC is doing renewing and building new business for organizations internationally. Henrik is a seasoned board member who currently supports good causes as an advisory board member of Accion East in USA and the Hanken School of Economics EMBA program in Europe. Connect with Henrik on LinkedIn: linkedin.com/in/htotterman.

Chapter 17
Becoming a Positive Leader: The Challenge of Change

Beatrice Bauer

Abstract Organizations are struggling with the development of an executive cadre that is truly competent in engaging subordinates and creating a positive working environment. The traditional leadership style of top down management is too slowly evolving into a collaborative approach that empowers employees and blurs the lines between boss and worker. Still too many leaders have not changed at all and seem to insist undaunted in their approach to managing people. Should we respect the global tendency to do more with less also in training or do we need to respect the real learning needs of our participants? Leadership development is a process of personal transformation. For the leaders we need today we cannot simply add skills to an existing portfolio but rather help them to develop a new Weltanschaung.

17.1 Introduction

Books on leadership come in all shapes and sizes, and with very creative titles. They are everywhere—in bookstores and even newsstands in train stations and airports. Business school programmes on leadership invent new topics and buzzwords every year to make learning how to lead more appealing. In the public sphere, the role of leaders in politics, their charisma and impact, is constantly at the centre of debate. I have done hundreds of interviews with leaders and the people who work for them, and seen the evidence provided by scientific research. Based on all this, and my direct professional experience, I can quite confidently say that we are probably talking about leadership so much because we have never faced a leadership crisis as serious as the one we are living in right now. Rulers have lost the trust of their citizens and electors, who are now leaning toward anti-elitism and want to reclaim their power. In companies of all sizes in every part of the world, followers are filled with doubt because they do not know if their leaders will be able to steer the proper course through this new and quite frightening *v.u.c.a. world* (volatile, uncertain, complex and ambiguous). In today's business reality, companies are expected to thrive against

B. Bauer (✉)
SDA Bocconi School of Management, Milan, Italy
e-mail: beatrice.bauer@unibocconi.it

© Springer Nature Switzerland AG 2020   245
N. Pfeffermann (ed.), *New Leadership in Strategy and Communication*,
https://doi.org/10.1007/978-3-030-19681-3_17

all odds while fighting to preserve their employees' jobs. I assume that one of the main reasons for this leadership crisis is that we have not been able to go beyond words and change the way leadership is done. It is astonishing, in fact, to see how little progress has been made in this direction. The behaviour of leaders in companies has varied very little over the past few decades. We have come to the point that the problems discussed in leadership training programmes today are very similar to the ones that participants would have brought up in the past.

We have tried with little success to convince organizations that people are truly important, not only because of the mission statement hanging on the wall or the last leadership model developed in order to implement change. If companies have to be far more agile, flexible and innovative in their approach to changing markets and societies, then the same flexibility, agility and innovativeness is needed for their people on all organizational levels, particularly at the top. If we want to create a new work environment that allows people to be more engaged and innovative, we must replace the traditional corporate organizational models with new ones. Out with centralized hierarchical structures and in with organizations where the power to decide and act is better distributed and based more on collaboration. For more than 20 years, we have been teaching that it is no longer appropriate to command and control our subordinates if we want to be competitive. We have been stressing that companies are far more likely to flourish if we adopt a participative and inclusive leadership style. This is what enables companies to create great teamwork and promote collaborative decision-making. Modern companies, though, while struggling to remain competitive, are ignoring the urgent need to make a profound change in their organizational structure, and to transform the traditional 'command and control' paradigm on which many are still based. The time has come for them to decide to invest in their leaders in order to help them to change their leadership style, which is as outdated as the system they belong to.

17.1.1 Traditional Leadership: It's Time to Change

We have seen that leadership does not evolve at the same speed as the change that companies experience. There has been far less focus on the role and competences of human resource managers, compared to the efforts dedicated to developing and updating their more technical skills. This is also obvious when we look at what companies typically ask for from managerial training institutions. The demand for leadership training is generally restricted to a short, intensive course that covers a wide variety of topics. If a company chooses to send a manager to a specific course focusing exclusively on leadership, this is often because the person in question has serious problems interacting with their workforce. In other words, the company is hoping for nothing short of a miracle. The commitment to just a few days of training obliges us, as teachers, to limit our intervention to short lectures, rather than helping leaders to translate new ideas into new behaviours. James Wilkinson, founder and for many years director of the Harvard Derek Bok Center for Teaching and Learning, once spoke a great truth at a conference: "A lecture transfers knowledge but can it transfer skills? The strength of information is first of all in generating a stimulus, not in shaping a response. It solicits further work, but does not in itself demand it. For all its focused energy and dramatic power, the lecture alone cannot meet all pedagogical needs" (Wilkinson 1982). This is particularly true if we think of the unprecedentedly broad spectrum of the skills that leaders need today.

Without true, deep-rooted leadership competences, managers simply try to do their best by imitating their boss's behaviour or the leadership style of someone they have randomly chosen as their own personal benchmark. Moreover, even if we admit that an important component of the success of a leader depends on intrinsic, personal factors that are difficult to change, the DIY approach can potentially be dangerous. It is easy for people to develop their own personal leadership styles, not in a conscious, articulated and flexible way, but instead as a response conditioned by the prevalent culture and their own personal experiences since childhood. In fact, this last factor plays a crucial role in personality development, particularly in the way people relate to others. In some cultures, family and education traditionally emphasise the importance of obeying norms and rules and respecting formal hierarchical power, rewarding those who accept and endorse conventional knowledge with no questions asked. Also, tradition repeats itself: an entrenched culture that establishes strong organizational routines for years eventually turns into a rigid belief system, resistant to change. If we are not lucky enough to learn from a boss with excellent leadership skills, or we are not immersed in a culture that is different from our own, we will easily impress upon our employees the values that have guided our own education: accepting traditional knowledge without question and inhibiting personal initiatives. We all need to be aware of how much each of us is a prisoner of our own life experience. We need to realize that we all tend to repeat acquired behaviours from the past, even if they are totally dysfunctional in the present. Recognizing all this is the first step to learning how to become a leader (Savickas et al. 2009).

Often leaders prefer to ignore or deny the roots of their leadership style. They are convinced that leading people (and themselves) is all about technical knowledge and rationality. But even the top CEOs and the highest-paid executives are human beings with a wide range of behaviours, cognitions and emotions. Most of the time, however, a restrictive artificial view of life in the workplace permeates their actions and words, reinforcing emotionally detached relationships. Do we, as management consultants and trainers, stick to this belief? After all, this attitude helps us to avoid any emotional involvement too, and allows us to illustrate complex leadership principles in a rational and concise way with a beautifully prepared Power Point. Even if technology can help us communicate effectively, what matters most in teaching is still the overall level of involvement. And a teacher's involvement in the classroom is closely correlated with his or her people skills. "We've looked through evaluations and found four things that virtually guarantee a high score," Wilkinson says. "Know your subject and be able to convey what you know; be enthusiastic about your subject; really care about whether the students learn your subject; and notice whether or not they learn. One of the surprising things for new teachers is that two of those four are about emotion, rather than focusing on intellectual content alone" (Wilkinson 2001).

17.1.2   The Cost of Traditional Leadership

What happens to traditional leaders if they successfully avoid promoting greater emancipation and empowerment of their workforce? What do their companies do when a person's commitment, personal initiative, autonomy and creativity become the critical factors for achieving success? The answer is obvious: they should be forced to change their leadership style. Why? Because it is impossible to expect innovation from their subordinates by pulling the same old strings, like a boss and his or her underlings. The problem for various management schools, consultants and trainers is that there is great resistance to this kind of change. In many cases, even when we demonstrate the limits and counter-productive effects of certain leadership styles, and back all this up with data, it is still not enough to convince people that they need to change.

When managing the complexity of work in a context characterized by global turbo-competition, the know-how of a single individual can no longer keep pace with the continuous evolution of the context where he or she needs to apply it. If the company fails to develop a more horizontal, shared approach to its problems, if it does not help its leaders play their role in an innovative way, the organization will force people who are well aware of their limited knowledge to feign competence in order to earn respect, sometimes with an extra touch of aggressiveness. In fact, bosses in positions of power who are aware of their limited leadership competences develop high levels of aggressiveness. As stated in the article, 'When the Boss Feels Inadequate: Power, Incompetence, and Aggression': "Power holders who do not feel personally competent are more likely than those who do feel competent to lash out against other people" (Fast and Chen 2009). In times of crisis and change, leaders are likely to find themselves at the head of a group of people who are worried about their future. Without the right competences, these are times when traditional leaders may not feel comfortable with people; they might prefer to withdraw from human relationships and focus only on solving company problems with less emotional content.

Instead, this same dominant position can give people who are less aware of their limits an exaggerated sense of their own importance, leading them to adopt an authoritarian leadership style. The problem with blind obedience is that it prevents us from seeing the real motivations of those who happily bow to the leader, who in turn gloats about that apparent full consensus. Forcing people to obey is useless if, in order to be successful, we need a group of workers with specific, complementary skills who are capable of using them autonomously. However, regardless of this, appeals often fall on deaf ears to delegate and to adopt a positive leadership model by giving followers space rather than orders. Once they have reached an executive position, leaders may feel sufficiently satisfied with themselves that they become convinced they no longer need to develop new leadership skills. They are not interested in listening to their workers' ideas or learning about their needs. Comprehensive analysis of self-evaluations by executives reveals a worrying lack of awareness of their abilities to engage people. In her book, "Act like a Leader, Think like a Leader", Herminia Ibarra

(2015) shows the data from a leadership evaluation performed by 3626 observers in comprehensive assessments. There is a large discrepancy between the positive self-assessment of leaders and the opinions of their workers and colleagues, in particular when it comes to their visioning the future, empowering and managing the team, and providing both feedback and reinforcement. Oblivious to the existence of this gap, or indifferent to the idea of taking responsibility for their behaviours, many leaders remain convinced that they do not need to worry about engaging their employees, or understanding their thoughts and feelings. With this rigid mind set, they generate stress and disengagement in their teams, to the point where they damage the companies they work for. Alternatively, worst case, they find themselves jobless and lacking the energy, motivation or time to gradually acquire new competences.

17.2 Self-awareness and the Motivation for Change

How can we help leaders boost their self-awareness and work on their dysfunctional behaviours without demoralizing them or damaging their self-confidence? Can we teach them to recognize their strengths and accept responsibility for their weaknesses? What is the best way to help them understand their resistance to change and show them where and how they can become more effective?

If we want to help traditional leaders to develop a new leadership style, we need to realize that their positions already require a good level of self-esteem. Teaching them therefore means helping them to "discover" a more functional way of being a leader in their own organization, rather than dictating what to do. Bosses or HR managers may send executives to leadership training because of a negative evaluation of their competences. However, accepting a training because of negative feedback, or feeling guilty and humiliated, are not the primary engines of change. Ironically, such experiences can even immobilize the person in question. Intrinsic motivation for change seems to arise when the person connects it with something of intrinsic value, something meaningful and cherished. Intrinsic motivation is engendered by an accepting, empowering environment that makes it safe for the person to explore the possibly painful present, and strive for what he or she wants and values (Miller and Rollnik 2002).

In creating a psychologically safe learning environment, the makeup of the class also plays an important part. In the past, groups were divided into hierarchical levels, as this was considered the key to creating an atmosphere more conducive to learning. Today, a group of executives that tackle a training course together can benefit from the diversity of the participants in terms of gender, age, company and cultural background. This can only happen if we succeed in making diversity visible. Creating a learning community that is relevant to every participant requires an initial investment. If we are able to highlight some positive characteristics of each participant, we can also see the positive influence of peer feedback on the learning process. Based on experience, we know that we are much better at observing the behaviours of others than we are at making an accurate self-assessment. So our peers can be

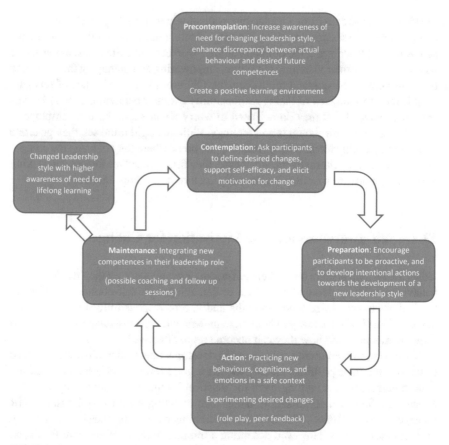

Fig. 17.1 The six stages of change in leadership training. Adapted from DiClemente and Prochaska (1998)

useful teachers by providing opinions and feedback when discussing personal case histories, for example.

In addition to putting leaders in a positive learning environment that rules out punishment or judgement, the pathway to change can only begin if we enhance their self-awareness. They need to recognize the current style of leadership they adopt, and each individual has to find the motivation to change. All too often we begin to promote behavioural change by providing good advice on which alternative behaviours should be enacted. However, participants may be either unaware of the problem behaviour or reluctant to accept responsibility for it. If they fail to see the problem and do not feel ready, willing and able to work on it, offering advice will only produce a negative reaction (Fig. 17.1).

An excellent aid for leading difficult participants through the process of change is the Transtheoretical Model (TTM). This model of intentional change offers an integrative framework for understanding the process of behavioural change, focusing on the decision-making process of the individual. TTM is based on the assumption that people do not change their behaviours quickly or decisively. Rather, a change in behaviour, especially habitual behaviour, occurs continuously through a cyclical process; in other words, change is viewed as a progression. The Stages of Change, a key component of the TTM, are the steps in the process people make as they change a certain type of behaviour. The initial *Precontemplation* stage is when the person is not aware of a problem or currently not considering change. Next comes *Contemplation*, where the individual seriously reflects on the reasons for or against change. *Preparation* follows, when planning and commitment solidify. Successful accomplishment of these initial stages leads to the individual taking *Action* to make the specific behavioural change. If this is fruitful, it leads to the fifth and final stage of change, *Maintenance*, in which the person works to preserve and sustain long-term change (DiClemente and Prochaska 1998). Adapting this model to the training stages in leadership development, we have highlighted six moments in the learning process.

17.2.1   The Precontemplation Stage

In the initial phase of a training course, it is useful to understand the stage participants are at in order to correctly evaluate their self-awareness and readiness for change. Executives in Precontemplation are either unaware of a problem behaviour or are unwilling or discouraged when it comes to changing it. They can be rather defensive about the targeted problem behaviour, and little convinced that the negative consequences of their leadership style outweigh the positive. A rational explanation will not alter, and may sometimes even intensify, the negative emotions linked to the training experience. Allowing participants the freedom to reflect and make their own decisions will facilitate a situation where the possibility of change can be explored in a non-threatening manner.

The key is to stress from the outset the idea that leadership is a result of a learning process influenced by the company cultures they have been exposed to throughout their working lives. This will help executives talk openly about how they have developed their leadership competences. So, typically, they will initially recall the beginnings of their career, the bosses they have worked for, the company values and rules that they have learned to respect which have contributed to their success. Identifying common experiences in their narratives and sharing what was considered a positive leadership attitude in the beginning of their careers is a very helpful exercise. This in-depth analysis of past working experiences helps to focus the attention of participants on outside causes for their actual leadership behaviour more than on their personal responses to these influences. In the beginning of a difficult personal process of change, it is not very useful to stress personal responsibility for a problem behaviour alone. "Life is the weaving together of many contextual influences that impact on it, as well as individuals' responses to such influences. Sometimes, to understand the 'mingled' nature of our career narrative we have to unpick and reweave it into a more cohesive yarn" (Watson and McMahon 2015).

So the first step is confronting experiences in past organizations and traditional leadership styles with the present and possible future development. This should help participants to link future challenges and changes their company will have to face to the consequently necessary changes in leadership. What challenges are they aware of? What challenges require new leadership skills? Which traditional leadership methods could even hinder successful change? Which of the leadership competences that are necessary for the future might be even diametrically opposed to those that have helped the company in the past to achieve its current success? Reasoning through these questions should shine a light on the discrepancy between the actual leadership behaviours and the competences needed to successfully help people to face present and future challenges.

It is best to leave the participants to find the answers to these questions on their own. Any brief analysis we as teachers could perform better and faster will translate into accomplishing the first phase of pre-contemplation more rapidly. However, this approach is counterproductive. A critical comment could push the participants in the opposite direction, making them feel the need to defend their traditional way of managing people. It is worthwhile to leave the leaders with the responsibility of reflecting on whether their past leadership style will be useful for guaranteeing a successful future for the company. Each group of executives can use their own experience to reconstruct the behaviours learned from and reinforced by the organization in which they began their career. Then they can differentiate these from the characteristics of the organization in which they are working now. What has changed over the years? Which hierarchical and bureaucratic attitudes and behaviours are still present in many senior leaders or traditional HR managers? What are the costs of ignoring the need for change?

It is also important to underscore the dramatic difference between the behaviour of a great leader from the past and the leadership style that is considered positive today. Helping leaders to recognize these differences is a necessary precursor to raising their awareness of the learning process involved. The dilemma with leadership development is that initially people may not understand the time and energy they need to invest to achieve the required emotional, social, or cognitive competences (Goleman et al. 2002b).

Having created a snapshot of the organizations of the past and present, the third scenario we should ask participants to describe is how they imagine the (fairly near) future of their company. We will see that, when talking about the future, what always emerges is a fairly realistic vision of the further drastic changes that every company must face. However, the company of the future cannot be defined using the traditional categories. The changes are rapid, "and require workers to develop skills and competences that differ substantially from the knowledge and abilities required in the occupations of the 20th century. Insecure workers in the information age must become lifelong learners who can use sophisticated technologies, embrace flexibility rather than stability, maintain employability and create their own opportunities" (Savickas et al. 2009).

17.2.2 The Contemplation Stage

The social learning theory developed by Bandura (1975) emphasises the importance of learning from others. So in this phase the trainer steps back and takes on the role of guaranteeing a psychologically safe environment. Allowing enough time for group discussion gives participants the chance to personally choose to acquire a series of new competences because they perceive them as being individually desirable and functional to future successes. This approach should encourage individuals to think imaginatively, to discover who they want and need to be in the future, and to explore all their personal potential with a positive sense of hope. As Boyatzis (2008) says, "the starting point in leadership development is the discovery of who the person wants to be." This discovery occurs the moment a new awareness emerges in the person's consciousness.

These first two stages of the learning process are also the most difficult. It may take time for a traditional leader to be ready to consider the need for change. Spending time eliciting intrinsic motivation through questioning, using the motivational interviewing approach (Miller and Rollnik 2002), is the best way to start the process in motion without creating resistance. A second positive aspect is that participants feel respected.[1]

What can we say about the role of the teacher? The rules of good teaching are as valuable for this type of intervention as they are for any workshop where the goal is to help people explore new behaviours, cognitions and emotions: good communication and listening skills, discussion teaching, empathy and respect for differences in expressing emotions, to name a few. The teacher's task is to release the potential of participants and to facilitate a natural change process, rather than imposing his or her ideas on "good leadership". The most critical and often underestimated aspect here is that the leadership style of the teacher as perceived in the classroom by the participants must be consistent with the underlying principles of the training course. Talking about engagement and the importance of listening and giving space to the ideas of subordinates, but then churning out a series of slides, jokes and examples, without providing the time or opportunity needed to consider them in depth, turns the participants into an entertained yet passive audience. This teaching style would totally contradict the sense of the term "participant". Unfortunately, education as entertainment is also the approach mostly favoured by people who are not really ready for change. Additionally, when we look at how leadership courses are marketed, we often find a long and attractive description of content, a promise to acquire competences that will turn the participant into Superman or Wonder Woman. But time limitations often fail to allow for real engagement and the mandatory focus on real learning. Neuroscientific research confirms what every student knows very well: "The first and most important step is to get people's undivided attention. The portion of the brain dedicated to learning and comprehension, the prefrontal cortex,

---

[1]Motivational interviewing is a collaborative—not a prescriptive—approach, in which a trainer evokes the person's own intrinsic motivation and the resources he or she can leverage to achieve it.

requires concentration to process new information. What matters most is the quality and quantity of attention paid to a new idea" (Rock and Schwartz 2007).

When we listen to different leaders analyse their interpersonal relationships in the company and their ideas on how to manage their workers, we have the chance to form a clear vision of their competences and adapt our training to the people and the problems that have emerged. On this subject, an international study on leadership skills (Weaver and Mitchel 2012) showed a truly demoralizing picture: (a) one out of three employees does not consider their boss efficient, (b) almost 40% of those interviewed feels offended and demotivated by their boss's actions, (c) 54% complains of a lack of engagement, or at least a lack of explanation about the decisions that concern them directly, and (d) 55% has considered leaving the company because of the boss. It is a safe bet that almost all of the respondents were careful not to say these things directly to their boss, feeding into the illusion of the boss's power, fuelled by that same silent, non-critical obedience we have already highlighted. The picture that emerges is that leaders in Europe still model themselves on examples from the past related to hierarchy, which are not effective in tackling the complex situations and social changes that have occurred over the last few years.

If, however, we analyse what the employees claim to be the most effective behaviours of the best bosses they have worked for, we can see that in the end they are not really as complex or unrealistic as we might expect. Employees say their best bosses:

1. recognized them appropriately;
2. supported them without taking over;
3. involved them in decisions;
4. listened to them;
5. took time to explain the reasons for their decisions;
6. took care to maintain their self-esteem.

These behaviours might seem simple, but adopting them involves a great deal of effort—it is not enough simply to improve company etiquette. Underlying all this is the demand for a leader who is above all a person with strong values based on transparency, respect and morals, one who is capable of encouraging people to participate and commit to their tasks, rather than expecting them to bow silently to the orders they get from the top.

17.3 Leadership and Engagement

The idea of empowerment started in the 1980s and first emerged, as always, in English-speaking business contexts. Empowerment means giving each worker the freedom, opportunity, motivation and authority to take the initiative and constructively solve the company's problems. Nevertheless, this concept—transferring the power to make decisions and act, strengthening competences and self-confidence—has been problematic for most leaders. The distance between the level of delegation used by the heads of our companies and the concept of empowerment was, and indeed still is, too far.

Since the 1980s a great number of articles and books have promoted a special focus on people as the premise for creating excellent, profitable companies, even in organizations with limited resources. This literature calls for strong engagement centring on more delegation, and management based on objectives as well as clear, pre-defined career paths, incentives and rewards. Increasing engagement also requires considerable effort by leaders to treat people like adult human beings, patiently forging relationships built on trust and collaboration. Unfortunately, in all these years none of the new management tools on offer has boosted the percentage of people who feel engaged, as we can see from the polls by Gallup (2017), an American company specialized in social and demographic research.

To measure the distance between what is practiced and what is preached, and to provide a few solid, practical indications on how to bridge this gap, it is useful to look at what scientific research in the field of psychology suggests. In fact, considerable progress has recently been made in this area. For several years now, psychologists have been studying what enables people to make strong commitments and perform at the highest levels, and what helps them develop their potential. From an analysis of the literature on engagement, what emerges is that a positive working environment that creates a positive emotional state motivates employees to invest themselves deeply in their work.

A positive approach to relationships along with a positive climate consistently offer various important benefits:

- It guarantees high energy levels and resilience, which helps build strong commitment, even in the face of difficulties.
- It develops states of mind characterized by strong enthusiasm and pride.
- It inspires employees and gives them a sense of challenge.
- It absorbs employees in their work, because they are fully concentrated and pleasantly engaged in what they are doing, so much so that time passes quickly and they even find themselves reluctant to take a break.

In a nutshell, an environment that helps people to be more confident, optimistic and effective engenders individuals and teams that are more likely to develop a strong sense of engagement and a good level of resilience, leading to positive results.

Every one of us has come across this type of positive experience in different situations, perhaps when we are playing a sport or doing a creative activity, or dedicating time to our favourite hobby. By becoming passionate about finding the solution to a problem or creating something meaningful, each one of us has experienced the feelings of surprise and amazement that come with discovering that we are capable of excellence. We can far exceed our expectations thanks to a strong emotional engagement. Unfortunately, though, the things we might succeed in achieving in our private sphere are not always easy to recreate in our working environment. This is primarily because our emotions are formally banned when we work for profit-oriented companies, which is actually detrimental to the fundamental role that our emotions play, above all in a work context. Instead, sometimes we even find an explicit attempt to curb people's enthusiasm, as if it were something transgressive, a threat to the leader. At the same time we ignore the irony of making employees fill out climate questionnaires, where assessing their emotions is considered critical for measuring the quality of their performance.

17.4   Engagement: Can We Do Better?

To put it simply, we can define engagement as a state that enables workers to become actively and passionately involved in company life, voluntarily doing more than their role requires. In other words, an engaged worker is willing to adopt positive or "free" behaviours that are fostered by emotional involvement rather than by an economic contract or dictated by a job description. These behaviours are also referred to as "above and beyond the call of duty" (ABCD), or 'positive deviance'—obtaining a commitment that exceeds expectations and achieving excellent performance from workers. This, on the other hand, requires the organization and top management to make an effort that deviates just as far from the status quo.

A recent 'State of the Global Workplace' report by Gallup (2017) identified the number of workers worldwide who describe themselves as "engaged", in other words, they work with passion and enthusiasm in their company and feel deeply involved in their job. The survey showed only 13% of workers are "engaged" while the remaining 87% are either "not engaged" or "actively disengaged" in their work. European workers, according to the poll, feel under-used and incapable of being at their best in the workplace. Europe scores far lower than the US in the ranking, as only 10% of the employees in European companies are engaged, compared to 30% among their American counterparts. Among the 18 European countries in the survey, this percentage is 17% in Norway and less than 10% in France, Italy and Spain.

The "not engaged" workers do not have a hostile, destructive attitude toward the company but instead simply fail to show any interest in company results, the work they perform or the customers they serve. They have learned over time to limit their efforts, and they believe that it is better to invest as little as possible in their job. Though they may have been in the past, they are no longer interested in seeking satisfaction from their work and they resist active engagement. We might think that this sort of detachment is only found in the lowest levels of the company population, but Gallup researchers claim the opposite is true, warning us that disengaged people can be found at all organizational levels.

There is also a group of employees who are defined as "actively disengaged" (in Europe the numbers range from 8% in Norway to 30% in Italy). These workers are seriously committed to reducing their participation in work and company life to a minimum, and find highly creative ways to proudly resist every possible form of engagement. The problem here is not simply that they are unproductive. Instead, they go so far as to destroy the positive energy of their colleagues and do everything possible to make the lives of everyone in the organization miserable, bosses and peers alike.

The researchers verified that there is a solid link between employee engagement and business performance, and not only in economic terms. They base this conclusion on trends in indicators such as customer portfolio, profitability, productivity, turnover, absenteeism, and quality. In other words, companies with a higher level of engagement among their employees also have the highest performance. But that is not all. The Gallup report also estimates that active disengagement costs the United States between 450 to 550 billion dollars per year. In Germany the same cost runs from 112 to 138 billion euro and in the UK the figures range from 52 to 70 billion pounds. A potential gold mine if we were only able to see it! Why are we so blind to this enormous waste of human potential?

If the workplace is becoming a place where human capital is consumed, or worse wasted, sooner or later companies will come to lack the very factor that all researchers agree is the only thing capable of responding most effectively to today's evolving markets. This is why we must try to change now in order to prepare for a better future. European companies must adopt new strategies for selecting and training their leaders, not only for their own good but also to meet different expectations that new generations have. How many times have we witnessed companies selecting and promoting leaders because of their excellent technical skills. "Not very good at managing people, but a quick training session will help." This is a comment we would very much like to never hear again!

We see that when people are engaged, this engagement derives from relationships that make them feel emotionally involved in their work and invested in the organization's objectives. They want to feel respected; they want to know that their work provides a significant contribution to the company, which in turn appreciates their efforts. Workers also want to work in an environment where they have good relations with other colleagues and with their managers. In particular, new generations of employees want to be sure that the company is interested in their personal development. They assess this interest in practical terms, based on feedback they want frequently, acknowledgement of their efforts by their bosses, and in the opportunity to participate in training and development programs. They also want a transparent relationship with their leader, which allows them to speak up, if necessary, when something is wrong. The autonomy and freedom granted to workers are fundamental elements for ensuring their well-being and guaranteeing they have the chance to express their potential (Baruffaldi 2019).

17.5   From Contemplation to Action: Defining Learning Needs and New Leadership Competences

The greatest challenge to an accurate self-image is for a person to see himself or herself as others do. Most executives perceive themselves as able to engage their subordinates, but the people around them may not offer truthful feedback on that score. In the second stage of the learning process, the Contemplation phase, a group discussion can help people accept direct and assertive feedback on their leadership competences and assist them in developing their own view of the change they desire.

17.5.1 Determination

Showing a group of executives a video where an incompetent boss is creating disengagement in his team (for example an educational video such as Max & Max by Covey (2004)), we can observe the behaviours and emotions typical of the 'Command and Control' leadership model. All the participants will find it easy to criticize the actions of an actor playing the part of a traditional authoritarian-style leader; they can plainly see why certain behaviours and attitudes can prove demotivating. Generally, watching such a scenario immediately sparks a discussion about how a good leader should behave, particularly if the company situation is critical and if it is losing customers (as in this video). This will reinforce and define even better the competences of a leader who needs strong engagement and involvement of people.

17.5.2 Leadership in Action

The next two steps of the learning process are more problematic. Once the participants have identified the most serious mistakes that cause subordinates to disengage, they are asked to imagine taking over this demotivated team and defining a leadership strategy for the first two months in this new position. A group discussion on diverse personal strategies usually generates a very lively debate. Successfully stepping into a new leadership role demands paying attention to every aspect of the situation. However, the proposed actions and strategies of traditional leaders only focus their attention on the objectives of their own department or team and on managing their employees, often in a very prescriptive way. The power of a leader is interpreted as a truly individual phenomenon, based on the power of the position. When asked to set out a strategy for the first 60 days, participants often completely neglect to include a broader focus on engaging bosses as well as colleagues from the other company departments. In their view, leadership is not based on the ability to create and influence a network of relations inside and outside the organization.

People who have an excessively individualistic (and narcissist) view of their own role often make another mistake: they forget to involve their superior in decisions. Because they feel self-sufficient, they may not see the need to forge ties with their boss based on mutual respect and understanding, or to share their success with colleagues. If leaders have been valued for years exclusively based on their own personal results, it is difficult for them to understand the importance of interpreting their role in a systemic organization. Teamwork, trust, empowerment and inclusion, group decisions and collaboration remain buzzwords that do not have any actual influence on their traditional values.

When through the class discussion the group has come up with a leadership strategy that takes into account all the critical aspects described above, the time has come to translate cognitions into actions by asking them to prepare a motivational speech. They must imagine that they now, as the new leader, are facing the group of passive, disengaged employees in the video, and they have to convince the team to get on board and change direction. The exercise requires them to explain their vision of the future in a clear, compelling way, detailing what they, as leaders, want to do and how they intend to engage their subordinates, how they see their collaboration in the future, and so forth. This apparently simple exercise highlights the difference between cognitively developing a good strategy with a clear understanding of the need to engage people, and the skills needed to translate it into coherent motivational communication.

Experimenting and testing the new behaviour in a safe setting gives participants the possibility to receive honest feedback on the attitude that comes across from their words. Is their speech compelling? Do they communicate in a way that conveys empowerment? Do they give the impression of being attentive to their team's needs and ideas? Do they sound supportive and open to discussion? Will they share future results with the team? Are they transparent and direct? Are they able to express their emotions? Again, it is best if feedback comes from their peers rather than the teacher, who must instead provide the participants with the right questions and summarize the statements they make, reinforcing a few principles of leadership. For instance, the instructor should emphasise the idea that lack of trust and a prescriptive approach limit the possibility of satisfying the basic psychological needs of subordinates; this in turn hinders individual engagement.

Indeed, based on the self-determination theory, we can predict that when individuals satisfy their basic psychological needs of autonomy (acting concordantly with their sense of self), relatedness (feeling connected to individuals and groups), and competence (feeling a sense of accomplishment from their own actions), they reach higher levels of subjective well-being (Deci and Richard 2000). The research suggests that, besides the negative effects on economic prosperity, specific cultural values (lack of trust, low respect, and high obedience) have a negative impact on the fulfilment of basic psychological needs. This is a highly insightful result, as satisfying these needs is considered crucial to achieving a high level of subjective well-being on the one hand and is also seen as a basic prerequisite for engagement on the other.[2]

---

[2]It is interesting to note that, "besides individual intrapsychic forces, the satisfaction of these needs also varies according to the characteristics of the social context. Among these characteristics, we consider, in the cultural dimension, the degree of limited morality, a feature of societies where obedience is highly valued, trust is low and respect is not recognized as one of the fundamental values to transmit to children. In these societies, trust and respect are bound to kinship-based relations and the individual's search for socio-economic opportunities is limited by the coercive power of the family (e.g. through the internalization of the obedience norm). As most of the socio-economic interactions are subordinated to the interests of closely-related persons, individuals achieve less self-determination. In particular, high obedience would hamper the satisfaction of autonomy and competence, while low trust and the scarce importance given to respect would limit the satisfaction of relatedness" (Conzo et al. 2017).

Changing one's leadership style means not only learning new behaviours but also experiencing new emotions. In order to make an effective motivational speech, executives must assert themselves, communicating their emotions together with their ideas. This also enables them to come to a deeper understanding of themselves, to find out what they feel (as opposed to what they think) about things, what they like and dislike, and what they want and do not want. Some people are able to express emotions appropriately and comfortably, while others struggle to find words for what they feel (Kets de Vries 2006). If an executive has worked for a lifetime in a company where emotions, especially for men, were strictly banned, and bosses were selected and promoted for their unemotional approach to problems and people, it will be quite challenging for them to let it go and allow themselves to experiment, even in a safe environment.

17.5.3 Maintaining New Behaviours

To help participants accept the risk of failure, when experimenting with a new leadership behaviour, they need to reinforce a positive view of themselves and the future evolution of their leadership. The strength-based approach (Niemiec 2017) can be very helpful to bolster and empower participants and to increase their self-efficacy to help them continue to practice the new behaviours. Learning and maintaining complex new behaviours cannot be achieved in a short training programme; it can start a positive change, but returning to the working environment there is a great possibility that if not reinforced by follow up sessions the old habits will soon take over. Business schools find it difficult to organize leadership training programmes over extended periods of time, despite the fact that the experience of developing and growing leaders in the context of a close-knit group is generally the most successful learning format. It helps participants to address personal difficulties in implementing change in a positive well-known environment.

An alternative possibility is to encourage leaders to take coaching sessions, immediately after the training experience, and help them understand how to use in future the help of an experienced coach.

17.6 Positive Leadership, a Possible Alternative?

Studies conducted over the last few years have proven what was already plain for everyone to see: working in an environment where relationships are positive is also beneficial to our physical and mental health. By the same token, working in a toxic climate causes serious damage to people and companies. Fostering the well-being and positive emotions in others and ourselves as the premise for an excellent performance is a field of research that seems miles away from the organized, practical rationality of our companies today. Knowing how to manage people appropriately so that they can feel positive emotions is an idea that is often laughed at, like the absurdity of Alice in Wonderland. Instead, companies that compete are more likely to resemble armies in battle, forging forward, steeped in rivers of tears and blood. Positive Leadership behaviour follows the lead of recently emerging Positive Psychology.[3]

Too often, leadership and management approaches in general have undertaken a negative perspective—trying to fix what is wrong with managers and employees, concentrating on weaknesses. By taking a positive approach, instead, researchers move from studying unhappy individuals with problems, to satisfied, optimist and talented people. In this regard, Seligman (2011) affirms that: "It's time for science to understand positive emotions, to back strength and virtue, to supply useful indications that will enable us to fulfil our potentials." He also states that hope, wisdom, creativity, courage, spirituality, responsibility, and perseverance have so far been ignored, and more attention is paid to negative impulses which are considered more authentic. In order to manage the employees in a company, leaders need to go beyond the predominant model consolidated over time, centred on obedience to power that is typically coercive, or in any case exercised in an authoritarian way. Instead, managing people must embrace positive attributes such as well-being, satisfaction and happiness.

In his recently published book, *Flourish: A Visionary New Understanding of Happiness and Well-being* (2011) Seligman uses five characteristics to expound on his scientific theory of well-being to explain what constitutes a good life and to outline what can be done to make it more satisfying in all fields, including work. Seligman uses the acronym PERMA to identify these five factors: positive emotions, engagement, relation, meaning and accomplishment.

PERMA
P Positive Emotions
E Engagement
R Relations
M Meaning
A Accomplishment

To better clarify this new way to understand the ingredients of positive leadership, it is important for us to briefly consider each element of PERMA individually.

17.6.1 Positive Emotions

The lack of negative emotions is not enough to ensure that an individual feels a long-lasting sense of well-being; indeed, research has shown that feeling positive emotions

---

[3]Positive Psychology refers to "the scientific study of positive human functioning and flourishing on multiple levels that include the biological, personal, relational, institutional, cultural, and global dimensions of life" Seligman and Csikzentmihalyi (2000). This psychological research and therapy emerged in the 1960s in the US, driven by theories and studies focusing on people's strengths and psychological capabilities.

plays a much more important role in achieving this goal. When we experience positive emotions, we tend to broaden our scope of attention. We are more likely to attend to many aspects of a given situation and look more at the big picture rather than the details of a circumstance (e.g. Wadlinger and Isaacowitz 2006). Positive affect also gives rise to an enlarged cognitive context (Isen et al. 1987) and helps us think outside the box, using more flexible cognitive categories. In addition, positive emotions create more stable and supportive relationships. Finally, when we experience positive emotions on a regular basis, such as gratitude, happiness or affection, this is essential to the well-being of each one of us, enhancing the subjective perception of our quality of life.

17.6.2 Engagement

People become happier and can fully develop their potentials if they are able to use their strengths. According to Mihaly Csikszentmihalyi (1990), if we allow ourselves to become engaged in something that we perceive as meaningful and important, we experience what this author calls "flow". Flow is characterized by intense concentration, loss of self-awareness, a feeling of being perfectly challenged (neither bored nor overwhelmed), and a sense that "time is flying". Flow is intrinsically rewarding; it can also help us achieve our goals (e.g., winning a game) or improve our skills (e.g., becoming a better chess player). Anyone can experience flow in different domains such as play, creativity, and work. Flow is achieved when the challenge of the situation meets one's personal abilities. A mismatch of challenge for someone with low skills results in a state of anxiety; insufficient challenge for someone highly skilled results in boredom.

17.6.3 Relations

As we already mentioned, positive relations are extremely significant for our physical and mental well-being. Being part of a social network, being able to rely on others and being able to help them in turn offers many people a huge sense of happiness. Additionally, receiving support from others and sharing our problems with them is one of the best antidotes to the inevitable low points of our lives. Research shows that being generous and looking after others has the more beneficial effect on the helper. Whether it is a relationship with a partner, with a group of colleagues at work, or a friendship, positive relations are one of the most important contributing factors in achieving satisfaction and well-being.

17.6.4 Accomplishment

People pursue accomplishment, competence, success and mastery as an end in itself in a variety of sectors, such as work, sport, leisure and so on. They even strive for success when it does not necessarily lead to positive emotions or significant relations. To achieve this we must define clear objectives in life and pursue them with commitment and dedication. Reaching our objectives and seeing ourselves achieve a goal that is important for us brings a deep sense of satisfaction, boosting our well-being and giving us a feeling of happiness, along with greater self-confidence.

17.6.5 Meaning

The importance of our work and the meaning that we attribute to it are largely responsible for determining the quality of our professional performances, our commitment and our satisfaction. It is therefore not surprising that for many employees, doing a meaningful job is more important than salary, working conditions or opportunities for promotion.

A meaningful job is defined as such when individuals perceive a real connection between their work and a wider life purpose. Meaning is also often associated with a sense of pride and accomplishment, with developing personal potential and with creativity. A sense of meaning and purpose can come from belonging to and serving a project perceived as larger than ourselves. Various institutions have the possibility to offer meaning to their members, such as religions, families, scientific communities, political parties or associations, but also working organizations and communities. If we can apply our strengths to a higher purpose we identify with and feel a sense of meaning, this is another big step towards leading a satisfying life. Over the last few years, we have witnessed a notable change in the values of industrialized countries. Young people have most of their material needs fulfilled; having seen professional success, they are aware of their own potential. These same young people are now finding the courage to leave well-paid jobs and seek a commitment that gives their lives meaning, perhaps accepting lower pay or working for non-profit organizations.

At the same time, however, we can also see a loss of meaning at all organizational levels and in a variety of roles. As David Graeber in his book "Bullshit Jobs" (2018) notes, people are not inherently lazy. "We work not just to pay the bills but because we want to contribute something meaningful to society. The psychological effect of spending our days on tasks we secretly believe don't need to be performed is profoundly damaging, a perception that is becoming increasingly more widespread even in medium and medium-to-high level qualified professions". For example, an excess of bureaucracy combined with the threat posed by technologies and artificial intelligence risk quashing the motivation of qualified doctors and teachers. These trends may make them lose the values and evaporate the expectations that had driven them with enthusiasm toward those noble professions. Even when we choose a job because it matches our potential and aligns with our values and objectives, we can lose our motivation entirely if the practical conditions in which we work change dramatically for the worse. The intolerable perception that we are not producing value for ourselves and for others creates a strong level of disengagement. If the boss tries to solve this problem of demotivation using an aversive approach based on punishment, this creates the ideal conditions for fostering active disengagement, even for those who were initially very highly motivated.

17.7 Learning a Positive Approach to Life and Work

One of the most effective ways of counteracting the general incredulity and resistance that automatically arises when we talk about positive leadership is taking participants on a company visit. It is important for traditional leaders to personally meet positive leaders at the companies where they work, to be able to hear them describe their experiences, ideas and plans in person and to ask them questions. This on-site experience allows learners to watch how a person can tackle even negative situations with a positive attitude and with encouragement. Observing the real-life application of the theoretical model is much more useful than a lecture. Seeing a leader surrounded by a positive climate and satisfied people can be energizing and enabling for the participants of a leadership course.

A valid example of a positive leader is undoubtedly Ali Reza Arabnia, president and CEO of Geico Taikisha (Gehrke 2018). This company, headquartered in Milan, is a leader in the field of technology for turnkey painting systems for the automotive sector. Arabnia is a leader who invested in his workers in a period of intense crisis. When faced with the dramatic trade-off between giving twelve more months of life to his company without investing in the people, or shortening its financial life by spending money to develop all his employees' professional skills and their ability to innovate, he chose the second alternative. And not just because his values always place the individual at the centre of everything! He was convinced that maintaining his human resources without letting anyone go (while other companies were sacking masses of workers to cut their costs) was an investment that would certainly have paid off when the market, worth hundreds of millions of euro, took off again. An entrepreneurial and managerial decision that paid off, as we can still see when we visit the company today in the smiles of all the employees at all levels. A positive leadership model was provided by a person who implemented an optimal solution to the problem of employee motivation and engagement, even at the peak of a crisis, sustaining a working environment that put people first, even before the company objectives. Ali Reza Arabnia overcame the crisis thanks to nothing more than a responsible way of perceiving the work and the management of his company. His story, the Geico story, is an excellent example of how to renew business models by focusing on people, putting their well-being and growth at the centre of attention, leaving behind the usual managerial schemes and trying to breathe life into innovation. After all, innovation is human and cultural before it is technological, as it always has been since the Renaissance.

Ali Reza Arabnia brought Geico out of financial hardship and gave it a competitive edge on the international markets, thanks to his unconventional choices founded on values, prioritizing people, and giving meaning to work for his employees. He designed the working environment on the model of a human being with a brain and a heart, which must first come together before anyone can create or innovate. The brain is the design offices, and the heart is what Arabnia calls "Laura's Garden of Thoughts." This space, dedicated to his wife, was conceived to be the place where everyone who works in the company can recharge physically and spiritually. In fact, there is a library, an art gallery, an auditorium for hosting thought-provoking cultural meetings and a wide variety of different plants, as many and as varied as the company's employees themselves. The company's innovation buzzes in the R&D centre, where the painting systems that the company produces move in real scale before the eyes of the delegations of clients who visit from all over the world. They initially seem curious and then they show tremendous enthusiasm. His idea of management has succeeded in striking a balance between the rational and emotional sides of the human brain, between the efficiency of a company and the satisfaction and motivation of its employees.

Today Geico sets the benchmark for Italian companies because it is successful in a global sector known to be over-populated and fiercely competitive. Yet this fierce competition never led the company to fire any workers, and has never even been used as an excuse by its leader to use fear as a tool for putting pressure on its people. In fact, quite the opposite is true: "For me fear is the thing that blocks people's creativity and initiative the most. Fear prevents us from deciding and the tension that we take home from a toxic working environment in the evening is not conducive to a good night's sleep. I don't understand how a leader can create fear in the belief that by doing so they will engage their staff more. I just don't get how a manager can manage a successful company by sowing fear."

The exceptional sensitivity, intelligence and moral leverage of the CEO and of those who have grown around him show that positive leadership can change people's level of engagement. What is more, by optimizing their potential, this approach makes it possible on one hand to have excellent performances (every Geico employee produces three times as much as the typical individual performance for the industry), and on the other people who are satisfied with their quality of life. Ali Reza Arabnia underscores his gratitude for the engagement of his staff, who helped him in times of crisis. The well-being of his employees is a genuine concern to him. "Anyone who has the luck and ability to achieve results has a duty to society. A company must not only make a profit, it must also take on social responsibility. After all, it's all about ethics in the end."

17.8   The Benefits of Positive Leadership

We do not need the backing of scientific studies to understand, based on our own experiences and those of our colleagues, that living in a psychologically toxic environment damages our health and compromises our professional performance levels. There is still much to be done in this field, starting from the commitment of everyone to create a working environment that is as positive as possible. However, those in leadership roles should be conscious of how much each of the aspects of PERMA can be influenced by their words and behaviours. The leader has a huge responsibility to create the adequate level of psychological and physical well-being required in his or her team so that the men and women who work for the company can fully express their potential.

When we find ourselves in situations of emotional discomfort, our cognitive abilities are significantly compromised. We attempt to shelter ourselves from the danger, by focusing on the details and trying to understand what is wrong in the situation and how to get ourselves out of it quickly. We do not have the adequate cognitive resources, at that point, to solve complex problems or carefully nurture a relationship. Yet many leaders at the head of organizations have paradoxically been chosen for their abrasive style, confirming the fact that being perceived as men or women who are tough enough with their workers still appears to be a quality that boosts a candidate's progress on the career ladder.

Positive Leadership is based on positive psychology and is designed to enable the best in people and inspire them to strive for remarkable results. Positive leadership is an integral part of managerial know-how. It does not replace the basic knowledge of the skills necessary to manage a company well (i.e. defining strategy and objectives, assessing one's employees, and providing feedback, etc.). Instead, it adds a positive vision of people and a different approach to solving problems. The image of the leader is also positive in the sense that it embodies a combination of optimism, sincerity, intellectual and moral integrity (Cameron 2012).

This leadership style requires a considerable evolution of the company vision and of its people. A leader must:

(1) focus on strengths and capabilities, setting aside the traditional deep-seated diffident approach to employees;
(2) believe that it is possible to develop people's personal passions and their intrinsic motivations;
(3) invest in the development of people so that they can spontaneously unleash their full potential and feel fulfilled while helping achieve the company's objectives;
(4) build more "positive capacity" using deliberate strategies to consistently create more positive experiences at work;
(5) concentrate on "positive deviance", in the awareness that in the world of global competition, only excellence wins.

At first glance, this might seem a rather trivially optimistic and sweetened definition of leadership. This is why we have referred to the principles underlying the research in psychology, neuroscience, and medicine on which this model of leadership is based. This approach is not the latest in a long line of marketing ploys used to sell a book, a course or a conference, but a serious, well-structured, proven and constructive model that should be adopted by those who, as leaders, feel responsible for excellent results in managing people and business.

The term "positive" refers to an optimistic vision that observes, develops and protects the positive elements that are available to us: the strengths of our people, their learning abilities, their motivations, etc. This is not a "positive" management style based on an unrealistic optimism that denies or purposely avoids seeing the negative, dysfunctional aspects in the organization. In this regard, it is worth remembering that "optimism, as a scientific construct, has been presented in different ways on the theoretical and operational levels. Among the many forms of optimism, one basic distinction, which must be made, is the difference between realistic and unrealistic optimism. Whereas the former performs important functions of adaptation, the latter, which leads to an optimistic distortion that is illusory and magical, has significant negative effects on the physical and psychological well-being" (Realdon and Anolli 2007).

Realistic optimism and positive vision are both characteristics that are not exactly easy to affirm, especially in times of crisis and hardship that unfortunately abound for many companies in this difficult time in history. When we are asked to "do more with less", it is hard to maintain an optimistic outlook and a positive leadership style. The political and social environment, not much more helpful, seems to send the message that there is one path to success: we only win if we behave aggressively, with cynicism and with no respect for others.

- Every day on television we see news of the umpteenth excess of brutality on the part of a political leader who resorts to the old "iron fist" approach to resolve delicate social problems and manage complex social and political situations.
- Expressions of hate and aggression are ten a penny on the social networks, transforming a medium that was supposed to unite us into a risky place for all users who have naively allowed themselves to be carried away by the desire to communicate and open up.
- Venues such as stadiums, concert halls or discos, communal spaces up until a short time ago where people could come together and socialise, have turned into high-risk areas where all kinds of violence can occur.
- In schools, teachers are having trouble defending themselves from violent adolescents and even in primary schools, places that should be synonymous with a world that is still sheltered and happy, bullying has spread like a modern plague.

The angry politicians, the haters and the bullies all seem to find support and consensus. Indeed, they may even be seen as effective models for managing interpersonal relationships, capable of earning the fearful respect of many. It is understandable then that companies adopt models that society deems successful. In such a climate, it is a challenge to introduce a leadership style that is functional for problem solving and beneficial to the people to be managed, but inconsistent with the dominant social models. It is crucial to make people aware of the extent to which the negativity (or positivity) of their social environment also unconsciously influences their behaviours. In a negative environment it is extremely problematic for the leader of a group to create and maintain a positive vision of the organization to which they belong for themselves and their workers, counteracting the pessimism and cynicism that surrounds them.

17.9 Conclusion

"Great leaders light up our passion and bring out the best in us. When attempting to explain why they are so effective, the keywords we use are strategy, vision or powerful ideas. The reality, though, is much more primitive. Great leadership works through emotions" (Goleman et al. 2002a). So a few words are enough to define the essence of leadership. The fact remains that in the short term, putting workers under pressure, threatening them and using fear tactics are strategies that get better results than a positive leadership approach. But, the fact that these behaviours are neither ethical nor useful to the company in the long run is easy for everyone to understand.

Individuals cannot be forced to improve their leadership competences, and they do bear ultimate responsibility for their development. One of the sincere hopes of every consultant and trainer is that more leaders and organizations will recognize the opportunity to both listen and give more to the people who work for them. Influencing the behaviour of traditional leaders is a more difficult and complex endeavour than we would expect. Profound professional and human dedication is needed if we want to be effective, along with adequate time and resources. As former Harvard dean Derek Bok once stated: "If you think education is expensive, try ignorance!"

Bibliography

Bandura, A. (1975). *Social learning and personality development*. New Jersey: Holt, Rinehart & Winston.

Baruffaldi, L. (2019). *Leading Millenials: Conoscere le nuove generazioni per costruire collaborazioni di successo in azienda*. EGEA

Boyatzis, R. E. (2008). Leadership development from a complexity perspective. *Consulting Psychology Journal: Practice and Research, 60*(4), 298–3139.

Cameron, K. (2012). *Positive leadership*. Berret-Koehler.

Conzo, P., Aassve, A., Fuochi, G., & Mencarini, L. (2017). The cultural foundations of happiness. *Journal of Economic Psychology, 62*, 268–283.

Covey, S. R. (2004). *The 8th habit*. Free Press, Video Max & Max. www.youtube.com/watch?v=bosxpMMSTpw.

Csikszentmihalyi, M. (1990). *Flow: The psychology of optimal experience*. Harper and Row.

de Vries, M. K. (2006). *The leader on the couch*. Wiley.

Deci, E. L., & Richard, M. R. (2000). The "what" and "why" of goal pursuits: Human needs and the self-determination of behavior. *Psychological Inquiry, 11*(4), 227–268.

DiClemente, C. C., & Prochaska, J. O. (1998). Toward a comprehensive, transtheoretical model of change: Stages of change and addictive behaviors. In W. R. Miller & N. Heather (Eds.), *Treating addictive behaviors* (2nd ed., pp. 3–24). New York: Plenum Press.

Fast, N. J., & Chen, S. (2009). When the boss feels inadequate: Power, incompetence, and aggression. *Psychological Science, 20*(11).

Gallup., (2017). *State of the Global Workforce*. Gallup Report: Gallup Press.

Gehrke, B. (2018). Verantwortungsvolle Führungskräfte in globalen Unternehmen. In V. B. Covarrubias, K. Thill, & J. Domnanovich (Eds.), *Personalmanagement. Internationale Perspektiven und Implikationen für die Praxis* (pp. 262–282). Wiesbaden: Springer Gabler.

Goleman, D., Boyatzis, R. E., & McKee, A. (2002a). *The new leaders: Transforming the art of leadership*. Little, Brown & Company.

Goleman, D., Boyatzis, R. E., & McKee, A. (2002b). *Primal leadership: Realizing the power of emotional intelligence*. Boston: Harvard Business School Press.

Ibarra, H. (2015). *Act like a leader, think like a leader*. Boston: Harvard Business Review Press.

Isen, A. M., Daubman, K. A., & Nowicki, G. P. (1987). Positive affect facilitates creative problem solving. *Journal of Personalit and Social Psychologyy, 52*(6), 1122–1131.

Miller, W. R., & Rollnik, S. (2002). Why do people change? In W. R. Miller & S. Rollnik (Eds.), *Motivational interviewing* (pp 3–12). Guilford Press.

Niemiec, R. M. (2017). *Character strengths interventions*. Hogrefe.

Realdon, O., & Anolli, L. (2007). Ottimismo disposizionale e coping emotivo. *Psicologia della Salute*, Franco Angeli, n1.

Rock, D., & Schwartz, J. (2007, October 4). Why neuroscience matters to executives. *Strategy + Business, Leading Ideas*.

Savickas, M. L., Nota, L., Rossier, J., Dauwalder, J. P., Duarte, M. E. Guichard, J., et al. (2009) Life designing: A paradigm for career construction in the 21st century. *Journal of Vocational Behavior*. https://doi.org/10.1016/j.jvb.2009.04.004.

Seligman, M. (2011). *Flourish: A visionary new understandings of happiness and well-being*. N. Brealey.

Seligman, M. P., & Csikzentmihalyi, M. (2000). Positive psychology: An introduction. *American Psychologist, 55*(1), 5–14.

Wadlinger, H. A., & Isaacowitz, D. M. (2006). Positive mood broadens visual attention to positive stimuli. *Motivation and Emotion, 30*(1), 87–99.

Watson, M., & McMahon, M. (2015). From narratives to action and a life design approach. In L. Nota & J. Rossier (Eds.), *Handbook of life design* (pp 75–86). Hogrefe.

Weaver, P., & Mitchel, S. (2012). Lessons from leaders from the people who matter. How employees around the world view their leaders. *Development Dimensions International*. http://www.ddiworld.com/ddi/media/trend-research/lessonsforleadersfromthepeoplewhomatter.

Wilkinson, J. (1982). The art and craft of teaching. *Harvard University Press*.

Wilkinson, J. (2001). Cited in Gudrais, E., Where pedagogy is "interesting". *Harvard Magazine*. (September–October).

Beatrice Bauer is Associate Professor of Practice—Leadership, Organization & HR, SDA Bocconi School of Management, Milano Italy, with thirty years' experience in the design and coordination of business courses for the development of positive leadership skills functional to the social and organizational changes taking place. In particular, given the multicultural background, she trained executives and senior managers in large multinational companies in Italy and abroad.

Part IV
New Leadership in Practice

Chapter 18
A New Model for Strategic Leadership in Healthcare: The A–G Model

Robert Pearl

Abstract Healthcare systems around the globe face the common challenges of rising medical costs, lagging quality and limited access. Improving medical care worldwide will require new leadership and innovative solutions. This article sets forth a brave leadership strategy designed to help clinicians overcome their fears of change and embrace the best solutions for their patients. The A-G model presented in this article gives leaders a powerful tool to help achieve meaningful transformation within a healthcare organization by allowing them to articulate a bold vision, generate the trust of their colleagues and demonstrate the courage needed to overcome the inevitable resistance they will encounter.

18.1 Introduction

Healthcare systems around the globe face a common and growing set of challenges. They include rising medical costs, growing unaffordability and shortcomings in the areas of quality, access and patient convenience. This chapter focuses on the U.S. healthcare system. Although American medicine is unique in some ways—which include a financing system wherein employers, not the government, cover more than half the population—approaches that have improved the structure and delivery of medical care in the United States can be directly applied to industrialized nations everywhere.

Despite its incredible scientific, medical and pharmaceutical advances in recent decades, the American healthcare system most closely resembles a nineteenth-century cottage industry. In every community, there is widespread fragmentation—doctors and hospitals operate in silos, disconnected from one another. Every service rendered is paid piecemeal; what healthcare providers call "fee-for-service." Technology—and, specifically, information technology—is held over from the last century. Doctors sit behind clunky computers powered by operating systems designed primarily for billing purposes and the most common way they transmit important

R. Pearl (✉)
Stanford University, Palo Alto, USA
e-mail: drrobertpearl@gmail.com

© Springer Nature Switzerland AG 2020
N. Pfeffermann (ed.), *New Leadership in Strategy and Communication*,
https://doi.org/10.1007/978-3-030-19681-3_18

medical information to colleagues in the community is via fax machine. Most important, there is little formal structure or leadership within medical settings, a reality that serves to perpetuate the problems already besetting American healthcare.

The consequences of this outmoded system are both predictable and deadly. Each year, hundreds of thousands of people die prematurely from preventable medical errors, omissions in preventive services (Lee et al. 2018) and the avoidable complications of chronic disease (Makary and Michael 2016). National healthcare expenditures in the United States now account for 18% of the Gross Domestic Product (GDP). With actuarial projections of $5.7 trillion in healthcare costs by 2026, economists fear that the expense of medical services could surpass the ability of the government, businesses and individuals to pay within 10 years, and might therefore lead to rationing of care as the lone remaining option (Tozzi and Tracer 2018). Furthermore, physicians are frustrated and burned out. One-third of doctors suffer from anxiety or depression, resulting in more than 400 suicides each year—among the highest rate of any U.S. profession (Andrew and Brenner 2018).

The need for strategic and operational improvements in healthcare has reached a fever pitch. Accomplishing them will demand new leadership and innovative approaches. To understand the challenges and opportunities of American medicine today, it is essential to begin with the evolution of medical practice since the mid-twentieth century.

18.2 Medical Care, Then and Now

Until the 1950s and '60s, doctors could do little to heal or help patients, save for prescribing basic antibiotics like penicillin, administrating immunizations against a few childhood diseases and performing a limited number of routine surgeries.

Due to the limited diagnostic and therapeutic approaches available, the cost of care delivery at the time was low—accounting for less than 5% of U.S. GDP. For most of the last century, nearly all patients who came to the doctor's office suffered from acute problems, such as pneumonia, influenza or appendicitis. Treatment of these diseases led either to full recoveries or quick death. The term "chronic disease" was nowhere to be found in medical books of the era because few people lived long enough to acquire them. In 1960, the average man died at 66 years old, 10 years younger than the average male today.

Bed rest was the treatment for heart attack, and most patients spent seven to 10 days in a hospital following a myocardial infarction, twice as long as they do now. After childbirth, mothers and their babies stayed a full week in the hospital compared to a day or two now. Back then, doctors had fewer than 100 drugs in their armamentarium. Procedures such as angioplasty—performed to unblock the coronary vessels to the heart—were still decades away from being perfected.

The health system was structured to meet the demands of that era's medical practice, which was far simpler and less sophisticated. The uncomplicated nature of care delivery at the time led doctors to work in solo practices and even make house calls. The absence of cutting-edge machines, expensive medications and complex procedures made a day in hospital relatively cheap, thereby allowing for extended inpatient stays. And with the exception of basic X-ray machines, simple blood tests and a rudimentary understanding of genetics, doctors relied primarily on their intuition and clinical experiences when diagnosing or treating a patient. Empathy and compassion were two of the most powerful remedies they offered, and patients were grateful for the care and concern they received.

All that began to change in the 1970s and '80s. Sophisticated MRIs, along with CT and PET scanners became powerful diagnostic tools, replacing much of the intuition doctors relied upon in the past. The history and physical exam, along with the highly personal doctor-patient relationship, progressively ebbed in importance as technology came to dominant medical practice. By the turn of the century, the human genome had been fully sequenced and analyzed. Specialty care eclipsed primary care as both the dominant force in healthcare and the most lucrative. With the introduction of ever-more expensive drugs and complex procedures, physicians were soon capable of extending human life far beyond what previously could have been imagined. As patients lived longer, chronic diseases like diabetes, heart failure and arthritis became more prevalent—an upward trend that continues today (Jones et al. 2012). It's estimated that by 2050, one-third of Americans will have diabetes (https://www.cdc.gov/media/pressrel/2010/r101022.html) (Fig. 18.1).

The challenges of medical practice are starkly different in the twenty-first century than in the twentieth. The physician's armamentarium is now stocked with more than 5000 available medications. Multimillion-dollar diagnostic and therapeutic machines can both find and treat cancer more effectively than before. Heart surgeries are routinely performed on patients over the age of 90 as people unable to get out of bed or feed themselves can be kept alive for years with machines and tubes. The electronic health record has made reams of data available at computer stations in hospitals, offices and even doctors' homes, so that physicians spend progressively less time at the patient's bedside, eroding the doctor-patient relationship.

One might think that broad access to these advances in the U.S. would make it a world leader in quality outcomes, but that promise has not been fulfilled. As a result, despite leading the world in healthcare spending, America ranks last among developed countries in almost every measure of clinical quality, from childhood mortality to life expectancy—with U.S. longevity even decreasing in recent years. Meanwhile, the American system's fundamental structure, reimbursement model and information technology systems have failed to evolve at the same pace of advancement as the rest of medical science.

The need for strong leadership is far greater today than in the last century. Back when the cost of care delivery was low, the goal of achieving operational excellence was less urgent and important. When chronic diseases were minimal, so was the need to implement systems of care that maximized coordination and collaboration among physicians. In fact, when intuition and personal experience dictated medical decision-making, there was little need for doctors to organize into teams or follow the limited research-based practices of the time.

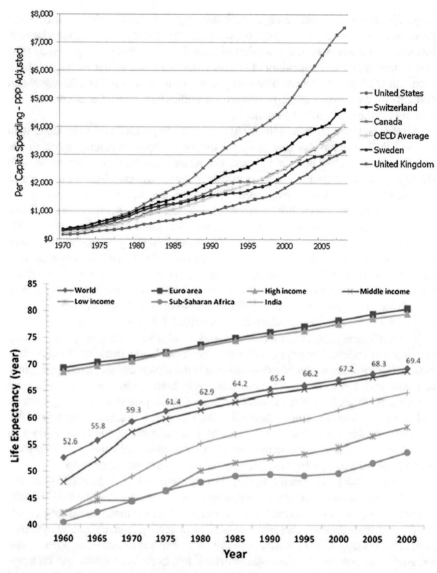

Fig. 18.1 Comparing the upward trajectories of healthcare costs and life expectancy around the world. Table 1 (https://www.kff.org/health-costs/issue-brief/snapshots-health-care-spending-in-the-united-states-selected-oecd-countries/) shows per capita spending for all OECD nations. Table 2 (https://en.wikipedia.org/wiki/Life_expectancy#/media/File:LifeExpectancy.png) shows life expectancy in years based on global location and income levels

Yesterday's approaches are inadequate for today's problems. Medical care has become so complex and expensive that the best outcomes are dependent on team-work, communication and evidence-based medical practice. The demands of the twenty-first century to lower costs, prevent medical errors and implement system-wide operational improvements will require more effective leadership.

And yet despite widespread discussion and agreement about the need to move from fee-for-service reimbursement to pay-for-value approaches, change is slow. Most healthcare leaders today are focused on maximizing the *quantity* of medical care delivered and the prices they can charge. Across the country, hospital administrators want to fill beds more than they want to find efficient, high-quality practices to empty them. Hospitals of today advertise the newest multimillion surgical robots and proton beam accelerators to attract highly remunerative patients, rather than questioning the clinical value in purchasing the machines in the first place (O'Neill and Scheinker 2018). Rather than developing a reputation for prevention and disease avoidance, doctors of today have staked their "brand" on offering miracle cures and life-saving surgeries for the sick and dying. If these problems were confined to the U.S., it would be possible, even logical, to blame the nation's unique culture, politics or private insurance system. But new data from around the world suggest that every industrialized nation is facing a similar healthcare-economic crisis. Addressing it will require skilled and innovative leadership everywhere.

18.3   Difficult Times Demand Bold Changes

If businesses in other industries worked similarly to American healthcare, the conse-quences would be disastrous (Parks 2016). It is impossible, for example, to imagine a capital-intense factory with expensive employees and machinery operating at 70% of capacity during the workweek. Yet, this is what happens in many community hospitals, operating rooms and radiology suites across the country (Pearl 2018). Or try to imagine that instead of hiring five skilled operators for particularly complex machines, you let everyone try their hand at each, rotating from one to the other throughout the week. None of the operators would gain maximum expertise and quality would suffer. And yet, that is how surgery in most specialties is performed in hospitals today—all the physicians in a single specialty doing a little bit of every-thing. Doing a procedure once a month creates variety for the surgeon but leads to poorer outcomes and more medical errors for the patient. Finally, try to think of a company whose employees spend huge amounts of time doing work that adds bil-lions of dollars of added costs, but no value for its customers. It might sound absurd but, in healthcare, 30% of what physicians do (tests, visits and procedures) has been proven to add no value for patients or improve clinical outcomes (Lallemand 2012).

18.3.1 Physician Resistance to Change

Inspiring and implementing change is difficult in any industry, but it is particularly problematic in healthcare. A major reason is the powerful role doctors play, accounting for 70% of all healthcare costs. It's not that the contributions of everyone else, including nurses, staff and pharmacists, are not crucial to patient outcomes. It's that without the support of physicians, any improvement effort is doomed to fail. Unfortunately, leaders who try to change physician practices meet resistance for a variety of reasons:

1. **Sense of entitlement.** During their years of training, physicians make huge sacrifices, often working 100 h a week while accumulating an average debt of $200,000 to earn their medical degrees. They spend the majority of their twenties and early thirties studying, training and working toward the mastery of their trade, often missing out on the more enjoyable experiences of their friends. After more than a decade of training, they feel entitled to practice as they choose.
2. **Knowledge differential.** Because physicians spend so many years training after graduation, they possess far more expertise than their patients, creating a "doctor knows best" mindset. Within healthcare's fragmentated environment, this mentality leads many doctors to believe their judgment and decisions are always right, even when the data offer evidence to the contrary.
3. **Fear of mistakes.** Although many doctors are overly confident, fear lurks inside all of them. In Silicon Valley, a failed venture can be perceived as a rite of passage if not a badge of honor. In medicine, where lives and professional reputations are on the line, failure is not an option. Mistakes result in pain and suffering for families and, in some cases, a malpractice suit and even the loss of a doctor's medical license. As a result, doctors are incredibly risk averse, particularly when it comes to embracing change.
4. **Outside threats.** For more than a decade now, doctors have felt as though their jobs and livelihood are under constant attack. Cost-cutting measures from both private insurers and the government have made physicians leery of "improvement programs." Getting anyone to see the wisdom of embracing change when it will negatively impact their income and their lives proves extremely difficult. Doctors are no exception.
5. **Personal Gain.**

Ultimately, it's difficult to shift dollars in the healthcare system from areas of relative ineffectiveness to ones with major impact. That's because individuals who will be negatively impacted consistently resist change. For example, studies show that investments in primary care and preventive medicine help patients avoid cancer, heart attacks and strokes far more successfully than investing in more complex, specialty care. However, any attempt to shift the dollars will be labeled by specialists as "rationing." They will lobby hard to stop it. Similarly, consolidating hospitals and clinical services to increase expertise and experience would maximize efficiency, improve clinical outcomes and lower costs, but attempting to shutter an underutilized facility or close a clinical service in a community will be contested by administrators, community leaders and the doctors who practice there.

Of course, physicians aren't the sole resisters to change. Hospital executives can be just as recalcitrant. They will lobby against any approach that moves patients out of their expensive inpatient OR suites into a community "surgicenter." They'll fight anything that limits their freedom to build additional hospital rooms in their facility when space already exists in a competitor's. Rather than looking for ways to increase efficiency and lower the cost of hospital care, hospitals and their leaders have increasingly sought to gain market power through mergers and the bulk purchasing of physician practices. With monopolistic control, they can raise prices and enrich their bottom lines with relative impunity. As a result, inpatient expenses in the U.S. have grown even more rapidly than overall medical inflation.

Over the past decade, medical care has become unaffordable for a growing percentage of the population, even for those with insurance. The out-of-pocket costs that patients are expected to pay in any given year now exceed the savings of 40% of Americans. In the United States, the number one reason families declare bankruptcy is the inability to pay medical bills. Although this is less of an issue for patients in countries with universal government-provided healthcare, economic challenges are making it harder for all developed nations to cover the medical needs of their citizens, thus diverting needed investments away from education, infrastructure and social welfare.

Opportunities to raise quality and lower costs abound. What is lacking are not ideas or proven solutions, but the leadership to translate opportunity into reality.

18.3.2 The Four Pillars of Healthcare Transformation

In healthcare, leadership requires vision of what is possible. The next generation of leaders will need to help their organizations transition from the failed structure, reimbursement and deficient technology left over from the last century, to solutions capable of solving the healthcare problems of the future. To that end, the "Four Pillars of Healthcare Transformation" can serve as a foundation for achieving what is possible (Salber 2017).

Together, these pillars facilitate solutions to the growing set of economic difficulties through improved quality of care and application of technological alternatives, rather than rationing. Each pillar seems straightforward, but all are complex. Understanding each is a crucial first step for any leader who will be leading the change process.

1. *Integration*. When physicians, both in primary and specialty care, work in a single medical group, collaboration and cooperation increase. Horizontal integration within a specialty encourages mutual support, implementation of best practices and added specialization. Vertical integration across primary, specialty,

and diagnostic care facilitates the rapid flow of information and supports innovative solutions to medical care.

2. *Capitation*. When physicians and hospitals are reimbursed on a prepaid basis—and rewarded for quality not volume—they have incentives to maximize prevention, eliminate medical errors and achieve greater operational efficiency. There are different ways American insurance companies are approaching prepayment. today. The first involves a single total-dollar payment to a hospital for a specific procedure, such as a total-joint replacement, to cover the total cost for the surgeon, anesthesiologist, operating room and post-operative care. Another approach is an annual payment to a group of doctors to cover all of the expenses that a cohort of patients with a common disease will incur (e.g., diabetes). Finally, in a fully capitated arrangement, a single per member per month fee is paid either to a medical group or an accountable care organization (ACO) for an entire population of patients and expected to cover the totality of their medical needs.

3. *Technology*. Information technologies in healthcare offer tremendous potential to lower cost, make care convenient and improve clinical outcomes, particularly for organizations that are integrated and capitated. Examples abound, including a single, comprehensive electronic health record system that provides the most up-to-date information on every patient regardless of where that individual is receiving care at the moment. Video and mobile technologies connect physicians with patients and with one another, inexpensively and regardless of geographic distance. The emerging field of data analytics can tell patients whether they are at risk of an impending medical problem, long-before providers recognize what is happening. And in the future, AI (Artificial Intelligence) will be capable of evaluating radiographs more accurately than clinicians, and sending the reports to doctors immediately, rather than hours or days later.

4. *Leadership*. As important as integration, capitation and technology are individually, the synergy of all three generates the greatest value for patients and providers alike. The following sections examine the types of positive outcomes that are possible through effective leadership, and describe the best approaches leaders apply to achieve them.

18.4   What Leaders Do

Leaders make positive change happen that otherwise would not. This process begins with strategic thinking that creates a clear vision of where they want their organization to go. Then it involves strategic action as they engage with the people they lead to align them around that vision and motivate them to move forward with confidence.

Strategy is often confused with the series of binders generated at this year's leadership offsite, stuffed with ideas and initiatives that are likely to be replaced in next year's binders. In reality, strategy should be thought of as a future-oriented process through which leaders decide where to position their organization to maximize the probability of success. How best to translate strategic thinking into action can vary

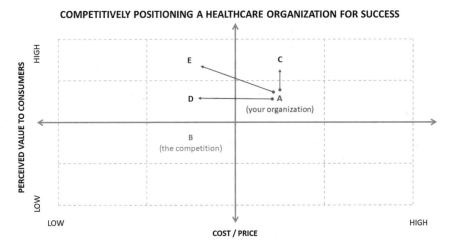

Fig. 18.2   Stratetic positioning in a competitive environment

year to year as new opportunities present themselves, but unless there are massive and unexpected shifts, the fundamental direction and competitive positioning should remain constant for most organizations.

In Fig. 18.2, we see a four-by-four matrix. One axis represents cost/price and, the other, perceived value in the mind of the consumer. The letter A represents where the leadership of the organization sees itself relative to its competitors (letter B). In this theoretical example, the organization shown is higher in perceived value (a combination of quality, access and personalized service), but also higher in cost than its competitors. The other boxes on the matrix represent future positioning opportunities: C (higher quality than today but at the same cost), D (same quality but at lower cost) and E (higher quality at lower cost). All are better competitive positions than exist today.

Of course, every organization would like to raise its perceived value as much as possible and lower its cost structure. But strategic thinking requires leaders to ask what is possible. How great are our competitive advantages? How fast can we implement technology, improve operations or change customer perceptions? How receptive will the people of our organization be to making the improvements required? In the end, leaders need to compare the relative risks of not improving fast enough (thus allowing the competition to leapfrog them) against creating unrealistic goals and failing to achieve them.

Having defined where leaders seek to reposition their organizations going forward, the final step is to identify the specific improvements they will implement to successfully achieve this new competitive positioning. Often, they realize that either (a) their plan is too aggressive and likely to fall short of full implementation or (b) it's overly conservative and risks losing market position. Phrased differently, leaders want a strategy that will "stretch" the organization without "breaking" its people.

Finding the right spot is a crucial skill in all industries, but particularly in healthcare, which is people-intensive. As simple as this four-by-four matrix looks, forces leaders to recognize and consider the magnitude of change required, the specific improvements needed and the impact the strategic plan will have on the people of the organization.

For relatively stable and static industries, the strategy of the past, one that made the organization success, often proves sufficient for the future. However, this is not the case for rapidly changing environments like healthcare. The rise of healthcare costs and the availability of new technology have created what can be titled a *strategic inflection point*. And the solutions required under that circumstance must be more radical than those that proved effective previously.

Strategic inflection points are times when innovative, new approaches can disrupt the winners of the past. When this happens, traditional companies like Borders, Kodak and Blockbuster go out of existence. Failure to change in these circumstances can prove deadly. The same will be true in healthcare. If doctors, hospitals and insurance companies don't radically alter their business models, they could find themselves replaced by other providers of care and coverage who are willing to make the improvements needed. Better care coordination and improved results happen through (a) integration of doctors and hospitals, (b) a reimbursement system that rewards superior outcomes rather than simply added volume and (c) modern technology offering patients greater convenience and more timely access at lower costs.

Developing the plan is only the first step. Next is translating it into reality. Ultimately, strategic thinking without action is powerless.

Leaders need to recognize the main reason change is so difficult has to do with how our minds perceive the magnitude of losses versus gains over time. The psychological literature demonstrates that our minds perceive a negative event in the short-run as far worse than an even bigger positive gain in the future. This imbalance explains why dieting is so hard and why only 50% of people are vaccinated each year against the flu. Leadership success in healthcare requires helping people reduce their fear of short-term loss. For doctors, giving up autonomy and personal control over decision-making is terrifying. That's true even when preserving such autonomy would prove even more problematic in the future than losing it. But change can be achieved through leadership. The rest of this chapter includes an approach that proved successful in implementing change for the 10,000 physicians and 36,000 staff who provide care to 5 million Kaiser Permanente members.

18.5 Meeting Healthcare's Biggest Challenges with Bold, Empathetic Leadership

The definition of a good healthcare leader is someone (a) who can create positive changes that would not happen on their own and (b) whose workforce—doctors, nurses, staff—would never wish to return to the approaches of the past once the improvements have been implemented.

The first step in aligning and motivating people is to understand their fears, including:

(1) Fear of the unknown.
(2) Fear the idea won't work.
(3) Fear they can't trust leadership.
(4) Fear others will see them as foolish for saying yes.
(5) Fear that too much will be asked of them.

The following model can be applied to a broad range of areas, from the doctor's office to the hospital to the patient's home. This model serves as a checklist to ensure leaders don't skip steps and helps them address the fears people have.

18.6   A New Model for Strategic Change

Aligning and motivating people is usually the hardest part of the change process. In other industries, some leaders try to drive change through what is called a burning platform—a crisis that's either natural or engineered and forces people to act. When crises are truly imminent, they can be effective motivators. But when they are not, people grow distrustful of leadership and will be even less willing to embrace change in the future.

Instead of leaning on catastrophe, effective leaders implement a more comprehensive set of steps that addresses people's fears and maximizes their trust. The following "A–G" model accomplishes that. Since all letters in the pneumonic need to be accomplished to overcome the five fears, the order is less important than the integrity and trustworthiness of the person leading the process.

To paint a clearer picture of this A–G pneumonic in action, the following example borrows from an approach that helped Kaiser Permanente reduce hospital utilization in half and inpatient costs by 30%. Although the A–G model itself works in any change-management situation, the specific actions required vary by circumstance. For this example, you should assume you are the leader of a 200-bed hospital in midsized market of the United States. A different set of steps would be needed for a small hospital in a rural area or a huge 500-bed facility in a highly populated urban setting. Regardless of the details, the leaders in each case would need to complete all the steps.

18.6.1   Introducing the Challenge

The opportunity for this particular transformative change begins with a little-known fact: Patients admitted to practically any hospital in the United States (or in the world, for that matter) on a Friday night will spend a day longer in the hospital than if they were admitted on a Tuesday night (Pearl 2013). The reason is simple. Care slows down on the weekends since practically every inpatient facility on the planet functions as a "five-day" hospital, not a "seven-day" hospital.

Of course, every hospital administrator and doctor will tell people that emergency care and urgent interventions take place around the clock. What they won't admit is how often patients with more routine needs are kept in the hospital and made to wait until Monday morning to have them addressed. The operating room, the interventional radiology suite, and most procedural areas tend to shut down on weekend for anything other than emergencies. Specialists and required support staff are typically "on call" but not onsite. Therefore, when a patient arrives in need of emergent intervention, all necessary doctors, nurses and technical staff are called into the hospital from wherever they may be to provide medical care—an expensive and inefficient process, reserved for patients whose needs can't wait.

Ask most doctors about the dangers of delaying care until Monday for more routine interventions and procedures, and they'll tell you hospitals are the safest places to be. This assertion is false (Saposnik et al. 2007). When diagnoses and interventions get delayed, even by a day, patients may be exposed to hospital-acquired infections—the third leading cause of death in hospitals. They also face a higher threat of suffering from a medical error, which kills 200,000 patients annually. And they are at risk of sleep deprivation and delirium with long-term detrimental health consequences (Freemantle et al. 2012).

This so-called "weekend effect" is not merely a U.S. problem. A British study found that patients who were admitted to a hospital on a Sunday faced a 16% higher risk of dying within a month than those admitted on weekdays. And Canadian scientists found that ischemic stroke sufferers admitted to hospitals over the weekend were more likely to die within seven days than those admitted during the week.

And it's not just quality of care that suffers either. Service levels (and satisfaction net-promoter scores) diminish as patients are forced to wait restlessly, spend anxiety-filled nights fearful of the worst possible outcomes. And of course, costs rise as nurses, dieticians and housekeepers are required to care for hospitalized people, regardless of whether they are actively being treated or awaiting intervention. Across the U.S. hospital costs account for over 30% of medical expenses and continue to rise faster than any other component of healthcare expenses, outside of drugs.

For all of these reasons, if the strategic plan includes simultaneously raising quality, lowering costs and increasing patient satisfaction, deciding to move from a five-day hospital to a seven-day facility might be one of the operational changes planned. But how best to achieve this outcome?

18.6.2 Applying the A–G Model

Most hospitals perform three basic functions. First, is emergency department care. The number of patients admitted here is similar on weekends and weekdays. Second, there are scheduled admissions, surgeries and procedures that will require hospital admission. In general, these are not done on weekends. In fact, facilities often front-load these functions from Monday to Thursday to keep the weekend census low. Finally, in many locations, routine outpatient procedures that require sedation (e.g., screening colonoscopy) are done at the local hospital on weekdays only. Only emergency colonoscopies to determine the exact cause and stop the bleeding are done Saturday and Sunday.

Perhaps your first thought in creating a seven-day hospital is that you would just add staff on Saturday and Sunday, similar to what you do midweek. This wouldn't work, since without the scheduled inpatient and outpatient procedures typically done on weekdays, there would not enough need to justify the added expense.

Another option would be to insist that on-call doctors, nurses and technical staff come to the hospital to provide the interventions the same day the need is identified. Once again, that solution would be inefficient and costly. That's because of the multiple times they would need to drive to the hospital, open a specialized room, and assemble the required supplies and instruments. Then, after they completed the procedure, they would need to put everything away, clean up and close the room, only to start once again a few minutes or a few hours later.

The best solution is to shift some of the elective work—the routine inpatient and outpatient procedures—from the weekdays to the weekend, reduce the Monday to Friday staffing and add people Saturday and Sunday. With this approach, there's no need to delay elective procedures until Monday since the clinicians and technical staff required would already be on hand and could add the patient on to that day's schedule with minimal extra cost or disruption.

Let's look at one example: a routine interventional procedure for an otherwise stable patient. This could be taking out the gallbladder or bringing additional blood supply to the heart. For the purpose of understanding the changes that would be needed, assume that there currently are 12 ORs in this hospital used for scheduled patients Monday to Friday and then none on the weekends (emergencies only). Shifting to a seven-day hospital would mean scheduling 11 rooms per day (Tuesday to Friday) and two each Saturday and Sunday. Note that both before and after the change, the weekly total is the same, 60 operative rooms for non-emergency procedures (Fig. 18.3).

On paper, the change appears fairly simple, logical and straightforward, so why has it not been universally adopted? The answer isn't scientific but, instead, personal. Most people want to spend weekends with their families. Doctors, nurses and technical staff are no different. As a result, they believe it is safe to keep patients with non-emergent problems in the hospital until it is more convenient to address them. As a leader wanting to lower cost without compromising quality or service, the picture looks different. To you, the delay is problematic and a major opportunity to make the care you offer more affordable and patient satisfying. But how do you get doctors, nurses and staff to see the necessity and make it happen?

Like most changes involving loss for those who must make them, this kind of transition proves difficult and cannot be achieved through fiat. Using the A–G model

Example of staffing a 5-day hospital for one month

MON	TUES	WED	THURS	FRI	SAT	SUN
12 ORs	12 ORs	12 ORs	12 ORs	12 ORs	0 ORs	0 ORs
12 ORs	12 ORs	12 ORs	12 ORs	12 ORs	0 ORs	0 ORs
12 ORs	12 ORs	12 ORs	12 ORs	12 ORs	0 ORs	0 ORs
12 ORs	12 ORs	12 ORs	12 ORs	12 ORs	0 ORs	0 ORs

Example of staffing a 7-day hospital for one month

MON	TUES	WED	THURS	FRI	SAT	SUN
12 ORs	11 ORs	11 ORs	11 ORs	11 ORs	2 ORs	2 ORs
12 ORs	11 ORs	11 ORs	11 ORs	11 ORs	2 ORs	2 ORs
12 ORs	11 ORs	11 ORs	11 ORs	11 ORs	2 ORs	2 ORs
12 ORs	11 ORs	11 ORs	11 ORs	11 ORs	2 ORs	2 ORs

Fig. 18.3 Moving from a five-day to a seven-day hospital

doesn't guarantee success, but it does increase the probability. Here's how it might be applied to moving from a five-day hospital to a seven-day hospital:

"A" is for Aspirational Vision. In healthcare, it's impossible to win over the brain, until you connect with the heart. As such, this truism must be front and center in any presentation you do or conversation you have. An aspirational vision needs to inspire, but if it doesn't include enough reality to be achievable, it's simply a dream. Improving quality, making care more convenient for patients and helping individuals avoid financial ruin are concepts that resonate with doctors and nurses and are possible. Improving a hospital's bottom line or meeting a regulatory requirement won't generate the same kind of enthusiasm. Your aspirational vision won't motivate everyone, but if it touches the hearts of enough people, you will begin the process of building a critical mass of believers.

At some point, everyone has experienced the anxiety of waiting in a hospital or doctor's office to find out whether a loved one will be okay. If hospitalized, we count the hours until the next test can be done, and lose sleep waiting to know if the diagnosis is cancer. Waiting from Saturday to Monday morning seems like a minimal delay to doctors but can feel like an eternity to patients and their families. Longer hospital stays on weekends add to the risk of complications and errors. Physicians learn in medical school to "First, do no harm." Delays in care do more than raise costs, they harm patients. Helping people recognize the lethal consequences of a five-day mentality is what is needed to generate a willingness for doctors to listen to the idea of a seven-day hospital.

Aspirational visions are not slogans, but rather destinations that others can see, feel and touch. Stories of personal experiences prove powerful when articulating your vision. They help people see the consequences of their resistance and the positive aspects of change. Most of us would resent delays in care if we were hospitalized. Then why should we want anything less for our patients? Once clinicians recognize this contradiction, they will be open to learning more.

Behaviors. Doctors are afraid that their leaders will ask too much of them or fail to understand how personally difficult this change will be. Hearing your plan to move from a 5-day to a 7-day hospital will activate a part of the brain called the Amygdala, which generates fear. Immediately, they will envision never seeing their partner or children again.

As a leader, you must anticipate this response and explain clearly what you are asking them to do differently in the future. Remember that behaviors are not attitudes. They are the visible actions people will need to take. For example, let's look back at the operating room. Assume the 12 elective rooms are staffed by 24 surgeons who spend half of their clinical time doing scheduled procedures Monday to Friday. Four OR rooms per weekend means that each physician would be asked to shift their patients from a weekday to a weekend once every three weeks. Operating an occasional Saturday or Sunday and having a free weekday will seem manageable to most physicians. Yes, it might mean they miss an occasional soccer game for their child, but it also would allow them to spend time at their kid's school. It's even possible some surgeons might prefer to work weekends, opting to ski or golf on weekdays when the slopes and courses are practically empty. By translating the idea of a "seven-day hospital" into the specific behaviors needed (one weekend day every three weeks), people begin to realize that their initial fears were overblown.

Context. Having engaged the heart and calmed the fear center of the brain, you must now win the intellectual argument. Remember doctors are accepted into medical schools because they study hard and ace the exams. They also love to engage in debate. As such, you may be asked, "Why not just continue to raise prices?" or "Why not buy a new multimillion-dollar diagnostic machine like the hospital across town and add revenue in that way?"

Hopefully, you considered all of these alternatives during your strategic analysis and concluded that greater efficiency is a safer path, more consistent with your facility's mission and an important step toward preparing your facility for the end of the price-inflation era. Providing context includes describing these external forces and explaining why the best time to act is now.

You might use Medicare, the source of most of your facility's revenue, as an example. The program pays American hospitals on a DRG (Diagnosis Related Grouping) basis, rather than by the day (per diem). Therefore, longer hospital stays not only delay patient recovery, but also cost more money without generating additional revenue. This combination limits the hospital's ability to purchase new technology and hire additional staff. And even for individuals whose insurance plan would pay more now, the businesses that purchase coverage are reaching the end of their financial ropes. They may begin excluding the most expensive hospitals and specialists from their networks and yours doesn't want to be left out.

Data. When it comes to their clinical practices, all doctors believe they are practicing at the top of their specialty. Of course, that's not mathematically possible. Data helps everyone more accurately assess their performance and learn from the best. Besides, physicians are scientists and, when confronted with the possibility of doing anything new, they insist on having the numbers to validate its effectiveness before they will commit.

The first step toward providing proof of concept is determining what you will measure. Comparing weekend length of stay for particular DRGs against weekdays is one possibility. Alternatively, you might decide to include this initiative with others and measure total inpatient days for similar sized Medicare populations.

What's even more important than the data itself is deciding in advance what you are going to do if the changes you outlined fail to produce the improvements required. One possibility is that you didn't do what needed to be done to support the operational transition. Maybe you failed to provide the nursing or pharmacy resources needed. If so, you need to acknowledge this and make the added investments required. But, assuming you have done so, you need to have planned what you will do next. If you simply continue to report no change in weekend length of stay compared to midweek, the message everyone will hear is, "Moving to a seven-day hospital isn't very important."

On the other hand, if the issue is that physicians in specific departments aren't performing the procedures on the hospitalized patients over the weekend, you will need to intervene to make it happen. For example, maybe you'll ask the nurses to immediately communicate the information to the nursing supervisor who can call you and tell you what is happening. Once you drive to the hospital and address the situation a few times, people will know you are committed to the success of the approach and will be unlikely to refuse to provide the added care in the future.

Engagement. No one will care how much you know until they know how much you care. Nowhere is this truer than in healthcare. Implementing change requires personally meeting with individuals and groups of medical care providers and, when necessary, intervening to ensure participation in supporting the change is broad and consistent. There is no other way to develop the trust needed and make sure your vision and ideas are understood. A study by the U.S. Dept. of Education demonstrated that it takes 17 exposures for students to learn a new word and integrate it into their vocabulary. It takes just as much repetition to master a new vision for the future and comprehend the actions needed to achieve it.

Trust is an essential ingredient in getting physicians to change their behaviors. They are afraid of what change could mean for their patients and for themselves. They worry that they will fail or look foolish to colleagues and friends alike. Trust in a leader is built over time. If your motivation is self-serving and not on behalf of patient outcomes, physicians will see it and their trust will wane. You can fool people once, but rarely a second time. Used-car salesmen can get away with it because they don't interact with the same customer more than once. Healthcare leaders, on the other hand, must maintain the highest integrity. And until they demonstrate it, those who must make the changes will withhold their trust. The only way to convince others of your sincerity and authenticity is through your actions.

If moving to a seven-day hospital is of major importance and central to your organization's strategy, then you need to look at your calendar, cancel some of your administrative meetings and carve out time to meet with clinicians individually and in small groups. Emails are good for delivering facts and data, but they do little to generate the trust needed to inspire meaningful change.

Remember that no one will be more committed to change than you are as the leader. As such, you need to be visible on weekends, thanking the physicians and staff and helping to smooth the transition. Being present builds trust in you, your word and your actions.

Faculty. The term faculty is usually applied to universities and describes the teaching or research staff that achieve the institution's mission. Here, it refers to the people on a healthcare leader's team who provide expertise and assistance in implementing the organization's strategy and fulfilling its mission.

Shifting from a five-day hospital to a seven-day one requires extensive clinical knowledge, financial analysis and operational redesign. None of us has enough time or expertise to do all of that alone. Finding others to help is an essential next step.

Clinical expertise is fundamental to medical training and physician culture. If you doubt the importance, go to a meeting involving doctors and ask them to make introductions. Everyone will include their specialty, even if the meeting has little to do with clinical practice. But expertise in one medical field rarely translates to credibility in another. If, for example, you are a practicing surgeon, an audience will accept your ability to opine on the operating room, but attendees will doubt you the moment you begin to talk about radiology.

Before trying to implement a complex program like the seven-day hospital, you will need to gain the buy-in of experts in each of the areas for which change is critical and, likely, challenging. Bring these experts with you to hospital-wide meetings and ask them to stand by your side when questions and problems arise. Moreover, successful change will require support from finance, HR, nursing, pharmacy, housekeeping and the support staff in each clinical area. A major reason change efforts of this magnitude fail is that leaders underestimate the number of people needed for success. Before you announce anything publicly, you will need to have engaged with many people and parties, individually and in small groups, listen to their thoughts and gain their confidence. Until then, you can't move forward.

Governance. Individual doctors in their own office answer to no one. Larger organizations (and even smaller, integrated medical groups) need a structure to make decisions, allocate resources and measure performance. Without this, the vision for the future is likely to remain just that—a vision. Governance has three parts, and all are important:

1. *Formal structure*. Often there is a board and medical staff committee that must affirm the direction and key parts of the plan. Frequently, leaders see this as the biggest hurdle, since these groups have clearly defined accountability and power. But leaders soon realize that the getting approval and support through the formal structure is the easiest part of the change process.

2. **Informal structure.** This group includes leaders who have massive influence, even without a high-ranking title. Every organization has "influencers" like these. They're the people doctors look to first before they'll consider supporting the change process. Leaders might falsely assume that gaining board/committee approval to move forward automatically means that physicians will follow. Ultimately, doctors believe they have the right to do what they think is best, no matter what they're told. When it comes to change initiatives, they usually look to their peers and the clinical experts, not the administration to tell them what to do. For these reasons, garnering the support of the informal leaders before moving forward is crucial.

3. **Incentive structure.** Finally, most organizations use financial incentives to motivate behavior. Leaders tend to view these visible "carrots and sticks" as most effective, but rarely is that the case. Financial incentives for performance can be powerful motivators. But, in healthcare, they rarely lead to the outcomes desired. Doctors are skilled. They got into medical school by acing tests. Financial incentives lead to change, but most often, not the ones leaders desired. Unintended consequences almost always are the result. For example, what happens if you pay people more to consult on weekends? Suddenly, doctors will begin recruiting requests for inpatient consultation from colleagues for patients for whom the additional opinion will add no value, but will be a waste of time and money. There are no shortcuts when implementing effective change in clinical practice. The purpose of the A–G model is to remind leaders of all the steps needed for success. Skip a step and the initiative will fail. Fail to provide a compelling aspirational vision and physicians won't hear the context, or care about the behaviors and data. Fail to engage as a leader and have a strong faculty, and the most powerful financial incentives will prove futile.

18.7 Measuring the Success of the Seven-Day Hospital at Kaiser Permanente

When The Permanente Medical Group undertook the move from a five- to seven-day model, it measured success based on the improvements in clinical outcomes and reductions in hospital utilization for the Medicare members it served. One reason for this choice was the availability of benchmarks across the country against which to measure performance. The second motivation was that patients in this 65+ age group are frequently utilizers of hospital services, so the impact of the change would be most significant.

In conjunction with this effort, the medical group decided to also raise quality and lower costs by maximizing preventive services and aggressively managing chronic diseases. Although determining the success of this approach involved a detailed and comprehensive analysis of clinical outcomes and costs, the easiest way to understand the resulting changes in hospital utilization is to examine the number of inpatient days per year per 1000 Medicare patients. To do that, you take the total number of hospital days for everyone covered through Medicare, divide by the total number of enrollees and multiply by 1000. That number is approximately 1400 days across the United States. Since the patients who enroll in Kaiser Permanente get all of their care through the medical group and multiple Kaiser Permanente hospitals, total inpatient utilization for the organization's hundreds of thousands of Medicare members could be directly measured and compared to external benchmarks.

Within five years of implementing the seven-day hospital and the various programs to make The Permanente Medical Group the nation's leader in quality outcomes, utilization in Kaiser Permanente was down to 700 days per 1000 Medicare members per year, half the national average. Of this 50% reduction, half could be attributed to the quality improvements, and half came from the shift to a seven-day hospital. Neither outcome would have happened without applying the full A–G model. And the clinical results included 40% fewer heart attacks, strokes and cancer, a decrease in mortality from these diseases by 30–50% (compared to national numbers) and a major cost reduction. Thanks to these changes, the organization could make further investments in medical care, fund capital improvements and lower prices for patients and purchasers.

18.8 The Next Generation of Healthcare Leaders

The healthcare-delivery organizations that lead in quality outcomes, patient convenience and affordability have already adopted and embraced each of the elements of the "Four Pillars." So why then is it so hard to institute similar changes everywhere?

The answer is that every change process involves "winners and losers." As a result, motivating everyone to move forward (in the same direction) proves difficult. The next generation of healthcare leaders can use A–G model to help inspire and implement effective change. And they need to remember the importance of three vital organs: the heart, brain and spine.

18.8.1 The Heart

Leaders must demonstrate and evoke passion and show compassion. Meeting regulatory requirements for quality and making care more affordable for populations of patients don't generate the same passion in doctors as saving a life in the operating room or performing a complex operative procedure. And yet these approaches to preventing disease in the first place or avoiding further complications in patients with chronic diseases have a far greater impact on mortality and life expectancy than heroic interventions.

Outcome data demonstrates that by controlling blood pressure effectively for patients, the incidence of stroke will drop by 40% for the entire total population served, compared to a similar number of people whose blood pressure remains high. But the deaths avoided will happen at some ill-defined time in the future and without knowing exactly whose life was saved. In contrast, pull the clot causing the stroke out of a major blood vessel to the brain, and the physician knows precisely whose life was saved. In addition, that doctor earns the family's gratitude forever.

The best leaders understand this challenge and overcome it by engaging the heart of doctors, nurses and staff. If you are trying to get people to wash their hands to prevent hospital-acquired infections (a growing threat in hospitals across the country), you can't just show an instructional video and expect to see change. People may forget statistics, but they will remember forever being in the same room with a man as he tells the story of his wife's death from an infection she developed inside your hospital. And their spines will stiffen when they're reminded of a difficult truth: No doctors think they're capable of transmitting the bacteria that kills a patient. But when they fail to wash their hands, that's exactly what they're capable of doing.

In my 18 years as CEO of The Permanente Medical Group, one of the most powerful experiences I can remember came during what I thought was a standard departmental meeting. At the time, all Kaiser Permanente medical centers were implementing a comprehensive EHR that added work to the reception and office staff. These individuals were already busy with their primary jobs, such as registering people or recording their weight and blood pressure measurements. Now they had to also check the medical record to see whether patients were due for breast cancer screening (mammography) or colon cancer assessment (stool sample kit).

Wanting to hear how the department was doing, I took a seat in the back row. The department chief began the meeting by pointing to a staff member sitting in the front row. He explained that this woman had identified a patient who was overdue for a mammogram and, in addition, went above and beyond in helping her get screened the same day. Even though the patient came into have her eyes evaluated, the staff member's actions led to an early detection of breast cancer, which was treated before the malignancy had spread.

The staff member, in front of her entire department and with her husband and children in attendance, received the organization's first-ever "I Saved A Life" award. Outside of the birth of her children, this was the happiest moment of her life. The next day, all of the medical-office departments were abuzz, each staff member hoping for similar recognition. The overall result of this program and others like it vaulted the screening rates for breast cancer among the patients treated by The Permanente Medical Group to the top of the nation, according to the National Committee for Quality Improvement.

18.8.2   The Head

As pointed out in the A–G model explanation, clinicians are scientific and demand data. A powerful example of how data can be used to improve clinical outcomes is in the effective treatment of sepsis. Early and aggressive treatment of this potentially lethal, system-wide infection could save as many as 70,000 American lives each year. However, effective sepsis treatment is very intrusive. And, despite saving numerous lives, the treatment can, on occasion, lead to the death of a few individuals who otherwise might have lived.

For doctors, not all deaths are the same. When someone dies from something a specific physician did, the doctor perceives that death as far more tragic than when multiple other patients die with no one to blame. For this reason, physicians have resisted early, aggressive intervention.

Dr. Diane Craig, a hospital-based physician at Kaiser Permanente, decided to address this problem. She studied the literature and met with leading national experts. Then she met with the emergency department physicians and presented the data on the number of lives that could be saved. She identified the specific steps needed whenever someone might be at risk and the time frame for doing so each time. She created "sepsis" teams to be summoned when a patient at risk was identified, similar to what is done for patients following a cardiac arrest. And she embedded the protocol into the health system's electronic medical record to ensure it was followed every time. The result was a dramatic reduction in hospital-wide mortality from sepsis—down below half of the national average.

18.8.3   The Spine

Changing clinical practice requires courage. The consequences of making a mistake are immense for physicians including nights of loss sleep and potential malpractice suits. Compared to risking being blamed, an unnecessary or redundant test feels inconsequential. Telling parents their child doesn't need an antibiotic is far more difficult than writing a prescription. Next generation leaders need to use the A–G model to make change happen and harness the power of the group to bring the most reticent along. Unfortunately, these approaches don't always lead to change. When stagnation happens, leaders have to step up and "do the right thing." And that takes spine.

An example of great courage came from Dr. Sharon Levine, one of the Associate Executive Directors at TPMG. She led our pharmacy efforts and accomplished remarkable work with the various "Chiefs of Quality and Therapeutics," especially in the areas of reducing inappropriate antibiotic use and prescribing more cost-effective generic medications for patients when appropriate. An area of concern for her was the pernicious impact that drug companies had on physician prescribing behaviors. From a strategic perspective, if we wanted our patients to trust the decisions of physicians, we had to insulate them from any nefarious financial incentives drug companies utilized to incent prescribing of unnecessarily expensive drugs when an equally or better low-priced alternative existed.

Even today, a decade after confirming the negative impact of drug-company dollars on medical practice, many physicians fail to accept that free lunches and lavish dinners have any effect on their drug ordering. And when confronted with the data that shows how much more often they deviate from the prescribing habits of their colleagues, they're certain the explanation is that their patients are different. Any implication that their prescribing habits are influenced by financial payment are rejected.

Sharron proposed, and our board accepted, a complete prohibition on taking anything from a drug company, including a mug or a pen, recognizing the slippery slope that can result when exceptions are made. She told me she was going forward, even though in her mind it was a potentially career-ending journey. A decade later, her policy remains the gold-standard against conflicts-of-interest. After its implementation, only two out of the 6000 physicians in the medical group at the time left the organization and she remained a highly respected and trusted leader until her retirement.

Leaders must be smart, able to analyze problems and skilled at communication. Without these abilities they rarely are chosen for leadership roles. But individuals wanting to be highly effective next-generation healthcare leaders will need to do more than that. They will be required to be skilled at creating a powerful vision for the future, aligning people around it and motivating them to move forward. They will apply the A–G model to making change happen and, through their efforts, move their organization closer to the four pillars. Each will understand the importance of the heart and mind and possess the spine needed to drive change when the right thing needs to be done.

Administration is the ability to make the things happen that people expect—from paying employees on time to following rules and regulations that govern the industry. In contrast, leadership is the ability to make things happen that otherwise would not.

Leadership in healthcare is difficult but, when done well, it's incredibly rewarding. Developing a strategy and an implementation plan that saves lives and makes medical care more affordable is a unique privilege. It's incredibly satisfying to watch an idea begin with people saying, "It can't happen," and then "It could happen," and, eventually, "It had to happen." Finally, "I'm glad it happened."

In healthcare, the best leaders don't act for personal gain, but on behalf of the patients to whom they are accountable and the doctors, nurses and staff they lead. In the end, if leaders don't experience pushback, they're not doing anything important. The measure of their leadership ability is whether they can overcome the resistance and make people grateful they did. When that outcome happens, the hours spent and dedication required prove well worth it.

18.9   Homework: Applying the A–G Model to Other Healthcare Challenges

Several years ago, I came across a sign hanging on the wall of a public health building that read "Quality. Service. Cost." in big, bold letters. And below that, in small print, "Pick any two." This all-too-common assumption reflected the mentality of twentieth-century healthcare, with its intrinsic belief that higher quality and better service could not be achieved without greater financial investments. As the example of the seven-day hospital demonstrates, this is simply not true.

Make no mistake, some businesses can be successful catering to customers with expensive taste. Examples include Lexus, Apple and Ritz Carlton. Even in healthcare, some brand-name hospitals have been able to command higher fees for the privilege of being cared for in their prestigious institutions. But in healthcare, leaders should strive to offer the highest quality in the most convenient ways at an affordable price for every patient. That is medicine's mission. As such, the next generation of healthcare leaders will need to address all three: "Quality. Service. Cost."

Doing so will be challenging, but there are far more opportunities than people assume. Here are four examples. Success in each will require powerful leadership, bold thinking and bravery. As a next-generation leader, ask yourself how you would apply the A–G model to achieve the following:

1. *Limit the number of physicians and hospitals doing procedures in each community to raise quality, increase patient confidence and lower costs*. The clearest predictors of superior outcomes and performance in medicine are volume and specialization. The most important question a patient can ask a surgeon is this: "How many of these operations did you do last year?" Getting this information can be difficult, but organizations like the Leapfrog Group are making that data available, as are some state-wide databases. To improve clinical outcomes, we need fewer specialists with great experience and expertise. We will also need to close low-volume surgery programs and refer more patients to high-volume facilities. It will be a great test of even the best leader to convince doctors and local communities to accept these realities.

2. *Focus on prevention*. It's no secret that it's far better to prevent a heart attack, stroke or cancer is better than to treat it. Unfortunately, that is not what the culture of medicine values or the achieves today. For example, hypertension (elevated blood pressure) is the most common etiology of stroke and contributes to heart attacks and kidney failure, and yet across the U.S. it is controlled only 55% of the time. When you compare those numbers to the best medical groups, which can achieve success rates of 90%, the call for action is clear. A major part of the solution is to include specialists in measuring blood pressure and coordinating with the patient's personal physician on modifying treatment. Getting everyone to focus on this major area of opportunity proves difficult, since reimbursement for doing so is minimal in comparison to doing more procedures.

3. *Implement effective technology*. Much of today's medical technology is expensive and no better than the traditional, manual alternatives. Multimillion-dollar robots and proton-beam accelerators have been shown to add little value in terms of clinical outcomes for most patients. At the same time, there is technology that can reduce cost, improve access and raise quality through earlier intervention. A powerful example involves video visits (often called "telemedicine"), which allow patients to get a consultation and participate in follow-up visits conveniently without having to miss work or school. Fewer than 10% of doctors offer

these today, and less than 1% of medical care is provided this way. The reasons are complex, but mostly involve the doctor's concerns, not the patient's. In fact, data demonstrates that patients are even more satisfied with these virtual visits, than in-person ones. In many countries, including the United States, these visits are not reimbursed by insurers, so physicians are loath to offer them. And even when they are paid, it can be more convenient and lucrative for doctors to ask the patient to come to their offices. Finally, many physicians are not "tech-savvy" and resist anything that is device- or application-dependent. In the future, as much as 30% of what is done in doctors' offices could be accomplished effectively virtually. Making that happen is a major opportunity for the next generation of healthcare leaders.

4. **Eliminate valueless interventions.** Much of what physicians do today adds little value, according to research published in peer-reviewed medical journals. This is true for orthopedics relative to knee arthroscopy and for angioplasty and stenting in patients with stable heart problems. The same failure to improve patient outcomes goes for complex back surgery in individuals with pain as their only symptom and even ordering an MRI to evaluate most cases of back pain in the first place is unnecessary. The same phenomenon of futile care can be seen in prescription of brand-name drugs, when identical generics exist and the over-use of antibiotics for viral infections for which they are completely ineffective and risk complications from taking them. And even treatment of some cancers, particularly prostate for patients with low risk of spread, has been shown to diminish the quality of life and fails on average to prolong it. Of course, that's not how income-generating physicians and hospitals perceive the situation. Often, they respond with intuitive, not scientific, justifications. They'll use phrase like "in my experience," or "you never can tell," or "I remember a patient who…" Even when the outcomes are better for patients and society overall, those changes that bring down individual incomes are trickiest to implement.

Leadership is a privilege. As these examples demonstrate, doing it well is difficult and time-consuming. Developing a strategy and a clear vision for the future is the first step. Helping the people of an organization overcome their fears and move forward to make change happen on behalf of patients is the goal. The A–G model offers specific steps that helps leaders be successful in achieving their vision. Like all skills, effective application takes practice and time. But as you will discover, nowhere is being a leader more rewarding than in healthcare.

Bibliography

Andrew, L., & Brenner, B. (2018). Physician suicide. *Medscape.*

Centers for Disease Control. (2017). National Health Expenditure (NHE) Data. *CMS.gov.*

Freemantle, N., Richardson, M., Wood, J., Ray, D., & Khosla, S. (2012). Weekend hospitalization and additional risk of death: An analysis of inpatient data. *Journal of the Royal Society of Medicine, 105*(2), 74–84.

Jones, D., Podolsky, S., & Greene, J. (2012). The burden of disease and the changing task of medicine. *New England Journal of Medicine, 366*, 2333–2338.

Lallemand, N. (2012). Reducing waste in health care. *Health Affairs Health Policy Brief.*

Lee, J., Jensen, C., & Levin, T. (2018). Long-term risk of colorectal cancer and related deaths after a colonoscopy with normal findings. *JAMA Internal Medicine, 179*(2), 153–160.

Makary, M., & Michael, D. (2016). Medical error—The third leading cause of death in the US. *BMJ, 353*, i2139.

Parks, T. (2016). Analysis of health care spending: Where do the dollars go? *AMA Policy Research Perspective.*

Pearl, R. (2013). *Why being hospitalized on a weekend costs more lives, health care dollars.* Forbes.

Pearl, R. (2018). *Saving America's hospitals: It's time to stop wasting time and lives.* Forbes.

O'Neill, D., & Scheinker, D. (2018). *Wasted health spending: Who's picking up the tab?* Health Affairs Blog.

Saposnik, G., Baibergenova, A., Bayer, N., & Hachinski, V. (2007). Weekends: A dangerous time for having a stroke? *Stroke, 38*(4), 1211–1215.

Salber, P. (2017). *Dr. Robert Pearl's 4 pillars of healthcare transformation.* The Doctor Weighs In.

Tozzi, J., & Tracer, Z. (2018). *Sky-high deductibles broke the U.S. health insurance system.* Bloomberg.

Dr. Robert Pearl is the former CEO of The Permanente Medical Group, the nation's largest medical group. In this role, he led 10,000 physicians, 38,000 staff and was responsible for the nationally recognized medical care of 5 million Kaiser Permanente members on the west and east coasts. In 2017 he authored "Mistreated: Why We think We're Getting Good Healthcare—And Why We're Usually Wrong," a Washington Post bestseller. He is a Forbes healthcare contributor and hosts the Fixing Healthcare podcast. He is a clinical professor of plastic surgery at Stanford University School of Medicine and is on the faculty of the Stanford Graduate School of Business, where he teaches courses on strategy, leadership, technology and healthcare policy.

Chapter 19
Skin Elements Ltd—The Importance of Knowledge Management in Commercialisation

Peter Malone, Tim Mazzarol and Sophie Reboud

Abstract This Chapter examines the case of Skin Elements Ltd., a Biotech start-up enterprise that successfully created, manufactured and commercialised an innovative skincare technology into a global market. The principal focus of this case study is on the role played by knowledge management and how this shaped the innovation strategy that saw the technology successfully commercialised and positioned within the global natural skin care market. The chapter examines the process of commercialisation, and demonstrates how knowledge management (KM), open innovation (OI), absorptive capacity (ACAP), and entrepreneurial operations management (EOM) played key roles in evolving the innovation strategy and commercialisation process. The chapter opens with an overview of the case study before introducing the concepts described above, and then enfolding the academic literature into the case to illustrate the relationships found. It then draws conclusions from the findings and lessons for research, policy and practice. The case draws from the personal experience of the lead author, who has provided first hand observations of the company's foundation and evolution over its first 12 years of operations.

19.1 Introduction

Skin Elements Ltd is a small, publicly listed biotechnology company headquartered in Perth, Western Australia, which produces a range of natural organic products under the *Soléo Organics* (sun care) and *McArthur* (skin care) brands. The company was founded in 2006, by inventor Leo Fung, financial specialist Craig Piercy, and serial

P. Malone · T. Mazzarol (✉)
UWA Business School, University of Western Australia, M263, 35 Stirling Highway, Crawley, WA 6009, Australia
e-mail: tim.mazzarol@uwa.edu.au

P. Malone
e-mail: peter.malone@bigpond.com

T. Mazzarol · S. Reboud
Burgundy School of Business, 29 Rue Sambin, 21000 Dijon, France
e-mail: Sophie.Reboud@bsb-education.com

© Springer Nature Switzerland AG 2020
N. Pfeffermann (ed.), *New Leadership in Strategy and Communication*,
https://doi.org/10.1007/978-3-030-19681-3_19

entrepreneur Peter Malone. By 2018, the company had publicly listed, undertaken a major corporate acquisition of the *McArthur* brand, expanded its brand distribution into 16 countries, launched an e-commerce strategy via an online sales channel, and launched a range of new products, including the *Elizabeth Jane* natural cosmetics skin care range. The commercialisation pathway followed by Skin Elements has evolved through four distinct phases as shown in Fig. 19.1, each of which are described in the following sub-sections.

19.1.1 Phase 1—Invention and Proof of Concept

This first phase took place between 2006 and 2009 and involved the initial R&D that created the formula for the *Soléo Organics* sun care product, through to the formal regulatory approvals for the product by the Federal Drug Administration (FDA) in the United States, and the Therapeutic Goods Administration (TGA) in Australia. As noted above, Skin Elements was born as a result of an alliance between three founders: the inventor, Leo Fung, who contributed his knowledge and expertise of naturopathic science; Craig Piercy, who brought his knowledge and expertise in financial management and capital raising; and Peter Malone, who contributed his knowledge and experience of successfully launching and scaling innovative entrepreneurial ventures.

The trio invested start-up capital of A$120,000 and spent the first year together in a micro office as they examined and tested the concept. This involved working across a number of laboratories in search of the right ingredients and manufacturing partners. Everything was being tested as part of the R&D proof of concept process. What was being developed was the first new sunscreen formula in forty years, and would need to meet and exceed all the attributes of the existing products available in the marketplace.

By early 2007 the first tubes of the *Soléo Organics* sunscreen were in test markets and certification programs. The partnership evolved to a private company structure to better reflect the ownership and accounting for further investment that would eventually be needed. The commercialisation process relied on the collective knowledge and expertise of the team, as well as their ability to access third party assistance. As the firm's Executive Chairman, Peter provided the overall leadership and strategic direction, while Craig, as Company Secretary, focused on operations. In turn, Leo focused on R&D and new product development (NPD).

Leo's research created a base formula around the use of natural zinc metal, a proven block to ultra violet (UV) radiation, and a selection of natural organic ingredients that provided cohesion and substance for the mixture when applied to human skin. A small laboratory was set up in an industrial estate within the Perth suburb of Canning Vale, in which the initial R&D was undertaken, and where the firm's offices were co-located. The founders maintained a close working relationship, retaining commercial confidentiality over their formulas, while also engaging selected third-party scientific testing services in the eastern city of Melbourne.

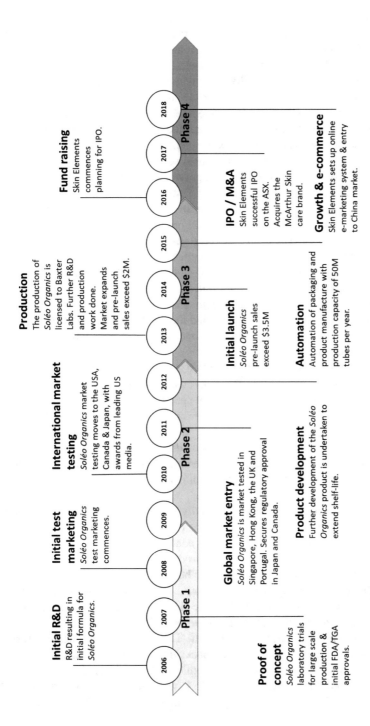

Fig. 19.1 Timeline of the skin elements case study

The primary focus of the R&D at this early stage was to identify and isolate the active ingredients, and create a stable formula that could be produced in volume. Due to the need for trade secrecy, independent test labs were selected carefully and only parts of the entire formula were revealed to each laboratory. Even then, such testing took place under strict confidentiality using non-disclosure agreements (NDA). The key active ingredient in the formula centred around zinc metal, and potential sources of the ingredients were examined from five international and domestic suppliers. A final supplier was selected, in the form of a scientist based in Melbourne, who had worked for the prestigious Commonwealth Scientific and Industrial Research Organisation (CSIRO), which is Australia's national science agency. He had commenced the manufacture of micronized zinc, which provided an important element within the formula. Zinc particles of one thousandth of a millimetre, invisible to the human eye yet larger than the pores in the skin, offered an alternative sunscreen formula when compared to existing formulas in commercial production.

At the end of the first year, from a technical perspective, the new product exceeded all criteria set in relation to the chemistry benchmarks, and was on track to deliver a significant improvement to the prevailing industry standards. Further modelling was undertaken on what the market attributes that the product would be required to meet and exceed. However, financial modelling and projections of future expenditure requirements undertaken by Craig, identified the need to secure additional funding to support the ongoing R&D and proof-of-concept work. A budget of $300,000 was signed off by the team.

Throughout the second year, Peter and Leo travelled extensively across Australia visiting suppliers and chemists to listen and examine the 'state of the art' in new natural ingredients, and obtain a clear set of specifications for both the product and the market. For example, Tea tree oil was examined as a natural preservative, but was not proceeded with even though it had strong preservative qualities. A barrier to its adoption was the feedback from focus groups that indicated its aroma was considered to be too over powering. A final combination of sixteen natural ingredients was selected to form the recipe of the formula and provide a platform for further development.

The appearance of the cream was a heavily debated issue. White or skin colour or both options. The consistency of the cream was critical to success. It needed to be easy to apply and non-greasy, yet able to adhere to the skin for a minimum three hours of water resistance. Other essential product attributes were a sun protection factor (SPF) of at least 30+ and the cream to be hyper allergenic.

Packaging was also going to be critical for the product to be able to stand out in an already overcrowded market space where the chemical ingredients of the incumbent products were the best of the out of patent synthetic formulations. A new incumbent would have to be able to differentiate itself. This saw the group explore various concepts, with Peter convinced that the packaging needed to clearly differentiate the product from the standard 'synthetic chemical tubes.' The product's name *Soléo Organics* reflected the organic natural shift. It was decided all products needed to be presented in a box telling the story on the ingredients, and the opportunity for customers to move to safer ingredients. Not only did the product need to standout, it also needed to create a point of difference.

At this stage the Skin Elements team began the process of signing off on the product specification and securing the intellectual property (IP) rights to the formula. An important milestone for the commercialisation process was securing approval from the TGA and FDA in order to allow the product to be sold for human use. This required a significant amount of testing. As this work progressed, the project team began to hear that rumours were spreading about a new sunscreen formulation based on natural ingredients was being tested for national release, which began to generate public interest.

The process of securing patent protection was therefore initiated covering Australia and all countries that would be targeted for future sales, and a full IP rights review was undertaken. However, an alternative IP protection program was examined, which adopted a trade secrecy approach using the concept of the 'secret formula.' This was considered to offer better protection for the product, which could potentially be replicated by competitors once the formula was disclosed as would occur with a patent. The success of *Coca Cola* and their "Formula X" was viewed as an example. This use of trade secrecy was considered to be especially relevant if the product was to be sold into markets where the governments and legal systems do not provide support to patent infringement. As a result, the company decided to opt for trade secrecy in relation to the protection of the *Soléo Organics* formula.

19.1.2 Phase 2—Market Testing and International Expansion

By early 2009 the Skin Elements program had reached the point where the 'voice of the customer' needed to be heard. The *Soléo* product also needed to be signed off by regulatory agencies within international markets to verify that it met their drug testing regimes and commercial criteria. The company also needed to assess its funding requirements to enable it to have sufficient financial resources to move forward. Leo was busy ensuring that the product was ready for market insertion, and Craig examined the opportunities for the company to raise capital to fund the next stage of its growth, with ambitious plans for entry into the multi-billion-dollar sunscreen market. This identified the need for a minimum capital investment of A$1 million, and all three founders liquidated assets to generate the funds needed to commence the work.

However, while the product was deemed ready for large-scale production, there remained many unanswered questions about consumer willingness to adopt a new, natural sun care product from an unknown Australian brand. What was becoming evident from the firm's existing market research was that there were growing concerns amongst consumers over the use of chemicals and their cumulative effects. This was identified as the key selling point for the *Soléo Organics* brand. Global research studies suggested that consumers were becoming concerned over the long-term health impacts of chemicals and other contaminants, in food and personal use products such as sunscreens and cosmetics. This was of particular importance in market segments such as baby or infant skin care products, or for older people with sensitive skin. These markets were not only large, but also less price sensitive, and willing to pay a premium for a product that was safe to use.

With the product ready to go, and eyes on the global market, Peter flew to the United States and met with a range of contacts he had made during previous projects and business ventures. His journey took him from New York to Los Angeles and ultimately to San Clemente in Orange County in Southern California. It was here that the hotbed for market testing was to commence. This provided a market environment that was not only affluent and open to innovation, but with a lifestyle that provided a substantial testing ground for sun care products. According to Peter, it was akin to the "Formula 1" racing circuit. Sun exposure in water sports was becoming a problem particularly for the international world surfing events which had been picked up by the new lifestyle companies of the early 2000s. Major surfing brands such as *Quiksilver*, *O'Neil*, *Hurley*, *BodyGlove* and *Ocean Pacific* were all on the west coast and heavily backed by mainstream investment funds. These were identified as the potential key to successful market entry into the north American market.

Skin Elements set up an office in San Clemente, hired a local team, and commenced its world marketing program. Meetings with the industry key players attended by Peter, Leo and Craig either individually or as a team, followed as the *Soléo* product was released into the USA. From California to Hawaii *Soléo* was soon visible in the surf industry. This marketing program ran for eighteen months and saw the product receive critical approval. Seventy-five thousand tubes were sold and/or given away as samples for ready to try customers. Selected retail outlets were given quantities to sell such as *Wholefoods* and *CVS Pharmacies*. Visibility was high and the company even secured the *Da Hui* or "Black Shorts" surfing gang of Hawaii to promote the product across the islands. The program saw the 2010 Hawaiian Triple Crown Surfing championships held at Waimea Bay promoted by Skin Elements with all other sponsors that year removed. No fees were charged to the company other than providing the new natural sunscreen. The significance of this decision by the Hawaiians didn't resonate until later. The need for safe cover was seen as critical and they were keen to make the statement.

The feedback from mainstream America was also compelling and the Skin Elements team regrouped back in Australia to assess their market research lessons. What was clear was that the *Soléo Organics* brand had been rated the number one sunscreen out of a total of 1728 brands sold in the United States by the prestigious Environmental Working Group (EWG). Further, the *Washington Post* had awarded *Soléo* the best new sunscreen, and *Elle Magazine* had given *Soléo* the award of best new cosmetic product of the year. The high-performance market segment had been won, and America was giving *Soléo* the clearance to move into mainstream segments. As early innovators it demonstrated the opportunity available to Skin Elements.

Encouraged by these positive responses in the United States, Skin Elements contracted a Perth marketing and advertising agency Marketforce to undertake consumer studies using focus groups to identify the key strengths of the *Soléo* brand across different market segments. This process lasted three months and led to findings that one of the most receptive consumers was the 'young mother', who sought a safe, but effective sunscreen suitable for babies and children. The natural formula of *Soléo Organics* sunscreen was viewed by these consumers as very important. Not only was the product safe for topical use on the skin, but also safe if ingested. Even a one-week old baby would be safe if it accidently ingested the *Soléo* product. The baby-care market segment was therefore identified as a key potential market, and one that was both large and relatively insensitive to price.

The Skin Elements team therefore developed plans for future market testing to fully assess the product's application to global skin types and fully assess and validate the formulation. The product was sent to a Japanese distributor that had seen the product in California and who saw the opportunity for the product with his customers. Leo wanted confirmation on the product's compatibility with Asian skin types and this appeared to be a good way to collect this data. A container of product left for Tokyo and within six months the skin assessment had come up positive. Further trials followed in Hong Kong, China, Singapore and Indonesia—all confirming acceptability of the *Soléo* formula.

This market testing also necessitated a parallel program of registration and licensing with the relevant drug administrations within selected Asian countries. An early submission to the Australian Therapeutic Goods Administration (TGA) had now come through as approved and this saw the Ministry of Health in Japan accepting the tests that had been undertaken in Australia. With this success in Asia, Peter and Leo travelled to the United Kingdom (UK) and began to develop the opening up of that market, followed shortly thereafter with entry into Europe. However, initial problems were encountered with the European Union (EU) regulatory agencies over the use of zinc as an active UV radiation block and this led to the product's entry being initially restricted for distribution in that market.

The cause of this problem proved to be largely one of bureaucracy. It had been over fifty years since natural minerals had been used in sunscreens and with the evolution of food and drug production licensing in Europe these had been long deleted from the codes. To address this challenge, Skin Elements linked up with a local EU manufacturer BASF from Germany, which was one of the five international companies pioneering the new micronized zinc products. Together Skin Elements and BASF brought this mineral back onto the register of effective UV radiation blockers. This subsequently paved the way for Soléo to move forward and saw the test marketing commence in the UK.

In selecting their foreign market entry strategy into the UK, the Skin Elements team searched for a partner able to provide both good initial market penetration and the opportunity to secure additional market research data. The *Fresh and Wild* store (later acquired by the Wholefoods Group of the USA) in the London suburb of Chelsea was identified as a prime candidate. Associated with the famous Chelsea Flower Show, the store was a major attraction for shoppers seeking natural and safe products. Peter and Leo travelled to London and met with representatives from *Planet Blue*, a company founded in London in 2005 by two Australians from Adelaide who were representing new Australian companies. Furthermore, at that time, their focus was on marketing new food and skin care products to the UK. Following negotiations, this company was appointed to undertake test marketing of the *Soléo Organics* product range in the UK.

A public relations (PR) person was also hired to gain exposure in the mainstream newspapers and presentations were undertaken for the press who were briefed on the product to assess the strength of the market. However, it soon became apparent that the UK was not a sophisticated sunscreen market, and in fact there were not an array of products offered compared to the USA, or for that matter, Australia and Japan. In Britain, a sunscreen was something that you purchased if you were heading to Spain or the Caribbean on holidays to deal with the sun. The sophistication of UV radiation was not something that the UK had really encountered. With the findings of this test marketing, Skin Elements decided to view the UK market as something of a 'green fields' opportunity.

19.1.3   Phase 3—Scaling Up for Full Product Launch

By 2012–2013, and encouraged by their market testing in Asia, North America and Europe, the Skin Elements team commenced planning for a full-scale production and launch of the *Soléo Organics* sun care product range. They spent around six months reviewing all the data that had been collected over the previous three years of market testing and assessment. It was decided that the company was now ready to move from proof of concept and market testing into mass production for a global market. However, although the product and market issues were secured, Skin Elements needed reliable production facilities and a significant injection of capital to fund the planned growth. As an interim step the Skin Elements team raised another A$1.5 million from within its own resources and a small group of private investors.

From a production perspective small batch product runs could be continued, but this was not sufficient to support large-scale market distribution. In 2013 an agreement was reached with Baxter Laboratories, a TGA accredited pharmaceuticals manufacturer in Melbourne, to sub-contract the production of the *Soléo* sunscreen product. Baxter Labs offered a range of services for skin care, sunscreen and topical pharmaceutical products. Under the production agreement the laboratory would produce 100,000 tubes of sunscreen in two 50,000 production runs. A 4-tonne vessel was used to mix the formula for the product.

This was a significant increase over previous production runs, which had involved only relatively small batches of 10,000 tubes. However, this scaling-up of production would offer an economy of scale that saved around 15% per-unit over the cost of the smaller production runs. All seemed set to go, with the ingredients delivered and the production runs commenced as planned, with each tube of sunscreen being filled within 30 min. Things seemed on track as far as production was concerned.

However, shortly after each tube came off the production line an inspection showed that all the ingredients had set hard like concrete! This threw the Skin Elements product launch into chaos and cost around \$250,000. As Peter described it, this was "an unmitigated disaster for the company." Leo quickly assembled a team from Baxter Laboratories and launched an investigation into what had gone wrong. This revealed that the product formula had been over mixed, in a manner similar to when cream turns into butter. Yet, in this case into a solid substance that was unable to be removed from the tubes. As Peter explained:

> As this is pure research and development there was no manufacturing recipe for scaling up. Everything is under development until solved. The process for scale up manufacturing settled on moving to the 1.5 tonne vessels (20,000 tubes).

Using this smaller batch production run approach resolved the immediate problem with the manufacturer of the *Soléo* sunscreen. However, this was not the end of the company's production difficulties.

Having resolved the problems in production in Australia, the first batch of 50,000 tubes for the UK market launch was shipped to London in late 2013. Within weeks of arrival the product was being distributed around the country through the Holland and Barrett Health stores. However, almost as soon as the product hit the street the company was hearing commentary coming back from the stores that the cream was 'scratchy' on the face especially if one rubbed the cream in hard. Customers were concerned, but had stopped short of actual complaints. Something wasn't right.

In response, Peter and Leo announced an immediate halt to distribution to the stores until an analysis could be undertaken on what had occurred. A full review was undertaken within Baxter Laboratories and it was here that the problem was identified. The wax in the mixture had, during the cooling down of the cream prior to filling into the tubes, precipitated out of the mixture back into its crystalline structure. This was generating the rough feeling when rubbing the cream onto peoples' faces. This resulted in yet another change to the way the formula was to be manufactured. The solution to the manufacturing process was to re-heat the mixture to its production temperature, and to ensure all the tubes were refilled prior to the mixture cooling down. In this way the wax remained in its liquid state for the life of the product. Despite these production difficulties Peter felt that the company had learned some valuable lessons that provided real benefits in the long-term, as he explained:

> Again, the procedure for manufacturing natural organic sunscreen ingredients was nowhere to be found. The formula was the company's and Skin Elements was the first in producing a mixture from these organic ingredients. It was part of the learning curve. And, it became part of the formula's security. The ingredients were not the critical part of the product – it was the manufacturing process. And this was to provide a major marketing edge. The company began to capitalise on this fact.

After the UK test program was completed the company gained valuable feedback and instituted a full review of manufacturing with the Baxter Laboratories production system. Skin Elements had invented a new formulation. It did not behave like synthetic chemical formulations and it was not something that existing laboratory chemists were experts in manufacturing. As part of this program the Skin Elements team examined every ingredient and its impact on the formula and every part of the packaging and its impact on the final product. Every item was to be sourced from only the certified suppliers and manufacturers. The manufacturing bill of materials was sealed and the process for manufacturing locked down. Each batch manufactured contributes to the body of knowledge on manufacturing the *Soléo* natural organic formula. All incremental improvements now are captured as part of each production batch. New approaches and systems are carefully managed before integrating into the manufacturing process. The formula is now finally robust for international commercialisation. What had been produced was not only a product innovation, but a process innovation.

19.1.4   Commercialisation, Financing and Growth

By 2016 Skin Elements had a decade of experience and had seen the product and its business model pivot through at least six cycles as it managed the flow of information between the market, the R&D/NPD process and the manufacturing operations. As a small company, Skin Elements relied heavily on the quality and integrity of its key suppliers and distributors. Peter had appointed four distributors to commence marketing the *Soléo Organics* product range within the Australian and New Zealand market. The market reaction was positive, and the product quickly became well-positioned within the retail health goods segment of the market, securing a place as a benchmark product. Sales were continuing throughout the USA via existing channels and through Asia where the initial test marketing programs had been operated. The company now seemed ready for a full-scale international launch.

It was clear that additional capital would be required, and Craig assessed the options for a final round of private financing. However, Peter felt that the future anticipated growth the company was forecasting would require a stronger capital base. Based on sales volume the estimated value of Skin Elements was approaching A$5 million, with the primary asset the *Soléo* sunscreen formula. It was decided to issue a prospectus and seek a public listing on the Australian Stock Exchange (ASX). Over the course of 2016 the company issued its prospectus under which it sought a minimum of A$3.5 million. The prospectus closed over-subscribed at A$3.7 million, which saw the Company listed on the ASX in early January 2017.

Just after the company had listed it was presented with an opportunity to acquire another Australian based skin care company *McArthur Skincare Pty Ltd*, which produced a range of innovative skin creams and soaps that used papaya (paw paw) extract as the foundation of their formula. The range of products produced by *McArthur* were complementary to the *Soléo Organics* sun care range, and it was deemed to offer a way to strengthen the product range of Skin Elements. The natural and organic formulas of these sunscreen and skin care products offered a strong market positioning for the Skin Elements brand within national and international markets.

By the end of 2018 Skin Elements was targeting three key global markets for its product range within North America, Europe and Asia. It had invested in a new e-commerce and e-marketing package designed to support its global market expansion. This focuses on the company's *SKINLIFE* program, involving a direct retail focus through on-line retailing and utilizing the SKINLIFE brand through selected retail stand-alone outlets. In addition, the company had entered into final negotiations with a health and medical group based in China, with plans to launch into the Chinese market with an initial A$20 million sales order. In addition, negotiations were continuing in Europe and the USA for similar sized orders over the course of 2019.

19.2   Analysis of the Case

The Skin Elements case provides an example of the successful commercialisation of an innovative product by a small to medium enterprise (SME). It was specifically selected because it offers a longitudinal case example, and because one of the co-authors of this chapter was a founder and principal actor within the story. In the following analysis we will examine the case and enfold the literature as we do.

19.2.1   Problems Facing SMEs in Commercialisation

Commercialisation is relatively poorly defined within the academic literature, despite the fact that it is a widely used term within that literature. It is generally associated with the process of taking new products, processes or services to market (Chakravorti 2004). Yencken and Gillin (2006) used the Scottish Enterprise definition of commercialisation as:

> The process of converting science and technology, new research or an invention into a marketable product or industrial process. (p. 215)

This conceptualisation of commercialisation as the successful launch and market adoption of an innovation, usually through securing good profits and returns to investment, is a consistent theme within the research literature (Chakravorti 2004). Further, the success of the commercialisation process is critical to the overall success of the entire innovation process and the competitiveness of the business, thereby making it an important area for research (Akgun et al. 2004; Pellikka and Lauronen 2007). However, as a concept, commercialisation has a range of meanings within the academic literature and encompasses a range of interrelated activities, or processes, that include the invention, early and late stage product development, proof of concept, new product launch and the subsequent marketing and distribution of the finished product (Ernst 2002; Ozer 2004; Yahaya and Nooh 2007). Despite its importance, the commercialisation process has been relatively poorly researched (Adams et al. 2006). This is a specific issue in relation to SMEs where there is not only limited research information, but unique conditions when compared to the environment experienced by large firms, which typically have superior resources, skills and knowledge (van Hemert et al. 2013; Mazzarol et al. 2014).

Pellikka and Virtanen (2009) identified at least four areas in which small firms engaged in technology commercialisation are likely to experience problems: (i) the commercialisation environment; (ii) marketing; (iii) financing; and (iv) management. Each of these will be discussed in relation to the Skin Elements case.

The first of these, the *commercialisation environment*, relates to the ability of the SME to gain access to the necessary support services needed for commercialisation (Kelley and Rice 2002; Malecki 1997; Dodgson 2000). In addition, the ability of the SME to access the necessary infrastructure (e.g. incubators, laboratories), to allow it to develop its technology and new products (Klofsten and Jones-Evans 1996; Autio and Klofsten 1998; Heydebreck et al. 2000). Finally, there is the firm's ability to secure the necessary resources for R&D and new product development (NPD) (Abetti et al. 1988; Dodgson 2000).

A second area is the need for the SME to develop the necessary competencies and resources to allow it to succeed in relation to *marketing* (Pellikka and Virtanen 2009; Kang et al. 2013; De Zubielqui et al. 2014). Marketing is critical for making assessments of the likely adoption rate of the new product, service or process (Jolly 1997; Ziamou 2002; Sedighadell and Kachquie 2013). This helps to gain clear insights into the needs and wants of the customer, and identify the value proposition that the customer is likely to respond to (Ford and Saren 2001; Huang et al. 2002; Ozer 2003). Also important is the ability to get the timing of any marketing, sales and promotion activities right so that they coincide with the NPD and production activities (Mohr 2001; Ford and Saren 2001; Pellikka and Lauronen 2007). Finally, there is a need for the SME to maintain close relations with their lead customers and/or end users to obtain real-time feedback on the product and address any problems (Athaide et al. 1996).

The third area that affects SMEs in relation to commercialisation is *financing* (Pellikka and Virtanen 2009). The ability of an SME to secure sufficient financial resources support the commercialisation of a new product, particularly within a national or global market, is a major challenge. Research into the *financing gap* for SMEs suggests that this problem is particularly focused on the innovative firms that are typically found in technology sectors, with "new business models and high growth prospects" (OECD 2006). NPD and commercialisation of innovative technologies generally demand significant investment in R&D, marketing and sales, and securing such funding is often difficult for small firms (Hoffman et al. 1998). The commercialisation process will place increasing costs on the SME and demand a substantial amount of working capital with which to support the R&D, NPD, production and marketing efforts required, as well as above average rates of profit from sales activity (Davidsson et al. 2009; OECD 2010, 2016).

The fourth area affecting commercialisation within SMEs is *management*, which relates to the ability of the firm's leadership team to manage resources, coordinate projects, undertake the often-complex processes of R&D, NPD, production, marketing and sales, which also need to be managed concurrently (OECD 2018; Do et al. 2018). In this area the overall success of the commercialisation process is likely to be found. Success may depend not just on the firm's R&D competencies, but also its ability to manage knowledge, learn and adapt to turbulent market environments, and strengthen their organisational capabilities (Park and Ryu 2015). The success of an SME seeking to commercialise its intellectual property (IP), within an open technology market, may also be diminished from a return to investment perspective given the information asymmetries that are typically found between such firms and their potential buyers (Padula et al. 2015).

Table 19.1 provides a summary of these four common problems that face SMEs engaged in commercialisation, and how Skin Elements addressed each of these challenges over the four phases of the firm's evolution 12-year journey from start-up to

Table 19.1 Summary of skin elements ability to address commercialisation challenges

Problems facing SMEs	Resolution of problems
Commercialisation environment: Access to support services, infrastructure, regulatory approvals, R&D and proof-of-concept	*Phase 1*: • Networking with many stakeholders in early Phase 1 R&D to identify the formula *Phase 2*: • Initial problems of securing EU regulatory approval resolved via collaboration with BASF in Germany *Phase 3*: • Strategic partnership with Baxter Laboratories secured to provide long term production facilities for scale-up *Phase 4*: • Enhancing resources via supplier and distributor channels
Marketing: Voice of Customer (VOC) test marketing, confirmation of customer value proposition (CVP), brand positioning, securing market access and sales distribution, and coordinating NPD, production and product launch activities	*Phase 1*: • Extensive 'out of the office' networking with key suppliers and chemists to identify 'state-of-art' product specifications *Phase 2*: • Test marketing in the USA (California and Hawaii) within surfing community, highlighting CVP of safe product • VOC test marketing in Australia identifying mother and child as key market segment • Test marketing in Asia to further assess VOC and CVP • Partnership with Planet Blue for UK market entry *Phase 3*: • Further test-marketing in UK highlights problems with 'scratchy' formula, which led to further refinement of the production process *Phase 4*: • Securing four primary distributors for the product in Australia and New Zealand • Development of market access into China, EU and USA • Creation of online e-marketing and e-commerce platform

(continued)

Table 19.1 (continued)

Problems facing SMEs	Resolution of problems
Financing: Securing start-up capital, ensuring that capital raising is able to keep pace with the firm's commercialisation strategy	*Phase 1*: • Initial start-up capital was 'bootstrapped' *Phase 2*: • Further 'bootstrapping' by founders • Systematic analysis of funding needs as the R&D, NPD and market development process unfolded *Phase 3*: • Fund raising of A$1.5m via 'bootstrapping' and private investors *Phase 4*: • IPO listing raises A$3.7m
Management: Ability to manage resources, coordinate projects, knowledge, IP rights and multiple stakeholders	*Phase 1*: • Early use of NDA confidentiality and trade secrecy with stakeholders in Phase 1 • Systematic approach to concurrent product R&D, NPD, IP rights and regulatory approvals, marketing R&D and capital raising *Phase 2*: • Management of TGA, FDA and EU regulatory approvals, in conjunction with product and market development *Phase 3*: • Proactive response to initial production problems by Baxter Laboratories turning disaster into IP rights benefit from new process innovation discovery *Phase 4*: • Coordination of post-IPO investor relations, global market expansion, production and logistics, online business model and acquisition of *McArthur Skin Care* as complementary business

global expansion. It is worth noting that the company was developing a new and potentially disruptive innovation, targeted at a mature, global industry, dominated by major corporations. If this weren't enough, the project team was continuously breaking new ground in terms of how the product would be designed, how it would be produced, how its IP rights would be protected, and how it would be branded, positioned and distributed.

The learning process within the three founders varied depending on their previous knowledge and experience. Leo, while knowledgeable in the field of science and applied chemistry, had limited previous experience with new venture creation and commercialisation. By comparison, Peter, saw the business from a broader perspective than most engaging in technology commercialisation for the first time. With previous history in successful technology start-up enterprises he realised the need for 'friends' in the environment. As he explained:

> Running from day to day with most things unknown the Skin Elements team needed to be strong of mind and want to succeed. Craig was not without experience in this sector and Leo was driven by a will to find a better solution to synthetic sunscreen formulations due to a skin cancer crisis in his extended family. I also understood the time periods that the team would be working to, my previous experience predicting a decade long drive to success would be required. This fact was signed off and imbedded into the team before commencement. I knew from experience that giving up was too often the result of start-up ventures due to the commercialisation environment.

This enabled the Skin Elements team to systematically manage the *commercialisation environment* to identify key contacts and potential alliance partners (e.g. Baxter Laboratories, BASF, Planet Blue) able to help the firm achieve its goals.

In relation to *marketing*, the Skin Elements team was well aware of the need for getting the marketing right. They were developing a very disruptive product to the prior art in their market. The team sought out from an early stage to determine the key customer value proposition in going natural and organic. The answer was safety. Pesticides and plastics from the petroleum discoveries half a century ago had seen many concerns surfacing. Public opinion was changing in relation to the perceived value of something that was useful on the one hand, but was now being questioned as injurious to health. As Peter explained:

> Our focus groups confirmed these concerns. As a way of gaining a real time feel for this shift in mood a global test program was commenced. The early innovators were the water sports where the problems of chemicals on the body were better understood. They were soon eclipsed by the young mothers who without exception globally were conscious of the need to cover the skin of their babies from the dangers of UV light including sunburn. In a test marketing program in Slovenia young mothers were queuing up to access a tube of Soléo from the main pharmacy in Ljubljana (the capital) after the product went on sale. Though the cost of a Soléo tube was twice the cost as compared to most other countries (due in the main from government import charges) this did not stop the demand for the product increasing every month during the market testing programme. A young mother gave up a coffee a week to provide her child safety from UV radiation.

The firm's approach to *financing* was also well-considered and systematic. As outlined in the case, an important aspect of the success Skin Elements had in securing financing was the ability of the founders to 'bootstrap' the venture in its early years, and then the skills and networks to raise additional private investment capital and ultimately take the company successfully through an IPO. As Peter explained:

> Our ability to meet funding costs from the team members through our own resources initially allowed the project to move rapidly in the development of the 'raw' formula. And as the project grew Craig and I had an understanding on how to scale up the venture. I also had contacts that became part of the seed capital as the venture began to gain momentum. And finally, the company IPO'd with the listing driven internally by the team. I was able to steer the company through a path that delivered the capital. And now, conscious as we are of the need for the company to achieve significant sales quickly, the team have targeted three key international sectors that for different reasons offer above average sales growth.

Finally, the overall *management* demonstrated by the Skin Elements team during the commercialisation process can be characterised as one of systematic flexibility and adaptiveness. An important aspect of the success Skin Elements had in its commercialisation was the previous experience of the company's senior management team in launching innovative ventures and commercialising innovations. This allowed them to meet and address the many challenges that emerged during the commercialisation process. As Peter explained:

> The team had been briefed on the need to back their own judgement and not give up against the odds. And this was the basis on which the three members of the project commenced. This proven track record by the three founders was a determining factor in the company's commercialisation success. Turbulent environments became everyday life and this became the norm. And as the company signed off on each stage of the program the investor assessments were able to be communicated directly from the team.

19.2.2 Knowledge Management in SMEs

As a concept, *knowledge management* (KM) refers to the way in which an organisation captures, stores, analyses and disseminates information, intellectual property, skills, competencies and knowledge, at the individual, group and enterprise level, so as to secure a competitive advantage (Civi 2000). According to Hedlund (1994) knowledge is found within organisations in at least three general forms. The first of these is *cognitive knowledge*, which is the theories and mental constructs that exist within the minds of the firm's employees and are used to guide thought and action. The second form is that of *skills*, which refer to the learned knowledge that comes from having the people within the organisation apply their cognitive knowledge and develop their own specific routines and rituals to achieve their desired goals. Finally, there is *embodied knowledge*, which refers to the products and services that the organisation produces, with each representing a manifestation of the cognitive knowledge and skills that transform the firm's resources into outcomes.

A critical aspect of KM is the relationship between *tacit* and *explicit* knowledge (Polanyi 1962). *Tacit knowledge* is that found within the individual, which represents their experience and wisdom. *Explicit knowledge* is that which has been codified into text, numbers, models or diagrams, and can be readily transfer from one individual to another. Thus, when an invention is codified into a patent and formally registered within a patent office, or literary works are copyrighted and published, this knowledge becomes transferable and moves from knowledge and intellectual capital or assets, to legally tradable intellectual property (Williams and Bukowitz 2001).

Research into KM within large organisations has focused on the need to establish systems for facilitating this process of transfer to and from tacit to explicit knowledge. For example, Nonaka and Takeuchi (1995), in their study of innovation within large Japanese corporations, developed the SECI model of knowledge transfer, which recognised the process as moving around four domains: (i) *socialisation*—converting tacit to tacit knowledge via interpersonal communication and social interaction;

(ii) *externalisation*—conversion of tacit to explicit knowledge via the formal codification of tacit knowledge via documentation and dissemination; (iii) *combination*—conversion of explicit to explicit knowledge through analysis of secondary sources, modelling, sorting and manipulation; and (iv) *internalisation*—conversion of explicit to tacit knowledge via learning and experience (Dimov 2007).

For knowledge to move effectively throughout an organisation Hedlund (1994) suggests that it must progress through three stages: (i) *articulation and internalisation*—where tacit knowledge is codified into explicit knowledge; (ii) *extension and appropriation*—where the codified explicit knowledge is distributed; and (iii) *assimilation and dissemination*—where the explicit knowledge is embedded into the organisation via formal training, procedures and coaching.

Within SMEs, the process of KM is generally highly idiosyncratic and informal in nature, typically centred around the knowledge and expertise of a small number of key people, who focus primarily within the *socialisation* domain of the SECI model. However, as the enterprise grows in size and complexity, more formal KM systems are needed, or are imposed by third-parties (e.g. investors, customers) to protect against the loss of key personnel (Bagshaw 2000). The formal management of KM within SMEs is therefore relatively poorly understood within the academic literature (Durst and Edvardsson 2012; Cerchione et al. 2016). What is identified within the literature is that most SMEs do not possess or use formal KM systems (Nunes et al. 2006), and where they exist, they generally are poorly aligned with the firm's corporate strategy (Pillania 2008). Many SMEs are aware of KM and its potential to add value (Radzevicience 2008), but lack the resources, personnel and expertise to implement KM systems (Keogh et al. 2005). However, when SMEs do make active use of KM systems, they typically experience benefits to long-term sustainable growth (Salojärvi et al. 2005).

Figure 19.2 illustrates the Nonaka and Takeuchi (1995) SECI model and how it applied to the KM process within Skin Elements. In relation to the SECI model Skin Elements operated for the early period within the *socialisation* domain where knowledge transfer essentially was between the three partners. This shifted to *externalisation* after the company appointed Baxter Laboratories, but interestingly only after a lengthy period following their appointment. During the first two years as the company was still adjusting the *Soléo* formula, the exchange of knowledge between Skin Elements and Baxter Laboratories was essentially socialisation (tacit-tacit), and required regular visits by the team from Perth to Melbourne (at a distance of 2721 km or 3½ h flying time). It only changed fundamentally at the time the company moved to commence its IPO.

As the company began the international test marketing program and then continuing with its activities to the present-day Skin Elements progressively found it needed to move to a more codified and explicit knowledge management system. This was driven to a certain extent by the need to bring in additional staff who had no history with the team. Likewise, the development of formal procedures has been coupled

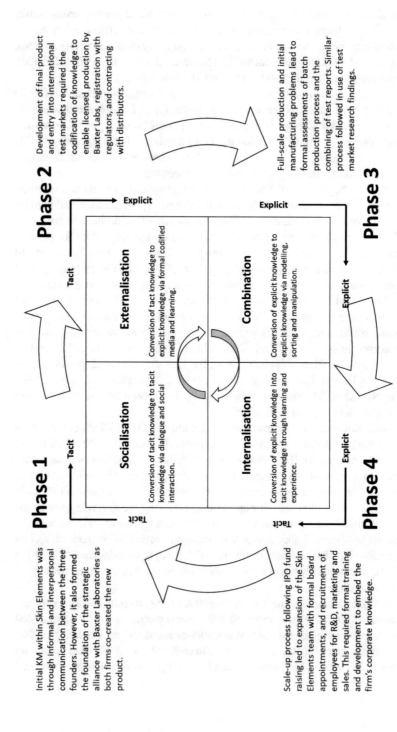

Fig. 19.2 SECI model applied to the skin elements case

with the need for regular reporting and this process has seen procedures developed for fast tracking work programs.

A specific example of this process was the evolution of the relationship between Skin Elements and Baxter Laboratories. As outlined in case study, Baxter Laboratories was first brought into the project after the Skin Elements team had undertaken a comprehensive assessment across Australia during Phase 1, to examine state-of-art knowledge in relation to the development of the product specifications, and to identify a long-term strategic partner to support the company's long-term ambition of scaling up the production of the *Soléo* product.

At the time of the initial contact between the two companies, Baxter Laboratories was also newly established and therefore open for discussion over a potentially new and innovative project. When Peter and Leo first met with the senior managers of Baxter Laboratories in Melbourne, the office and laboratory was still unfurnished. As Peter explained:

> At the time Leo and I had our first project meeting the furniture had not arrived and the meeting was held sitting on crates in a brand-new laboratory in Boronia in Victoria just weeks after it had been built! Skin Elements was among the first customers and the building was only running at 5% capacity. Today the Laboratories are three times larger and capacity is over 80% on the expanded site which is 50 million tubes per year. And taking the brief to commence the production of a privately owned natural organic sunscreen formula was still viewed with some trepidation. This was still the world of synthetic chemicals full stop.

However, the business relationship between Skin Elements and Baxter Laboratories was not just between businesses, it was between the founders and shareholders of two young, entrepreneurial companies. The founders of Baxter Laboratories, Craig Baxter and his brother Brent, forged an alliance with the founders of Skin Elements, which was founded on a common purpose and sense of collaboration in an exciting and potentially valuable new product commercialisation opportunity. The interpersonal communications and *socialisation* (tacit to tacit) exchange of knowledge was a necessary and important aspect of this engagement. As Peter recalled:

> The knowledge management process was tacit-to-tacit for good reason. Everything was new and time was precious. Meetings were face to face, secrecy was high and trips were often just for the day. Depart Perth at 6am arrive Melbourne at midday. Meeting 1 pm to 6 pm and back to Perth at 8 pm arriving at 9 pm. But when the crises occurred everyone knew their role and their place and what they had to do. Things went into overdrive without the need for procedures and instructions, and solutions were born.

However, after the product formula had been stabilised and subsequent production runs had become more orderly and routine, the KM process changed. The manufacturing process for the *Soléo* product and how to do a production run, was codified within a strict trade secrecy agreement between Skin Elements and Baxter Laboratories. This set clear, formal processes that included the procurement of the micronized zinc from a selected manufacturer and not from one of the Baxter Laboratories regular suppliers. This evolution of the KM process between Skin Elements and Baxter Laboratories also followed the knowledge transfer process described by Hedlund (1994), flowing through the three-stages from articulation and internalisation, then through extension and appropriation, to assimilation and dissemination.

What seems to drive the adoption and use of KM systems are the firm's strategy, purpose, organisational culture, the leadership and support of the senior management, and the engagement of employees willing to share knowledge (Wong and Aspinwall 2004a, 2005; Shelton 2001). Sparrow (2005) found that in relation to KM systems adoption, most SMEs are either unengaged, focused on retaining knowledge ownership within a few key people, willing to share knowledge via learning and co-production on a project-by-project basis, or systematically engaged in formal KM practices. Despite this, most SMEs are informal in relation to KM (Hutchinson and Quintas 2008; Edvardsson 2009), relying on simple ICT support tools and knowledge exchange via interpersonal communication (Evangelista et al. 2010). Nevertheless, how well KM is embraced by SMEs and used within innovation and commercialisation processes is contingent on the firm's leadership and culture (Gray 2006).

Despite these limitations there is a general consensus that the ability to access external knowledge and bring it into the SME, and then make use of this knowledge is important (Chen et al. 2006). The adoption of KM is generally found within SMEs that are focused on innovation than those that are cost-driven (Levy et al. 2003). Further, it can be a source of competitive success for innovative SMEs (Perez-Araos et al. 2007; Harris 2008; Alegre et al. 2013), in particular those engaging in international growth strategies (Fletcher and Prashantham 2011). For SMEs seeking to adopt KM systems at least five steps should be followed: (i) develop a clear structure to organise the process; (ii) identify different types of knowledge within the firm; (iii) include any KM related processes or activities that can manipulate knowledge; (iv) identify forces that can affect KM outcomes; and (v) develop a balance between the use of ICT systems and social engagement (Wong and Aspinwall 2004b, 2006).

Initially as a team of three running a start-up program the KM system was limited. However, over time the company operated under a OI structure seeking links with specialists and third part contractors to work on the creation of the technology. This saw a change in KM controls. Ultimately the company completed the *Soléo Organics* formulation and subsequently re worked the papaya formulations and now *Elizabeth Jane* natural cosmetics. These are all filed under documentation and access is secure. Now as a public company post-IPO the communication is very formal given the laws of the ASX regulators. Peter explained the influence of this formalisation with the KM system of Skin Elements:

This has seen a change in how knowledge is managed within the firm. No longer are they idiosyncratic and social. Once the production process transferred to a formal plan with Baxter Laboratories the way communication and knowledge transfer was undertaken also followed suit. Orders became now formal, procedures are cast in stone for staff to follow and flexibility has now been removed from how communication and knowledge is transferred between the companies. Skin Elements delivers orders, Baxter Laboratories produces, packs and despatches to a clear specification and all under a GMP manufacturing agreement. Today there is an alignment of strategy, purpose and culture together with the leadership and staff of both companies. Both Baxter Laboratories and Skin Elements continue to review knowledge exchange processes and now access third party information as part of a continual knowledge management program.

19.2.3 Open Innovation, Absorptive Capacity and Operational Entrepreneurship

Closely related to KM are the concepts of *open innovation* (OI) and *absorptive capacity* (ACAP). The first of these (OI) was introduced by Chesbrough (2003) following his examination of how NPD and commercialisation processes worked within leading high-technology firms in the United States. The notion of *open* rather than *closed* approaches to innovation reflects the dichotomy between traditional in-house management of R&D, NPD and commercialisation, with all activities undertaken within the firm's own research laboratories and production plants. However, the increasing pace of competition and the emergence of new digital technologies, resulted in a change from a 'closed' to an 'open' approach to innovation management by large companies. Driving this change was the recognition that speed to market and competitive success was contingent on enhancing the flow of knowledge and ideas not only within the firm, but between it and a wider network of customers, suppliers and third-party actors (Chesbrough 2006). KM lies at the heart of this process, which has been identified as critical to success in the commercialisation of radical innovation. As noted by Greco et al. (2016) in a study of companies in the European Union:

> **Firms seeking to engage in open innovation**, …should enter into collaborative agreements with a few knowledge-intensive partners, ensuring frequent interactions that may favour the transfer of knowledge across organizational boundaries. (Greco et al. 2016, 514)

How formal such engagements are is likely to depend on the nature of the project and the firm's strategic goals (van de Vrande et al. 2009; Garriga et al. 2013). While OI has benefits, it has been viewed as a function of R&D management with less attention given to its place in general management and economics (West et al. 2014). For firm's seeking to engage in OI, a more flexible mindset within the management team is required, in particular in relation to how IP rights are managed (Gambardella and Panico 2014). It is also important for all employees to become involved, and to understand that they can and should develop and maintain commercially valuable relationships with outsiders (Carayannis and Meissner 2017). However, OI is not without its challenges, and requires a strong commitment from the top management, supported by processes and ICT systems that enhance KM processes and other tools for innovation and commercialisation (Parida et al. 2012; 2014).

Skin Elements saw the need to form strategic alliances to help facilitate its commercialisation program from an early stage. As a result, it embarked on an OI program rather than the alternative of creating a large internal project team. After the breakthrough with the creation of the *Soléo* formula, the Skin Elements team linked with a group of three key technology partners. Initially a specialist medical advisor was contracted on signing off on the base formulation. Peter and Leo travelled to Melbourne to utilise the services of his dermatology labs and understanding of human skin as an expert cosmetic surgeon. The team worked in and out of his rooms in Toorak Victoria for the final six months of the development phase. This was followed by a concurrent tie up with a consulting chemist in Sydney, and a naturopathic specialist in practice in Brisbane. The key to the rapid prototyping was the use of the expert panel to round off the product development process. As explained by Peter:

> Within these open innovation arrangements with partners the ability to solve problems constructively enabled Skin Elements to leverage our time and resources to maximum purpose. This mutual value cocreated at the end strengthened all parties.

Another important aspect of both KM and OI is ACAP, which relates to the firm's ability to effectively manage the flows of information and knowledge from a range of external actors (e.g. customers, suppliers, competitors and R&D partners), bring this into the firm, and integrate it into the commercialisation process (Huang and Rice 2009). ACAP is not just about how the organisation acquires and assimilates information and knowledge, but how it exploits it and how rapidly it can do so (Zahra and George 2002). Of prime importance is the firm's ability to connect and communicate with external actors and then between internal actors, facilitating meaningful knowledge exchange in the process (Cohen and Levinthal 1990). ACAP not only plays a key role in OI, it also helps to foster enhanced innovativeness and financial performance within firms (West and Bogers 2014). Within SMEs, the small size and close proximity of the firm's management and project teams means that the R&D, NPD, marketing and sales activities involved in commercialisation are usually undertaken by the same people, or at least with all members involved. Formal systems are uncommon and information flows largely interpersonal and informal (Sparrow 2001; Bougrain and Haudeville 2002).

Skin Elements exemplified this process within the product development and commercialisation undertaken in the creation of the *Soléo* product. As explained by Peter:

> The closeness of the small team in the technology development phase saw the mode of communication able to be better managed. This delivered faster and more immediate decision making. The ACAP of the team was high and allowed for the exploitation of the opportunity quickly. Management systems were largely informal with communication and information flows between all more informal and interpersonal. When the formula initially failed Leo was able to assemble the team at short notice and solve the problem. The ACAP was there to do this. Similar issues occurred during the test marketing phase and again the ACAP of the company was able to deliver.

Shepherd and Patzelt (2017) suggest that ACAP and the application of systematic NPD processes such as *StageGate*® (Cooper and Edgett 2005; Cooper and

Kleinschmidt 1993, 1995), comprise a potential theoretical concept of *operational entrepreneurship*, which they define as follows:

> *Operational entrepreneurship* can be defined as the selection and management of transformation processes for recognizing, evaluating, and exploiting opportunities for potential value creation. (Shepherd and Patzelt 2017, p. 122)

This concept focuses on the use of *transformation processes* to assist the firm's management team to reduce uncertainty and risk in the commercialisation process by applying a series of decision making frameworks and tools to help validate assumptions prior to making ever increasing levels of investment. A wide-range of techniques and methods have been developed that lie outside the scope of this chapter. However, a common feature of these transformation processes is that they seek to assess the customer and market perception of value in a new product or service, as an integral part of any R&D, NPD commercialisation process.

York and Danes (2014) reviewed three of the best known processes: (i) the *Stage-Gate*® NPD process (Cooper 2019); (ii) the *Fuzzy Front-End* (FFE) process (Koen et al. 2002); and (iii) the *Customer Development* (CD) process (Blank and Dorf 2012). Their analysis identified similarities and differences between these three approaches. Of them *StageGate*® is the most structured and formal, but this reflects its origins in large north American manufacturing companies. *StageGate*® has evolved over time since its launch in the 1980s is now offered in a variety of new forms for smaller projects, rapid development or technology development (Cooper 2019).

The FFE process is not a complete NPD system, but more an experimental approach designed to undertake initial customer and market assessments in the early stages of the NPD process. The CD process is developed from the work of Ries (2011) and the *Lean Start-Up* concept, which focuses on creating business models that are built on understanding the value proposition that can be made to customers, validating this through the creation of *minimum viable products* that enable early market testing, and then either preserving with future investment, or *pivoting* around one of a large number of pivots to reinvent the product concept and select one of three *growth engines* upon which to build the business growth strategy. A key attribute of the CD/Lean Start-Up process is the need to build any new products or services around what the customer/end-user views as representing value, rather than seeking to push technology onto a customer.

The key to the rapid prototyping and the use of the expert panel described above, was the ability to round off the product in a very short period as compared to alternative internal approaches. This allowed the Skin Elements team to commence the assessment of the market of the key value drivers of the natural organic sunscreen over the state of the existing market offering more effectively. The *operational entrepreneurship* transformation processes utilised by Skin Elements saw the team working through the application of FFE and CD principles as it progressed through the product identification and specification process, then the assessment of the needs of the customer and exploration of opportunities in selected international markets.

For example, as part of specification development for *Soléo Organics* the assessment of the customer needs commenced via an FFE process (Koen et al. 2002) that saw Peter and Leo meet with and test the product in a number of targeted markets and market segments. This was part of the initial product discovery process. The Skin Elements team then progressed from this period to the more formal market insertion program in Phase 2, where they followed the basic principles associated with CD (Blank and Dorf 2012), and Lean Start-Up (Ries 2011). It should be noted that these techniques were not applied in a formal manner, although the project team did participate in commercialisation education programs run by a local university during 2015–2017 where they addressed these techniques through the course. This included the *StageGate®* NPD process (Cooper 2019), which became a more common approach within the firm from Phase 3 onwards. As Peter explained:

> In the early phases, our ability to assess individual markets parallel, and pivot the technology offering for the market, resulted in a change to the selection of ingredients that in the laboratory appeared to meet what was required. Only by moving through a full market insertion of progressive versions of the Soléo Organics did the company develop the product that met market acceptance globally.

19.3 Conclusion and Lessons Learnt

The Skin Elements case provides a study of the evolution of an innovative SME from initial start-up to public listing and international market entry. It addresses a gap in the literature on how KM is understood and applied within SMEs (Durst and Edvardsson 2012; Cerchione et al. 2016). The case also highlights the importance of having within the foundation team, a good cross-section of experience, skills and knowledge. In the case of the three founders, Leo brought his scientific knowledge, Craig, his knowledge of finance and company operations, and Peter, his knowledge and networks in the foundation of entrepreneurial ventures and the strategic leadership required to navigate the challenging process of commercialisation.

As the case shows, the Skin Elements team successfully overcame each of the major problems that Pellikka and Virtanen (2009) suggest are common challenges facing SMEs engaged in commercialisation. Their ability to do this was made possible by their willingness to embrace OI practices and look for outsiders who could provide them with the necessary knowledge, skills and resources that they might otherwise have had to create in-house at much greater cost and time. The success of their concurrent securing of product and production systems, and market entry and distribution channels, reflected the team's ACAP, KM system and use of *operational entrepreneurship* techniques (Shepherd and Patzelt 2017). These also evolved and became more sophisticated as the company grew and developed its internal systems. The case suggests that SMEs can apply KM, OI, ACAP and formal NPD processes just as effectively as large firms. It also addresses the way effective interpersonal communications is the backbone of managing KM within such a company, building trust and understanding between the members of the project team, and their strategic alliance partners. This shows the importance of OI strategies for innovative SMEs (Wynarczyk et al. 2013).

The case is not without its limitations. It offers only one example of a case that was somewhat atypical due to the range of skills, knowledge and expertise that the three founders brought together. Future research should examine multiple cases of SMEs engaged in the commercialisation of innovative technologies and examine how they evolve their strategies in the successful creation of entrepreneurial innovation value (Malone et al. 2015).

Finally, for managers and entrepreneurs seeking to commercialise new innovations, the case highlights the importance of selecting the right cross-section of people, with the right skills, knowledge and expertise to address the major challenges that face SMEs. Having a clear roadmap for the concurrent evolution of the product technology, market development, capital raising to support growth, and expansion of the company is another key lesson. Finally, the case illustrates the vital importance of building and maintaining strategic alliances that can not only provide the SME with the resources and market access that it needs, but also access to new knowledge that can provide information that will prove essential to the team's learning and subsequent strategic decision making as they navigate through the uncertain waters of the commercialisation process.

References

Abetti, P. A., Lemaistre, C. W., & Wacholder, M. H. (1988). The role of Rensselaer Polytechnic Institute: Technopolis development in a mature industrial area. In R. W. Smilor, G. Kozmetsky, & D. V. Gibson (Eds.), *Creating the Technopolis* (pp. 125–144). Cambridge, MA: Ballinger Publishing.

Adams, R., Bessant, J., & Phelps, R. (2006). Innovation management measurement: A review. *International Journal of Management Reviews, 8*(1), 21–47.

Akgun, A. E., Lynn, G. S., & Byrne, J. C. (2004). Taking the guess work out of new product development: How successful high-tech companies get that way. *Journal of Business Strategy, 25*(4), 41–46.

Alegre, J., Sengupta, K., & Lapiedra, R. (2013). Knowledge management and innovation performance in a high-tech SMEs industry. *International Small Business Journal, 31*(4), 454–470.

Athaide, G. A., Meyers, P. W., & Wilemon, D. L. (1996). Seller-buyer interactions during the commercialisation of technological process innovations. *Journal of Product Innovation Management, 13*(5), 406–421.

Autio, E., & Klofsten, M. (1998). A comparative study of two European business incubators. *Journal of Small Business Management, 36*(1), 30–43.

Bagshaw, M. (2000). Why knowledge management is here to stay. *Industrial and Commercial Training, 32*(5), 179–182.

Blank, S., & Dorf, B. (2012). *The start-up owner's manual: The step-by-step guide for building a great company*. Pescadero, CA: K&S Ranch Publishing.

Bougrain, F., & Haudeville, B. (2002). Innovation, collaboration and SMEs internal research capacities. *Research Policy, 31*(5), 735–747.

Carayannis, E. G., & Meissner, D. (2017). Glocal targeted open innovation: Challenges, opportunities and implications for theory, policy and practice. *Journal of Technology Transfer, 42*(2), 236–252.

Cerchione, R., Esposito, E., & Spadaro, M. (2016). A literature review on knowledge management in SMEs. *Knowledge Management Research & Practice, 14*(2), 169–177.

Chakravorti, B. (2004). The new rules for bringing innovations to market. *Harvard Business Review, 82*(3), 58–67.

Chen, S., Duan, Y., Edwards, J. S., & Lehaney, B. (2006). Toward understanding inter-organizational knowledge transfer needs in SMEs: Insight from a UK investigation. *Journal of Knowledge Management, 10*(3), 6–23.

Chesbrough, H. W. (2003). *Open innovation: The new imperative for creating and profiting from technology*. Boston, MA: Harvard Business School Press.

Chesbrough, H. W. (2006). Open innovation: a new paradigm for understanding industrial innovation. In: H. Chesbrough, W. Vanhaverbeke, J. West (Eds.), *Open innovation: Researching a new paradigm* (pp. 1–12). Oxford: Oxford University Press.

Civi, E. (2000). Knowledge management as a competitive asset: A review. *Marketing Intelligence & Planning, 18*(4), 166–174.

Cohen, W., & Levinthal, D. (1990). Absorptive capacity: A new perspective on learning and innovation. *Administrative Science Quarterly, 35*(1), 128–152.

Cooper, R. G. (2019). The drivers of success in new-product development. *Industrial Marketing Management, 76*(1), 36–47.

Cooper, R. G., & Edgett, S. J. (2005). *Lean, rapid and profitable new product development*. Canada: Product Development Institute Inc.

Cooper, R. G., & Kleinschmidt, E. J. (1993). Screening new products for potential winners. *Long Range Planning, 26*(6), 74–81.

Cooper, R. G., & Kleinschmidt, E. J. (1995). Benchmarking the firm's critical success factors in new product development. *Journal of Product Innovation Management, 12*(5), 374–391.

Davidsson, P., Steffans, P., & Fitzsimmons, J. (2009). Growing profitable or growing from profits: Putting the horse in front of the cart? *Journal of Business Venturing, 24*(4), 388–406.

De Zubielqui, G. C., Lindsay, N., & O'Connor, A. (2014). How product, operations, and marketing sources of ideas influence innovation and entrepreneurial performance in Australian SMEs. *International Journal of Innovation Management, 18*(2), 1–25.

Dimov, D. (2007). From opportunity insight to opportunity intention: The importance of person-situation learning match. *Entrepreneurship Theory and Practice, 31*(4), 561–583.

Do, T. H., Mazzarol, T., Soutar, G. N., Volery, T., & Reboud, S. (2018). Organisational factors, anticipated rents and commercialisation in SMEs. *International Journal of Innovation Management, 22*(2), 1–30.

Dodgson, M. (2000). *The management of technological innovation. An international and strategic approach*. Oxford: Oxford University Press.

Durst, S., & Edvardsson, I. R. (2012). Knowledge management in SMEs: A literature review. *Journal of Knowledge Management, 16*(6), 879–903.

Edvardsson, I. R. (2009). Is knowledge management losing ground? Developments among Icelandic SMEs. *Knowledge Management Research & Practice, 7*(1), 91–99.

Ernst, H. (2002). Success factors of new product development: A review of the empirical literature. *International Journal of Management Reviews, 4*(1), 1–40.

Evangelista, P., Esposito, E., Lauro, V., & Raffa, M. (2010). The adoption of knowledge management systems in small firms. *Electronic Journal of Knowledge Management, 8*(1), 33–42.

Fletcher, M., & Prashantham, S. (2011). Knowledge assimilation processes of rapidly internationalising firms. Longitudinal case studies of Scottish SMEs. *Journal of Small Business and Enterprise Development, 18*(3), 475–501.

Ford, D., & Saren, M. (2001). *Managing and marketing technology*. London: Cengage Learning Business Press.

Gambardella, A., & Panico, C. (2014). On the management of open innovation. *Research Policy, 43*(5), 903–913.

Garriga, H., Von Krogh, G., & Spaeth, S. (2013). How constraints and knowledge impact open innovation. *Strategic Management Journal, 34*(9), 1134–1144.

Gray, C. (2006). Absorptive capacity, knowledge management and innovation in entrepreneurial small firms. *International Journal of Entrepreneurial Behaviour & Research, 12*(6), 345–360.

Greco, M., Grimaldi, M., & Livio, C. (2016). An analysis of the open innovation effect on firm performance. *European Management Journal, 34*(5), 501–516.

Harris, R. J. (2008). Developing a collaborative learning environment through technology enhanced education (TE3) support. *Education + Training, 50*(8/9), 674–686.

Hedlund, G. (1994). A model of knowledge management and the N-form corporation. *Strategic Management Journal, 15*(2), 73–90.

Heydebreck, P., Klofsten, M., & Maier, J. (2000). Innovation support for new technology-based firms: The Swedish Teknopol approach. *R&D Management, 30*(1), 1–12.

Hoffman, K., Parejo, M., Bessant, J., & Perren, L. (1998). Small firms, R&D, technology and innovation in the UK: A literature review. *Technovation, 18*(1), 39–55.

Hossain, M. (2015). A review of literature on open innovation in small and medium-sized enterprises. *Journal of Global Entrepreneurship Research, 5*(1), 1–12.

Huang, F., & Rice, J. (2009). The role of absorptive capacity in facilitating "open innovation" outcomes: A study of Australian SMEs in the manufacturing sector. *International Journal of Innovation Management, 13*(2), 201–220.

Huang, X., Soutar, G. N., & Brown, A. (2002). New product development processes in small to medium-sized enterprises: Some Australian evidence. *Journal of Small Business Management, 40*(1), 27–42.

Hutchinson, V., & Quintas, P. (2008). Do SMEs do knowledge management? Or simply manage what they know? *International Small Business Journal, 26*(2), 131–154.

Jolly, V. K. (1997). *Commercializing new technologies: Getting from mind to market.* Boston, MA: Harvard Business School Press.

Kang, J., Gwon, S. H., Kim, S., & Cho, K. (2013). Determinants of successful technology commercialization: Implication for Korean government-sponsored SMEs. *Asian Journal of Technology Innovation, 21*(1), 72–85.

Kelley, D. J., & Rice, M. P. (2002). Advantage beyond founding the strategic use of technologies. *Journal of Business Venturing, 17*(1), 41–57.

Keogh, W., Mulvie, A., & Cooper, S. (2005). The identification and application of knowledge capital within small firms. *Journal of Small Business and Enterprise Development, 12*(1), 76–91.

Klofsten, M., & Jones-Evans, D. (1996). Stimulation of technology-based small firms—A case study of university-industry cooperation. *Technovation, 16*(4), 187–193.

Koen, P. A., Ajamian, G. M., Boyce, S., Clamen, A., Fisher, E., Fountoulakis, S., et al. (2002). Fuzzy front end: Effective methods, tools and techniques. In P. Belliveau, A. Griffin, & S. Somermeyer (Eds.), *The PDMA toolbook 1 for new product development* (Chapter 1). New York: Wiley.

Levy, M., Loebbecke, C., & Powell, P. (2003). SMEs, co-opetition and knowledge sharing: The role of information systems. *European Journal of Information Systems, 12*(1), 3–17.

Malecki, E. J. (1997). *Technology and economic development: The dynamics of local, regional and national competitiveness* (2nd ed.). Boston, MA, USA: Addison-Wesley Longman Publishing Co., Inc.

Malone, P., Mazzarol, T., & Reboud, S. (2015). Understanding commercialization in entrepreneurial firms: A case example. In *International Council of Small Business (ICSB) 60th Annual Conference 2015*, 6–9 June, Dubai, UAE.

Mazzarol, T., Clark, D., Gough, N, Olson, P., & Reboud, S. (2014). Commercialisation in SMEs: Case studies from Australia, New Zealand and the United States. In B. Kotey, T. Mazzarol, D. Clark, D. Foley, & T. McKeown (Eds.), *Meeting the globalisation challenge: Smart and innovative SMEs in a globally competitive environment.* Melbourne: Tilde University Press.

Mohr, J. (2001). *Marketing of high-technology products and innovations.* Englewood Cliffs, NJ: Prentice-Hall.

Nonaka, I., & Takeuchi, I. (1995). *The knowledge creating company: How Japanese companies create the dynamics of innovation.* New York/Oxford: Oxford University Press.

Nunes, M. B., Annansingh, F., Eaglestone, B., & Wakefield, R. (2006). Knowledge management issues in knowledge-intensive SMEs. *Journal of Documentation, 62*(1), 101–119.

OECD. (2006). *The SME financing gap: Theory and evidence* (Vol. 1). Paris: Organisation for Economic Co-operation and Development. www.oecd.org/bookshop.

OECD. (2010). *High-growth enterprises: What governments can do to make a difference*. Paris: Organisation for Economic Co-operation and Development.

OECD. (2016). *Financing SMEs and entrepreneurs 2016: An OECD scoreboard*. Paris: Organisation for Economic Co-operation and Development (OECD) Publishing.

OECD. (2018). *Promoting innovation in established SMEs*. Paris: Organisation for Economic Cooperation and Development (OECD). www.oecd.org.

Ozer, M. (2003). Process implications of the use of the internet in new product development: A conceptual analysis. *Industrial Marketing Management, 32*(6), 517–530.

Ozer, M. (2004). Managing the selection process for new product ideas. *Research Technology Management, 47*(4), 10–11.

Padula, G., Novelli, E., & Conti, R. (2015). SMEs inventive performance and profitability in the markets for technology. *Technovation, 41–42*(2015), 38–50.

Parida, V., Oghazi, P., & Ericson, Å. (2014). Realization of open innovation: A case study in the manufacturing industry. *Journal of Promotion Management, 20*(3), 372–389.

Parida, V., Westerberg, M., & Frishammar, J. L. (2012). Inbound open innovation activities in high-tech SMEs: The impact on innovation performance. *Journal of Small Business Management, 50*(2), 283–309.

Park, T., & Ryu, D. (2015). Drivers of technology commercialization and performance in SMEs. *Management Decision, 53*(2), 338–353.

Pellikka, J., & Lauronen, J. (2007). Fostering commercialization of innovation in small high technology firms. *International Journal of Technoentrepreneurship, 1*(1), 92–108.

Pellikka, J., & Virtanen, M. (2009). Problems of commercialisation in small technology-based firms. *International Journal of Entrepreneurship and Innovation Management, 9*(3), 267–284.

Perez-Araos, A., Barber, K. D., Munive-Hernandez, J. E., & Eldridge, S. (2007). Designing a knowledge management tool to support knowledge sharing networks. *Journal of Manufacturing Technology Management, 18*(2), 153–168.

Pillania, R. K. (2008). Strategic issues in knowledge management in small and medium enterprises. *Knowledge Management Research & Practice, 6*(4), 334–338.

Pillania, R. K. (2009). Demystifying knowledge management. *Business Strategy Series, 10*(2), 96–99.

Polanyi, M. (1962). *Personal knowledge: Towards a post critical philosophy*. London: Routledge.

Radzevicience, D. (2008). Developing small and medium enterprises using knowledge management frameworks. *Aslib Proceedings, 60*(6), 672–685.

Ries, E. (2011). *The lean start-up: How constant innovation creates radically successful businesses*. London: Portfolio Penguin Books.

Salojärvi, S., Furu, P., & Sveiby, K.-E. (2005). Knowledge management and growth in Finnish SMEs. *Journal of Knowledge Management, 9*(2), 103–122.

Sedighadell, S., & Kachquie, R. (2013). Managerial factors influencing success of new product development. *International Journal of Innovation Management, 17*(5), 1–23.

Shelton, R. (2001). Helping a small business owner to share knowledge. *Human Resource Development International, 4*(4), 429–450.

Shepherd, D., & Patzelt, H. (2017). *Trailblazing in entrepreneurship: Creating new paths for understanding the field*. Cham, Switzerland: Palgrave Macmillan.

Sparrow, J. (2001). Knowledge management in small firms. *Knowledge and Process Management, 8*(1), 3–16.

Sparrow, J. (2005). Classification of different knowledge management development approaches of SMEs. *Knowledge Management Research & Practice, 3*(3), 136–145.

Theyel, N. (2013). Extending open innovation throughout the value chain by small and medium-sized manufacturers. *International Small Business Journal, 31*(3), 256–274.

van de Vrande, V., de Jong, J. P. J., Vanhaverbeke, W., & de Rochemont, M. (2009). Open innovation in SMEs: Trends, motives and management challenges. *Technovation, 29*(6), 423–437.

van Hemert, P., Nijkamp, P., & Masurel, E. (2013). From innovation to commercialization through networks and agglomerations: Analysis of sources of innovation, innovation capabilities and performance of Dutch SMEs. *The Annals of Regional Science, 50*(2), 425–452.

West, J., & Bogers, M. (2014). Leveraging external sources of innovation: A review of research on open innovation. *Journal of Product Innovation Management, 31*(4), 814–831.

West, J., Salter, A., Vanhaverbeke, W., & Chesbrough, H. (2014). Open innovation: The next decade. *Research Policy, 43*(5), 805–811.

Williams, R. L., & Bukowitz, W. R. (2001). The Yin and Yang of intellectual capital management: The impact of ownership on realizing value from intellectual capital. *Journal of Intellectual Capital, 2*(2), 96–108.

Wong, K. Y., & Aspinwall, E. (2004a). Characterizing knowledge management in the small business environment. *Journal of Knowledge Management, 8*(3), 44–61.

Wong, K. Y., & Aspinwall, E. (2004b). Knowledge management implementation frameworks: A review. *Knowledge and Process Management, 11*(2), 93–104.

Wong, K. Y., & Aspinwall, E. (2005). An empirical study of the important factors for knowledge-management adoption in the SME sector. *Journal of Knowledge Management, 9*(3), 64–82.

Wong, K. Y., & Aspinwall, E. (2006). Development of a knowledge management initiative and system: A case study. *Expert Systems with Applications, 30*(4), 633–641.

Wynarczyk, P., Piperopoulos, P., & Mcadam, M. (2013). Open innovation in small and medium-sized enterprises: An overview. *International Small Business Journal, 31*(3), 240–255.

Yahaya, S.-Y., & Nooh, A.-B. (2007). New product development management issues and decision-making approaches. *Management Decision, 45*(7), 1123–1142.

Yencken, J., & Gillin, M. (2006). A longitudinal comparative study of university research commercialisation performance: Australia, UK and USA. *Innovation, 8*(3), 214–227.

York, J., & Danes, J. (2014). Customer development, innovation and decision-making biases in the lean start-up. *Journal of Small Business Strategy, 24*(2), 21–39.

Zahra, S., & George, G. (2002). Absorptive capacity: A review, re-conceptualization and extension. *Academy of Management Review, 27*(2), 185–203.

Ziamou, P. L. (2002). Commercializing new technologies: Consumers' response to a new interface. *The Journal of Product Innovation Management, 19*(5), 365–374.

Peter Malone is an experienced entrepreneur currently completing doctoral studies at the University of Western Australia. His research focuses on developing a model of Entrepreneurial Innovation Value (EIV). He has 10 years' experience as a researcher and consultant in the fields of innovation and entrepreneurship and has successfully launched a number of start-up firms in technology industries. He has taught at the University of Western Australia, delivered presentations at Curtin University and the University of Notre Dame Australia on innovation and holds an MBA from the University of Western Australia and a Bachelor of Architecture from Curtin University. Peter has delivered several papers on EIV internationally and his research is at the forefront of understanding commercialisation value in entrepreneurial SMEs. Peter is the founder and Executive Chairman of Skin Elements Ltd, a start-up biotechnology company that has formed the basis of the case study presented in this edition. Peter is also involved in other innovative business ventures as a director and shareholder, and is the founding Chairman of the Commercialisation Studies Centre (CSC) Ltd., a not-for-profit organisation dedicated to enhancing knowledge and understanding of commercialisation across industry and the wider community.

Tim Mazzarol is a Winthrop Professor in Entrepreneurship, Innovation, Marketing and Strategy at the University of Western Australia and an Affiliate Professor with the Burgundy School of Business, France. He has around 20 years of experience of working with small entrepreneurial firms as well as large corporations and government agencies. This includes strategic management, marketing, and support to commercialisation. He is the author of several books on entrepreneurship, small business management, and innovation. His research has been widely published internationally. He holds a Ph.D. in Management and an MBA with distinction from Curtin University of Technology and a Bachelor of Arts with Honours from Murdoch University, Western Australia. Tim is also a Qualified Professional Market Researcher (QMPR) as recognised by the Australian Market and Social Research Society (AMSRS), director of the Centre for Entrepreneurial Management and Innovation (CEMI), coordinator of the Co-operative Enterprise Research Unit at UWA, and a founder director and Company Secretary for the Commercialisation Studies Centre (CSC) Ltd., a not-for-profit organisation dedicated to enhancing knowledge and understanding of commercialisation across industry and the wider community. He has co-authored several text books on small business management, and entrepreneurship and innovation.

Sophie Reboud is Professor of Strategy and Management of Innovation at the Burgundy School of Business, France and an Honorary Research Fellow at the University of Western Australia. She has over ten years of experience as a researcher and consultant in the field of management and strategy of small firms. Originally trained as an agronomist she served as a research engineer for École Nationale Supérieure des Mines de Paris for five years, and completed her Ph.D. there. Sophie's research interests include the strategic management of innovation and technology. This includes firms in the food sector and low-tech industries with specific focus on intellectual property and strategy in small firms. Sophie is an affiliate fellow with the Centre for Entrepreneurial Management and Innovation (CEMI), and a collaborative researcher with the UWA Co-operative Enterprise Research Unit. She has co-authored several text books on small business management, and entrepreneurship and innovation.

Chapter 20
Group Work: Application and Performance Effectiveness in Musical Ensembles

Phillip Cartwright and Kadeshah Swearing

Abstract This research deeps and broadens understanding of group work and the effectiveness of achieving learning and performance benefits from collaborative task-based activities. The focus is on the applicability of group work applied to music ensembles of mid- to large sizes. Research shows that in order to achieve working as a group, the group must pass through two phases: one, cooperation (where they agree to disagree) and two, collaboration (where they iron out their differences and disagree to agree). Based upon results and observations thus far, the key to transition from phase one to phase two is emotional intelligence, as this is the most important skill for a conductor-leader.

20.1 Introduction

The purpose of this chapter is twofold. First, it is intended to deepen and broaden the understanding of group work and the effectiveness of achieving learning and performance benefits from collaborative task-based activities. Studies of artistic organizations are useful for improving the knowledge about how those organizations operate and perform as well as informing other disciplines and organizations. Second, the chapter is focused specifically on the applicability of group work applied to music ensembles of mid- to large sizes.

The topics of group work, collaborative learning and performance within organizations, including educational and arts institutions, are broad and complex. These topics are important for understanding organizational models and human interactions. They have implications for artistic success determined on the basis of a combination aesthetic and commercial measures. Motivating sources for this chapter are the recent

P. Cartwright (✉)
PSB Paris School of Business, 59, rue Nationale, 75013 Paris, France
e-mail: phillip.cartwright@gmail.com

K. Swearing
University College of the Cayman Islands, 168 Olympic Way, PO Box 702, George Town, Grand Cayman, Cayman Islands
e-mail: KSwearing@ucci.edu.ky

© Springer Nature Switzerland AG 2020   329
N. Pfeffermann (ed.), *New Leadership in Strategy and Communication*,
https://doi.org/10.1007/978-3-030-19681-3_20

work by Swearing (2019), *An Investigation and An Exploration into the use of Group Work by Lecturers in Tertiary Institutions across the Caribbean: A Mixed Methods Approach* (Swearing 2019); the edited volume of Rink et al. (2017), *Musicians in the Making: Pathways to Creative Performance* in particular, the second part of that volume focused on creative processes; and Lewis' dissertation (2012), *The Incomplete Conductor: Theorizing the conductor's role in orchestral interpretation in the light of shared leadership practices.* Swearing (2019) describes trends and ramifications associated with group work in tertiary institutions. Rink et al. (2017) volume includes articles by Wise et al. (2017), Ginsborg (2017) and Cottrell (2017). These authors address behaviors and performance issues in solitary practice and work of both small and large ensembles, respectively. Finally, Lewis (2012) provides considerable insight on leadership strategies supported from her perspective as a practicing conductor.

Broadly, this research falls under the meta-subject of the creative process. The academic literature is replete with articulate descriptions of the creative process. For example, in an effort to integrate the engineering design and cognitive psychology literature, Howard et al. (2008, p. 167) argue that "… the generic creative process model…contains the three stages of analysis, generation and evaluation".

A psychology-based interpretation of the creative process draws on divergent thinking which supports generating novel solutions to loosely defined problems (Gibson et al. 2009). Hagman (2005) sets out a narrower definition focused on musical practice and performance. According to Hagman,

> The creative process of the musician involves two phases, which are described below. The first, the *practice phase*, is psychologically dynamic and involves specific challenges to the musician's self-experience. It is during the practice stage that the musician is most creative (one exception to this might be the improvisations of jazz musicians). The second phase is that of the *public performance,* at which time the results of the creative effort are displayed. (Hagman 2005, p. 104)

Of course, taken in context, each of the many definitions claim validity. For this research, a simple definition will suffice. The creative process refers to a series of actions involving recognition of problems followed by the generation, development, and communication of new ideas.

Cohesiveness of the subject matter proposed for this work, thus falls under the banner of creative process and suggests logical relations between the subjects of group work, ensemble and performance. Following Swearing (2019), it is intended that group work be considered as a teaching strategy promoting academic achievement and socialization (Forslund and Chiriac 2012). It involves individuals who interact with one another and join together to achieve a common goal, and whose interactions are structured by a set of roles and norms (Johnson et al. 2014). This shared purpose provides a reason for members to invest and interact in the group, to share ideas and create a sense of unity.

The focus is primarily on music ensembles ranging in size from several member to full symphonies, although clearly, an ensemble might be as small as a duo.[1] According to the *Harvard Dictionary of Music* (2003) an ensemble is

A group of musicians performing together. One speaks (1) of a "good" or "bad" ensemble with reference to the balance and unification attained in the performance of a string quartet, etc.… *Harvard Dictionary of Music* (pp. 204–205)

Young and Coleman (1979, pp. 12–13) go a step further describing the musical ensemble as "…an unusual kind of social group whose mode of interaction involves a degree of intimacy and subtlety possibly not equaled by any other kind of group".

Finally, as to ensemble performance Keller et al. (2016, p. 2) submit, "… musical ensemble performance—as a social art form—entails multiple individuals pursuing shared AESTHETIC GOALS. These goals are typically realized through the nonverbal communication of information about MUSICAL STRUCTURE and EXPRESSIVE INTENTIONS to co-performers as well as audience members". Further, Ginsborg (2017, p. 156) makes clear the thought that *ensemble practice* and *performance* must be considered as the stages where collaboration related to technical, expressive and perhaps psychological challenges such as those of recognition and communication can take place.

Following this Introduction, Sect. 20.2 sets out additional context for the research and key ideas deemed imperative for this chapter. Section 20.3 focuses on particular organization in education and music. Section 20.4 presents conclusions and recommendations.

20.2   Group Work and Ensembles in Theory

20.2.1   *Group Work*

Swearing (2019), describes group work as a process of enquiry and discovery in which all participants are prepared to engage themselves and each other in some type of activity. It is a process of enquiry and discovery in which all participants are prepared to engage; group work is also a shared endeavour towards resolving problems or circumstances felt by participants to be real or urgent (Preston-Shoot 2007). In an educational setting, group work promotes students' collaboration. This increases students' achievement, persistence, and attitudes toward science, as well as promotes cognitive restructuring that leads to learning (Wilson et al. 2018). Group work encourages brainstorming among its members which leads to a more informative experience when dealing with complex and unfamiliar problems.

---

[1] Solitary practice or performance is excluded for the creative process or creativity. In fact, solitary practice is shown to be important for advancing technical and interpretative aspects of performance as well as for ensemble skills (Ginsborg 2017, p. 143).

As matters of clarification, first, in group work, there is a distinction that should be made due to a misconception between working *in* a group and working *as* a group. Students can form a group, be working individually as a part of the group, but there is no coming together for them to be able to work together as a group and this pretty much defeats the purpose of giving group work in the first place.

Second, Chiriac and Granström (2012) state that there are two approaches to learning in a group: cooperative learning and collaborative learning. While there is no universal definition for cooperative learning, it is mostly focused on ensuring that participants are working together (Davidson and Major 2014). This kind of learning is very similar to jigsaw puzzles where each piece fits a certain area of the whole picture or each person is working on an individual section. Sometimes there is absolutely no need for interaction between members of the group. Once each person completes his or her section and in a timely manner then the group assignment will be completed. Usually in this type of learning, one participant's output is not an input to any other participant, and so, participants can work individually and separately without actually working together. At the end of the day, however, each piece of the jigsaw must be placed together to create the big picture so there must be some kind of collaboration. This level of collaboration is minimal, which is why there is a preference for collaborative learning.

Collaborative learning, which is the co-construction of knowledge shared among members of the group, is the ultimate goal (Schoor et al. 2015). Therefore, collaborative learning focuses on working with each other toward the same goal. It is working toward the discovering, understanding, or production of knowledge as a whole (Davidson and Major 2014). It is akin to the cogs in a machine; the pieces are interconnected with each other, with their combined roles aiming for one specific purpose. Collaborative learning puts a greater emphasis on the process. It allows for appreciation of the input from others. It helps participants to deal with diversity and be able to accept that there are multiple perspectives or ideas. Undoubtedly, however, these two classifications of cooperative and collaborative learning constitute for the practice known as group work.

Roles exist within groups. Possibly the most distinct role an individual can have within a group is being the leader. In general, a group leader is an individual who is put in charge of the group and is responsible for motivating, managing and directing the group towards the agreed upon goals and objectives (Saylor Academy 2012). Leaders tend to have a very definite effect on groups they belong to. As important as their position is, so is their effect on the group. Often, groups with defined leaders in position tend to function at the pace of the leader—it is the leader that guides and shows initiative firstly among the group members. The potential of leaders to communicate and create a sense of *shared identity* is an important determinant of the probability that their efforts to energize, direct, and instill particular work-related behaviors within the group will be successful (Ellemers et al. 2004). If a group leader is efficient in directing and leading the others, the group is most likely to succeed in their common goal. However, if the group leader is inefficient, then the group may experience problems that will be detrimental to the success of the team.

The literature on leadership theories and models is vast. Ardichvili and Manderscheid (2008) have highlighted predominant contemporary theories: leader–member exchange (LMX) theory, situational leadership, transformational leadership, servant leadership, authentic leadership, and complexity theory. In summary, based upon Ardichvili and Manderscheid (2008, pp. 622–624), one can consider the following:

1. LMX theory or vertical dyad linkage (VDL) theory evolved from the examination of relationships between leader and followers. LMX theory contends that leaders develop separate "in-group" and "out-group" relationships with direct reports. A leader's relationship with their direct reports is viewed as a series of dyadic links. Direct reports become part of the in-group or the out-group based on how well they work with the leader and how well the leader works with them.
2. Situational leadership was developed by Hershey and Blanchard (1969), and is premised on the idea that different situations call for different types of leader action.
3. Transformational leadership postulates that there are three types of leadership behaviors: transformational leadership, transactional leadership, and laissez-faire leadership.

   a. Transactional leadership focuses on the exchanges of favors that occur between leaders and followers, and on reward or punishment for good or poor performance.
   b. Laissez-faire is a passive form of managing people, basically an attempt to leave people to their own devices and is, thus, considered to be "impoverished" leadership.
   c. Transformational leadership is characterized by the process whereby a leader engages with others and creates a connection that raises the level of motivation, commitment, and morality in both the leader and the follower.

4. Servant leadership is based on the distribution of power to followers. When leaders subscribe to stewardship or servant leadership principles, they work to serve their followers for the purpose of achieving organization objectives.
5. Authentic leadership relates to the leader's self-awareness, self-regulation, as well as moral perspective, and high ethical standards that guide decision-making and behavior. It emphasizes greater self-awareness and self-regulated positive behaviors on the part of leaders and associates.
6. Complexity theory considers leadership as part of a dynamic and evolving pattern of behaviors and complex interactions among various organizational players. The emphasis is placed on the interactions of people in organizations and the creation of patterns of behavior.

Autocratic leadership has been excluded, not because it is irrelevant, e.g., Cottrell's reference to Solti and Toscanini (Cottrell, p. 194), but because it is not a preferred leadership type among prevalent scholars in the field, nor most likely, musicians. Today the most likely models for ensemble or orchestral leadership are transactional or transformational leadership (Cottrell 2017, p. 194). Recall that transactional leadership refers to management practices focused on specific tasks or objectives, the achievement of which are tied to recognition and rewards. Transactional management is not typically recognized as appropriate for organizations where innovation or artistic achievements are prioritized.

Transformational leadership is identified by four characteristics: *idealized influence*, i.e., vision building that relates to the organization's goals; *inspirational motivation* wherein leaders communicate high expectations and important purposes; *intellectual stimulation* that promotes intelligence, rationality, and problem solving; and *individualized consideration* wherein leaders have a heightened sensitivity to individual followers (Robbins and Judge 2016; Bodenhausen and Curtis 2016; Lodders and Meijers 2017). Various literature has explored the effects and reliability of transformational leadership in organizations, but the presence of transformational leadership in an orchestra will be one focus of the study. Researcher Atik (1994) presented three stages of transformational interaction, namely:

1. the testing phase;
2. the working relationship;
3. the transformational stage that includes more sharing of responsibilities and diminished hierarchical boundaries.

Atik expressed, however, that these stages are not universally accepted by orchestral players. He claimed that while the musical community condemns the now "extinct" autocratic leadership style, they continue to want for the "strong and directive figure" who will tell and guide them toward the specific actions they need to execute in the minimum amount of time (Atik 1994). His point of conclusion regarding transformational leadership style, wherein he presented that it is a more difficult approach (i.e., superior-subordinate relationship is subject to quick changes; hierarchical boundaries are less fixed; control is not straightforward; participation issues, and delegation and decision-making lead to uncertainty) seemed troubling. However, when comparing the transformational leadership style with the charismatic leadership style, Atik stated:

> The one clear distinction between the two leadership styles is that follower autonomy, independence and, therefore, future development of the player is a conscious part of the leadership style, whether for the individual performance or for the long term. This would appear to confirm Bass's (1985) suggestion that, "the transformational leader has a developmental orientation towards followers". Thus there appears to be a greater emphasis on making music with players in the transformational, as compared to the personalized charismatic style. (Atik 1994, p. 27)

With this statement, Atik pointed out that there appears to be a greater emphasis on the transformational than the charismatic leadership style. In addition, the overall conclusion that he offered suggests that an interactive and dynamic perception of the relationship between superior and subordinate can make for a more effective style of leadership. His study also noted that there are possibilities of positive consequences of gradual relinquishment of control for the collective good. These points can qualify as characteristics of how a successful transformational leadership can be good for the orchestra in the long-run.

In the context of Boerner and Freiherr von Streit's (2005) study, the relationship between a transformational leadership style and the group mood or climate of an orchestra is examined. Boerner et al.'s research showed that a conductor's transformational leadership style in an orchestra typically, on its own, does not increase the artistic quality of a symphony orchestra. However, it is notable that if the transformational leadership style is accompanied by a highly cooperative group climate, then the result is an enhanced artistic quality. As such, it can be argued that the success of a transformational leader in an orchestra is dependent on the cooperative group climate of its symphony (Boerner and Freiherr von Streit 2005).

Not considered by Ardichvili and Manderscheid (2008) is the theory of distributed leadership. While not entirely independent of other theories or models, Lewis (2012) argues convincingly for its relevance in consideration of music ensembles. According to Bennett et al. (2003, p. 3), "Distributed leadership is not something 'done' by an individual 'to' others, or a set of individual actions through which people contribute to an organization…[it] is a group activity that works through and within relationships, rather than individual action".

Distributed leadership has been defined and re-defined by a number of researchers over several decades. It has also been compared to and sometimes interchanged with leadership-related contents, such as 'shared', 'collaborative', 'collective', 'emergent', and 'democratic', to name a few. Some researchers define distributed leadership as the practice of school leadership (Gronn 2002; Spillane et al. 2001, 2004). In particular, Gronn (2002) argued that a more expanded unit of analysis, wherein this unit of analysis may possibly encompass patterns, would better serve students and practitioners of leadership or varieties of distributed leadership (Gronn 2002). Furthermore, Spillane et al. (2004) constructed a framework of distributed leadership with the findings as follows:

1. It proposes theoretical grounding for studying day-to-day leadership practice, which then allows investigations of practice to more than just documenting lists of strategies that leaders use in their work;
2. It considers that leadership activity, at the level of the school, is the appropriate unit of analysis in studying leadership practice, rather than focusing at the level of an individual leader or a small group of leaders;
3. It identifies an "integrative model for thinking about the relations between the work of leaders and their social, material, and symbolic situation, one in which situation is a defining element in leadership practice";
4. And lastly, the distributed leadership perspective also indicates the need for more complex approaches to studying the expertise of leaders, in particular wherein expertise is not simply a function of a leader's thinking and mental schemata.

Note that numbers one, three, and four have implications for musical ensembles. Spillane (2005) also notes that distributed leadership is more about leadership practice rather than leaders or their roles, functions, routines, and structures; it is also a system of practice that consists of a collection of interacting components, such as leaders, followers, and situation (Spillane 2005). It can also be democratic or autocratic.

Researchers Fitzsimons et al. (2011) further explore distributed leadership (as well as shared leadership) by examining the existing literature on the concept. Their research aimed to highlight the gaps in the literature and to challenge and clarify existing theory (Fitzsimons et al. 2011). They suggest that because researchers claim that these leadership practices are emerging due to challenges, more attention must be paid to the systematic affect that such learning demands (Fitzsimons et al. 2011).

Additionally, Bolden (2011) notably observes three factors relating to distributed leadership—power and influence, organizational boundaries and context, and ethics and diversity—as well as three methodological and developmental challenges, namely ontology, research methods, and leadership development, reward and recognition (Bolden 2011). He further concludes that descriptive and normative perspectives must have, in addition, more critical accounts that recognize the "significance of DL in (re)constructing leader–follower identities, mobilizing collective engagement and challenging or reinforcing traditional forms of organization" (Bolden 2011). Yet, Bolden also notes that attempting to provide a definitive definition of distributed leadership would fail to capture the complexity of the field and may potentially foreclose a series of ongoing debates and discussions about it.

The distributed leadership model also provides a greater opportunity for the conductors to be able to become transformational leaders as it offers them more mobility in their decision-making processes. However, for these benefits to manifest, it is imperative that a sense of the distributive justice climate (DJC) is established. Distributive justice climate can be defined as the shared views of fairness that rewards and resources are distributed, which is an important indicator of how effective a group will be (Whitman et al. 2012). Thus, some groups, such as teams, tend to prefer leaders who share rewards in proportion to effort and productivity. Furthermore, it is vital to take note of not only the social but also the transitional aspects when encouraging successful completion of tasks (Cogliser et al. 2009).

Distributed leadership is not the same as the traditional sense of leadership not only because it is shared, but also because it focuses on the skills, traits and behaviors of the leaders. It consists of five (5) dimensions: context; culture; change; relationships; and activity, based on the context that external and internal influences are needed for change. Based on a study conducted by Jones et al. (2012), it was discovered that the following factors relied on academics, executive, and professional staff working in close collaboration: the provision of professional development to encourage shared or distributed leadership, the resourcing for collaborative activities, and working conditions to support individual participation. These factors include values and practices such as "trust; respect; recognition; collaboration; and commitment to reflective practice; associated with personal behaviors that includes the ability to: consider self-in-relation to others; support social interactions; engage in dialogue through learning conversations; and the opportunity to grow as leaders through connecting with others" (Jones et al. 2012).

This study defines leadership as the ability to influence a group toward the achievement of a vision or a set of goals and objectives (Robbins and Judge 2016). Such influence can either be formal, wherein the role is assigned by an external higher-ranking authority (i.e. orchestral leaders who decide who should be the conductor) or informal, wherein an individual within the group rises up to the position. Group leaders are supposed to engage in three sets of activities that encourage and promote collaborative learning in teams and groups, namely: (a) developing, promoting, and maintaining a learning climate within the team; (b) helping members develop and use particular learning tools, both individually and collectively; and acting as learning partners for group members at different stages of group development, learning, and performance (Zaccaro et al. 2008). One of the responsibilities of a group leader is his or her ability to manage diverse and sometimes problematic motivations of individual members, which is often the key to a group's success (Baumeister et al. 2016).

20.2.2 Ensembles, Orchestras and Performance

Elaboration on all dimensions of group work and ensemble behavior is simply not possible in a single chapter, however, there are some elements that stand out from others. A review of the literature related to musical ensembles, organization and performance points to the fact that recognition of ensemble size is critical for understanding participant requirements including not only proficiency with an instrument, but coordination, communication, the role of the individual and social factors (Goodman 2002, p. 153). The importance of ensemble size is clarified by the articles of Wise et al. (2017), Ginsborg (2017) and Cottrell (2017). These authors address solitary practice, and work of small and large ensembles, respectively. This chapter primarily concerns ensemble sizes from multiple person ensembles such as chamber orchestras to full symphonies, so a brief summary of key points is warranted.

There are different points of entry to the understanding group work in ensembles. One can, for example, begin by studying the role of a conductor and leadership. This approach is interesting, but it does not encompass the whole of the 'big picture' and it can easily take attention away from the broader concern for ensemble or orchestra members. Further, conducting, as recognized today, has origins in the middle and late nineteenth century. Prior to that era, musicians had other ways of communicating and coordinating. The academic works of Atik (1994), and more recently, that of Hunt et al. (2004) are of interest. In the realm of popular writing, the very interesting book by Mauceri (2017) offers a personal perspective on conductor-orchestra relationships.

Ginsborg (2017) devotes attention to groups of musicians engaging in verbal and nonverbal communication as well as listening (Sicca 2000). At the outset of rehearsal or practice, verbal communication is likely to dominate (Seddon and Biasuttii 2009) giving way to nonverbal communication as social and musical relationships become understood. As to nonverbal communication, musicians signal using body movements, eye contact and gestures (Goebl and Palmer 2009; Williamon and Davidson 2002). Mauceri (2017, p. 19) makes clear the importance of gesturing in chamber-music and orchestra sections. All in all, these sources suggest a strong sense of teamwork.

Group work and teamwork are not interchangeable, yet the relationship between the two is important. Teamwork, as defined by Cordery (2005), is the extent to which the members are *"truly reliant on each other's actions"* and is not to be confused with group work—which does not require a high degree of interdependency unlike a team. Nonetheless, this methodology has some similarities to group work, and this can be seen when we look at the symphonies. Similarly to group work, teamwork requires that the members communicate as much as possible. As stated by Robbins and Judge (2016), in order for individuals to be connected in such a way that they are able to work effectively and efficiently, communication plays an important role. It fosters group cohesion which in turn increases productivity. However, Katzenbach et al. (2013) argue that the key to a group's success lies not within the content of the discussions that the group members have, but rather the manner in which its members communicate. According to a study conducted by them, the results revealed that a group's communication patterns were as important as all other factors such as intelligence, personality, and talent combined. To emphasize this, Katzenbach et al. (2013) state that the researchers involved in the study could even predict which groups would be successful versus which ones would not, based on their communication patterns, without meeting the members. Thus, it is then logical to ask, what conversation patterns should a group exhibit in order for it to be successful. According to their study, it is highly likely that a group will be successful if they exhibit these characteristics:

1. Everyone talks and listens in roughly equal measure, keeping contributions short and sweet.
2. Members face one another, and their conversations and gestures are energetic.
3. Members connect directly with one another—not just with the leader.
4. Members carry on back-channel or side conversations with the other members.

In essence, Katzenbach et al. (2013) is saying that a group will not simply do well because the group members are accomplished or smart, but rather because its members understand how to properly communicate with each other in a way that follows a successful communication pattern.

Group members need to know how to communicate effectively. A group is successfully communicating when the "group flow," is present. A group flow is a collective experience of fluid harmony that is similar to the individual experience of flow that psychologist Mihaly Csikszentmihalyi observed in creative individuals. "Group flow increases when people feel autonomy, competence and relatedness. Many studies have shown that team autonomy is the top predictor of team performance" and emerges under the following 10 "flow enabling conditions". In particular, Sawyer (2007) identifies:

1. The group's goal both emerges from and guides its interactions.
2. Group members listen to one another so closely that they can respond to new ideas or actions as though they knew about them in advance.
3. Group members concentrate. They often use symbols and slogans to create a sense of shared identity.

4. They control "their actions and their environments".
5. Individual performance becomes less important than the group's shared creation.
6. All members have equal say.
7. They take time to get to know each other, which enable them to make good decisions quickly and easily.
8. Group members communicate with each other constantly, although often informally.
9. They find ways to move forward. Bugs, glitches and failures don't stop the process, but instead become opportunities to come up with new ideas.
10. They allow time for failure. In the art world, this means rehearsing; in the business world, it means assuming some projects will fail and learning from them.

Matters of leadership cannot be ignored. Visual narrative and auditive narrative researcher Koivunen (2006) argues that there are two types of leadership present in symphony orchestras. The first is the visual leadership narrative that is closely connected to the traditional leadership literature cited in the previous section. The second, the auditive leadership narrative has much in common with participative approaches to leadership (Koivunen 2006).

The visual leadership narrative examines the discussions of leadership that posits possessive individualism as a main theme. Possessive individualism has two central epistemological themes. First, the assumption that the knowing individual is understood as an entity (i.e., individual's knowledge and access to their own mind content is perceived as properties of entities, or as individual possessions). Second, these individual possessions are the ultimate origins of the design and control of internal nature and of external nature, including other people or groups i.e., individuals get chosen as leaders *because* of their specific and superior characteristics or traits (Dachler and Hosking 1995).

Traditional leadership discussions place emphasis on the singular individual. In the context of an orchestra, the emphasis is on the conductor as the leader. As the most visible in a group, the individuality of a person stands out and unerringly draws attention to him or herself. Thus, the similarities of the traditional leadership and visual culture with respect to the conductor are how the visual leadership narrative is formed.

> Leadership literature is an ode to individuals, and vision is a sense of individuality as well. Vision isolates, distances and separates the viewer from the object in a similar manner the leader is separated and distanced from the subordinates. Due to the distance, the leader and the viewer are not closely affected by what happens to the objects. Both leadership literature and the visual primacy expect clear and permanent results that can be observed, rechecked and controlled. The culmination of this is the literature on visionary leadership that praises vision as the highest value. (Koivunen 2006, p. 3)

Aside from the visual narrative approach, Koivunen also posits the narrative of auditive leadership. She states that there is a lot less in common between the writings of auditive culture and leadership literature than it was for visual culture. Koivunen further states that in most cases of traditional leadership, the hierarchy of importance is set on the leader's message; literature on communication emphasizes the significance of ordering and explaining rather than receiving and listening (Koivunen 2006). However, the case is different for the observed literature on shared leadership. Shared leadership presents a better dynamic in that individuals relate more to each other. As an example, Dachler and Hosking (1995) present the partnership model. In the context of partnership, networking is meaningful as it gives voice to the varying perspectives of individuals in a group. It therefore encourages the possibility of negotiation, as well as recognizing and respecting differences as different, but equal.

As such, shared leadership, when paired with auditive culture, does make sense as the importance of *listening* is emphasized. In the context of an orchestra, it is doubly more important to focus on what the conductors and players hear. Synchronicity and harmony is key to the success of the group. It is important to note, however, that Koivunen presents these two observations not to compete with each other but to supplement. As the auditive aspect of things is usually ignored in favor of the more dominating sense (the visual), then it is necessary to consider it.

> Visual culture promotes individuality, distance and endurance while the auditive culture cherishes collectivity, exposure, unity and temporality. The concern here is not to replace the visual model with the auditive culture but merely to draw attention to auditive aspects and to reach a healthy balance where all senses are equally appreciated. (Koivunen 2006, p. 15)

Performance is by no means tangential to the current line of research, although it is a very broad subject matter itself. Performance connects performing artists to listeners, and to use the words of Keller et al. (2016, p. 2), connects the "musical structure" and "expressive intentions" of the performers to the audience. Implicitly, the performer(s) and audience must be distinct. Briefly, adapting Godlovitch's model, live performance can be considered as:

> a datable sound sequence (that is, sonic event), • immediately caused by some human (-like) being, • the immediate output of some musical instrument, • intended to be caused at a specified time and place, and in a specified manner, • the exercise of skilled activity, • the outcome of appropriately creditworthy physical skill, • an instance of some identifiable musical work, • intended as an instance of such a work, • successful as a constraint-model of such a work, • intended as a constraint-model of such a work, intended for some third-party listener, • presented before some third-party listener, • listened to by some third-party listener exercising active concentrated attention. (Godlovitch 1998, p. 50)

In the end, listeners respond to performance, initially by way of audible acceptance or displeasure followed by experience sharing to peers. In turn, depending on the level of success attached to performance, there arises a level of demand for future invitations and performances. Thus, arguably, the ensemble or orchestral organization is tied to overall effectiveness and success of performing organizations.[2] Success is a function of audience approval or disdain.

---

[2]One can argue that even a dysfunctional organization might achieve success for a period of time based upon factors such as celebrity or past reputation. Over time, surely below expectations performances will have deleterious outcomes.

20.3   Conductors, Ensembles, and Orchestras

Contemporary research on music ensembles and symphony orchestras date back to Westby (1960), Faulkner (1973), and Parasuraman and Nachman (1987). More recent contributions to this research area are to be found in the studies of Castañer (1997), Koivunen (2003) and Koivunen and Wennes (2011). The earlier work focused primarily on leadership models, while the latter work considers leadership from the perspective of aesthetics and psychology.[3] Many studies have concentrated on inter-relationships between orchestra members and the conductor-orchestra nexus. Hunt et al. (2004) examined the conductor-orchestra relationship as an example of a creative organization, while Marotto et al. (2007) studied collaborative leadership or collective virtuosity in organizations with an application to orchestras.[4] Other examples in this line of research are Khodyakov (2014) and Atik (1994). A prevalent theme is that leadership styles across musical ensembles and symphonies are varied, and in fact, most leadership paradigms can be successfully applied to the orchestral setting. Further, leadership roles are often unclear in orchestras (Atik 1994; Allmendinger and Hackman 1996).

Two research studies explore organizational culture and leadership of very different organizations. The first is the thesis of Piqueé (2015) and subsequent article by Furu et al. (2015) focusing on the Berlin Philharmonic. The second analyzes the organization of The Happy Roll Elastic Ensemble (HREE), a community music ensemble supported by Tainan Culture Centre in Taiwan (Hsiehand and Kao 2012).

Widely recognized as the world's most democratic symphony orchestra (Furu et al. 2015, p. 7), the Berlin Philharmonic is unique in that decisions concerning the hiring of conductors and new musicians is made by the orchestra itself. Leadership is distributed across the orchestra, which as indicated by Furu et al. (2015, p. 7) poses challenges for any organization, but is especially challenging for a symphony orchestra holding to high standards and a sonic tradition. The Berlin Philharmonic is interesting for study both from a performing arts and business perspective. For purposes of this research, the interest is in organization and leadership.

The democratic organization of the Berlin Philharmonic has highly meaningful implications not only for the conductor-orchestra relationship, but there are implications for the interaction between orchestra members as well. While the Berlin Philharmonic is not the only orchestra to have adopted such participatory values, it is certainly a leading organization in the music world to have done so. Clearly the organization takes pride in its values as indicated on its website.

---

[3]Specifically, this research takes a relational constructivist approach. Botella and Herrero (2000) considers relational constructivism in detail.

[4]Collective virtuosity refers to the aesthetic experience in a group transformed by its performance.

When the Berliner Philharmoniker elect a new chief conductor it is always a turning point for the musicians and their public. Take the conductors who held the post during the period after the Second World War, for example: Wilhelm Furtwängler, Herbert von Karajan, Claudio Abbado and Sir Simon Rattle – each of them a distinct personality who had a lasting impact on the orchestra in terms of interpretation and programming. Kirill Petrenko will become the orchestra's next chief conductor in summer 2019. Accessed from https://www.berliner-philharmoniker.de/en/titelgeschichten/20182019/petrenko/.

The vast research related to leadership makes clear that the effects of organizational climate can be seen in employee motivation, employee development and retention, and employee performance (Holbeche 2006). Furu et al. (2015) are quite detailed in describing the organizational-administrative aspects of the Berlin Philharmonic. Suffice it to state that members of the orchestra are in control of the hiring process, including the artistic director and general manager. A two-thirds majority vote is required for acceptance of a new member, and once accepted, a new member must pass a two-year trial period. The election of the artistic director only needs a majority of votes. The orchestra plans its turns of performance by section, and it is particularly noteworthy that musicians decide where they will sit in a section for a given performance. The effects of such self-organizing behavior are of interest to study, perhaps in the context of complexity theory and participatory nature of interactive processes (Stacey et al. 2000). What is particular value of considering the Berlin Philharmonic is its successful adaptation to principles discussed in the context of distributed leadership and group work.

The Happy Roll Elastic Ensemble (HREE) is a community music ensemble supported by Tainan Culture Centre in Taiwan. With enjoyment and friendship as its primary goals, it aims to facilitate the joys of ensemble playing and the spirit of social networking. This article highlights the key aspects of HREE's development in its first two years (2009–2011), including musical and administrative decisions and operations.

HREE experienced issues during the course of development and adaptive actions were seen as necessary. The key problems were musical as well as issues related to rehearsal and administration (Hsiehand and Kao, p. 53). The group was established as self-managed and membership required payment of an annual fee set at 500 New Taiwan Dollar (less than 20 US dollars). Member meetings were held, and an annual election was held for the administrative director and board members. Each member was assigned an activity group and a sectional rehearsal group. The leaders of these activity groups were responsible for reporting on member attendance and on matters related to the general administration of the group. The section leaders ran small group rehearsals.

Forming an ensemble was challenging. The conductor was responsible for music cognition and performance of music. The music arrangers were his supporters. An important responsibility on the part of all of the members was to be both players as well as listeners. Disputes over matters of opinion were discussed openly and with civility. In fact, the collegiality within the group was considered as imperative for growth and progress. In short, the authors likened the ensemble to a family and the group was quite successful at finding consensus.

Glen Inanga, a concert soloist pianist, conductor, and music professor, considers the conductor as the key component of the orchestra. He says:

> The person conducting them has all the information of music that they're reading in front of him or her. And what they do is that they are in charge of making sure that they realize the vision for how that music should sound as they interpret the music that they're all looking at. So that **conductor**, in a sense, is the **key person**. That affects what you hear coming out of the orchestra. And they can play the same song, but different conductors, and you get different products. Different sound, different speeds, different nuances come out because they're focusing on so many different things.

He further compares the position of the conductor to a designer, wherein the performance of an orchestra is the design. The design would be the same idea, but it will be executed in many different ways because the designer (or the conductor) is not limited to only one way. This presents originality to the same music piece played by different orchestras. From this, we can argue then that the conductor intimately affects the orchestra. Recall that Boerner and Freiherr von Streit's (2005) research presents the idea that a conductor, on its own, does not increase the artistic quality of a symphony orchestra; rather, the existence of a good mood climate is necessary in order to achieve an enhanced artistic quality. Still, it cannot be denied that the conductor plays an important part in ensuring that the orchestra's mood and climate is in peak condition. In fact, Glen states that:

> And a lot of the times the conductor is the one who basically is in charge — at that level, at that expert level. These people can all play; the people are all, themselves, masters, graduates, whatever it is… They're all performing professionals in their own capacity. And they're all experts in what they do. You can't teach them to play the instrument. They teach other people.

And, as a conductor:

> You have to prove yourself, that you understand what you're doing, and what you're trying to do. […] A lot of the times you have to prove to them what you're trying to do, or try to get all of them to believe that what you're trying to tell them to do actually works. And trust that you know what you're doing.

The conductor is in charge of ensuring that the orchestra members are meshing well and performing harmoniously. As such, it is also the conductor's responsibility that the mood or climate of the group is going well. Thus, it can be argued that the elevated artistic quality of the orchestra depends on the success of the conductor's leadership.

With a cohesive group, the orchestra reaches a synergy (also referred to as emergence) wherein their performance is elevated. But how does the conductor guarantee that a group is cohesive and unified? Recall the distinction between cooperation and collaboration in group work that we identified before, based on Swearing's (2019) research. When asked which one works for an orchestra, Glen replies:

> I think a combination of both. I'll tell you why. Because each of the parts has been written in an orchestral kind setting whereby one thing fits into the other specifically. Everything is tied to a particular point. You play for a certain amount of time, you shut up for a certain amount of time, et cetera. Everything has to be in its rightful place. It's like a jigsaw puzzle. You have to do your part very well. It doesn't make sense to do your part well and the next person doesn't, because it compromises the overall picture of the jigsaw puzzle. So everything fits

well. But… It's very hard sometimes to just stay in your little tiny place and play. Because a lot of the times, the music that you're playing relies on your listening to what's happening, and responding sometimes to what's happening, to the part next to you.

He further elaborated, saying:

It fits like a glove in terms of… It's quite organic in that sense. So if you have somebody who is just telling you what to do, the music will sound very nicely put together, but in terms of the synergies and that sideways integration of the material, you will not get that, because you will lose out in that component. **The music will still sound good, but it will not be taken to the next level**. […] There are two ways of playing that same piece: you can either play exactly what's written on your part, and I play exactly what's written on my part. Once you've got that jigsaw puzzle, you can leave it there and people will be happy with the product, but you can take things to the next level whereby you are gluing things together and responding in a very organic way to what you're hearing from the other person. And then the music has a life of its own that shows vision in the sense that it's taken to the next level.

From his answer, we are then presented with the view that group cohesion, or achieving successful harmony in an orchestra, cannot be contained to either just collaboration or cooperation. In fact, we can argue that there are two levels to group work. The first level is cooperation, which we have already established beforehand. Here, as Swearing (2019) puts it, the members tend to 'agree to disagree' when faced with conflict. Glen's answer provides us with the meaningful insight that in the case of a symphony or an orchestra, musicians *can* play cooperatively—by which that means, they can play solely by themselves during their turn and still contribute to the output as a whole. They can play their own part without caring for the others. However, in order to reach the next level, as Glen puts it, the performers need to work together by "gluing things together and responding in a very organic way to what you're hearing from the other person". This statement then supports the collaborative side of group work, which is where the 'disagree to agree' aspect comes in. In order to achieve outstanding harmony, these players must collaborate with one another. As such, the types of group work applicable to an orchestra are therefore sequential. The first phase is cooperation, and then the group transitions to the second phase, which is collaboration. For a group to function to effectively and efficiently, collaboration is the ideal phase to be in. So the question then becomes: how do we transition from cooperation to collaboration? The answer is emotional intelligence.

The most important skill for a conductor to have is **emotional intelligence**. In an orchestra, you're not just managing people as is the case with regular groups. You're trying to influence expert people, which means that you have to manage people and their egos.

Thus, the use of emotional intelligence is a very important aspect of distributed leadership. Before, it was usually thought that the most important skill that a leader should have is mastery of whatever he or she is managing. However, over the years, the qualifications of leaders have developed far beyond just their skills or knowledge base. They must possess emotional intelligence, as it is particularly important. This is in line with Swearing's (2019) findings on the importance of emotional intelligence in group work. According to Goleman (1995), emotional intelligence is defined as "being able to rein in emotional impulse, to read another person's innermost feelings, and to handle relationships correctly and smoothly" (Goleman 1995).

Contrary to what people commonly understand, emotional intelligence is not restricted to "being nice" but knowing *when* to be nice and when to choose not to. Individuals who have the ability to sense and understand the emotions and moods of other people will make handling diversity in groups a little less challenging, and give them the ability to work with various people no matter their background (Swearing 2019). Since an orchestra is typically made up of a very diverse cast, effectively managing diversity at the level of masters is an important skill that a conductor should have. According to Glen, one of the most common problems of an orchestra conductor is getting the performers to see their vision. He gives an example of playing a very famous piece, Beethoven's Fifth Symphony, and explains that for a lot of musicians, this piece has been played for about "five hundred times". The real struggle comes because these expert musicians have their own way of doing and playing the piece, and so might question the conductor on his or her way of conducting that same piece. It also presents the challenge of getting these players in sync because each has played the piece in his or her own way. This is in line with Anderson's (2008) research about how shared leadership, despite being the most supportive amongst leadership models, can cause a power struggle between both sides (Anderson 2008). In this case, the performers and the conductors are all professionals. The performers tend to know the pieces that they're playing, which may sometimes lead to them being defensive when they are being instructed by someone who might not even specialize in their instrument. Glen explains that,

> Some of them [conductors] might feel insecure. It's quite intimidating for them because they're conducting people who know that piece inside out, and have played it probably even better than the conductors.

Thus, we can gather that emotional intelligence is what bridges the gap between cooperation and collaboration.

20.4   Conclusions

This chapter as focused upon group work and gains from collaborative initiatives. The applications of interest are musical ensembles. It has been argued that even though a group is two or more individuals, if two or more individuals are working on a task, this does not necessarily constitute group work as there is a difference between working *in* a group and working *as* a group. The focus of this research is to get musicians to work *as* a group. The most important part of any group is the group leader as research shows that the success of a group is dependent on its leader. The literature points toward distributed leadership as the most suitable type of leadership for conductors or ensemble leaders. Distributed leadership is a group activity that works through and within relationships, rather than individual action.

The research effort behind this chapter shows that in order to achieve working as a group, the group must pass through two phases: one, cooperation (where they agree to disagree) and two, collaboration (where they iron out their differences and disagree to agree). Based upon results and observations thus far, the key to transition from phase one to phase two is emotional intelligence, as this is the most important skill for a conductor-leader.

Bibliography

Allmendinger, J., & Hackman, J. R. (1996). Organizations in changing environments: The case of East German symphony orchestras. *Administrative Science Quarterly, 41,* 337–369.

Allmendinger, J., Hackman, J. R., & Lehman, E. V. (1996). Life and work in symphony orchestras. *The Musical Quarterly, 80*(2), 194–219.

Amit, R., & Zott, C. (2001). Value creation in e-business. *Strategic Management Journal, 22*(6/7), 493.

Anderson, K. D. (2008). Transformational teacher leadership: Decentring the search for transformational leadership. *Management in Education, 2*(2), 111.

Ardichvili, A., & Manderscheid, S. V. (2008). Emerging practices in leadership development, an introduction. *Advances in Developing Human Resources, 10*(5), 619–631.

Atik, Y. (1994). The conductor and the orchestra. *Leadership and Organization Development Journal, 15*(1), 22–28.

Baregheh, A., Rowley, J., & Sambrook, S. (2009). Towards a multidisciplinary definition of innovation. *Management Decision, 47*(8), 1323–1339.

Bartel, C. A., & Garud, R. (2009). The role of narratives in sustaining organizational innovation. *Organization Science, 20*(1), 107–117.

Bass, B. M. (1985). Leadership and performance beyond expectations. New York: Free Press.

Baumeister, R. F., Ainsworth, S. E., & Vohs, K. D. (2016). Are groups more or less than the sum of their members? The moderating role of individual identification. *Behavioral and Brain Sciences, 39,* 1–56.

Bennett, N., Wise, C., Woods, P. A., & Harvey, J. A. (2003). *Distributed leadership.* Nottingham: National College of School Leadership.

Berliner Philharmoniker. (2019). *The Berliner Philharmoniker online.* Last accessed March 2, 2019 at https://www.berliner-philharmoniker.de/en/.

Bettis, R. A., & Prahalad, C. K. (1995). The dominant logic: Retrospective and extension. *Strategic Management Journal, 16*(1), 5–14.

Blackwell, A., Phaal, R., Eppler, M. J., & Crilly, N. (2008). Strategy roadmaps: New forms, new practices. Paper presented at the Diagrams 2008. In *Fifth International Conference on the Theory and Application of Diagrams*, Munich.

Bodenhausen, C., & Curtis, C. (2016). Transformational leadership and employee involvement: Perspectives from millennial workforce entrants. *Journal of Quality Assurance in Hospitality & Tourism, 17*(3), 371–387.

Boerner, S. (2007). Promoting orchestral performance: The interplay between musicians' mood and a conductor's leadership style. *Philosophy of Music, 139.*

Boerner, S., & Freiherr von Streit, C. (2005). Transformational leadership and group climate: Empirical results from symphony orchestras. *Journal of Leadership and Organizational Studies, 12*(3), 31–41.

Bolden, R. (2011). Distributed leadership in organizations: A review of theory and research. *International Journal of Management Reviews, 13*(3), 251–269.

Botella, L., & Herrero, O. (2000). A relational constructivist approach to narrative therapy. *European Journal of Psychotherapy, Counselling and Health, 3*(3), 407–418.

Bresciani, S., & Eppler, M. J. (2009). The benefits of synchronous collaborative information visualization: Evidence from an experimental evaluation. *IEEE Transactions on Visualization and Computer Graphics, 15*(6), 1073–1080.

Carlile, P. R. (2002). A pragmatic view of knowledge and boundaries: Boundary objects in new product development. *Organization Science, 13*(4), 442–445.

Castañer, X. (1997). The tension between artistic leaders and management in arts organizations: The case of the Barcelona symphony orchestra. In M. Fitzgibbon & A. Kelly (Eds.), *From maestro to manager. Critical issues in arts and culture management* (pp. 379–416). Dublin: Oak Tree Press.

Chiriac, E. H., & Granström, K. (2012). Teachers' leadership and students' experience of group work. *Teachers and Teaching: Theory and Practice, 18*(3), 345–363.

Cogliser, C. C., Schriesheim, C. A., Scandura, T. A., & Gardner, W. L. (2009). Balance in leader and follower perceptions of leader–member exchange: Relationships with performance and work attitudes. *The Leadership Quarterly, 20*(3), 452–465.

Cohen, E. G., & Rachel, L. A. (2014). *Designing groupwork: Strategies for the heterogeneous classroom* (3rd ed.). New York: Teachers College Press.

Cordery, J. (2004). Another Case of the Emperor's New Clothes? *Journal of Occupational and Orgnizational Psychology, 77,* 481–484.

Cordery, J. (2005). Team work. In D. Holman, T. D. Wall, C. W. Clegg, P. Sparrow, & A. Howard (Eds.), *The essentials of the new workplace: A guide to the human impact of modern working practices* (pp. 99–110). Chichester: John Wiley.

Cottrell, S. (2017). The creative work of large ensembles. In J. Rink, H. Gaunt, & A. Williamon (Eds.), *Musicians in the making: Pathways to creative performance* (pp. 186–205). Oxford: Oxford University Press.

Dachler H.P., & Hosking D.M. (1995). *The primacy of relations in socially constructing organizational realities*. Avebury, Aldershot.

Davidson, N., & Major, C. H. (2014). Boundary crossings: Cooperative learning, collaborative learning, and problem-based learning. *Journal on Excellence in College Teaching, 25*(3–4), 7–55.

Deiglmayr, A., & Schalk, L. (2015). Weak versus strong knowledge interdependence: A comparison of two rationales for distributing information among learners in collaborative learning settings. *Learning and Instruction, 40,* 69–78.

Ellemers, N., De Gilder, D., & Haslam, S. A. (2004). Motivating individuals and groups at work: A social identity perspective on leadership and group performance. *Academy of Management Review, 29*(3), 459–478.

Faulkner, R. (1973). Orchestra interaction. Some features of communication and authority in an artistic organization. *The Sociological Quarterly, 14,* 147–157.

Fitzsimons, D., James, K. T., & Denyer, D. (2011). Alternative approaches for studying shared and distributed leadership. *International Journal of Management Review, 13*(3), 313–328.

Forslund, F., & Chirac, E. H. (2012). Group work management in the classroom. *Scandinavian Journal of Educational Research, 56*(5), 1–13.

Furu, P., Piqué, M. M., & Reckhenrich, J. (2015). Simon Rattle and the Berlin Philharmonic: Co-creating leadership and organizational culture. Last accessed February 23, 2019 at https://www.researchgate.net/publication/303285560_Simon_Rattle_and_the_Berlin_Philharmonic_Co-creating_leadership_and_organizational_culture.

Gibson, G., Folley, B. S., & Park, S. (2009). Enhanced divergent thinking and creativity in musicians: A behavioral and near-infrared spectroscopy study. *Brain and Cognition, 69*(1), 162–169.

Ginsborg, J. (2017). Small ensembles in rehearsal. In J. Rink, H. Gaunt, & A. Williamon (Eds.), *Musicians in the making: Pathways to creative performance* (pp. 164–184). Oxford: Oxford University Press.

Godlovitch, S. (1998). *Musical performance: A philosophical study*. London: Routledge.

Goebl, W., & Palmer, C. (2009). Synchronization of timing and motion among performing musicians. *Music Perception, 34*(3), 427–438.

Goleman, D. (1995). *Emotional intelligence*. New York: Bantam.

Goodman, E. (2002). Ensemble performance. In J. Rink (Ed.), *Musical performance: A guide to understanding* (pp. 153–167). Cambridge: Cambridge University Press.

Gronn, P. (2002). Distributed leadership as a unit of analysis. *The Leadership Quarterly, 13*(4), 423–451.

Hagman, G. A. (2005). The musician and the creative process. *Journal of the American Academy of Psychoanalysis and Dynamic Psychiatry, 33*(1), 97–117.

Hershey, P., & Blanchard, K. H. (1969). *Management of organizational behavior.* Englewood Cliffs, NJ: Prentice Hall.

Holbeche, L. (2006). *Understanding change: Theory, implementation and success.* Burlington: Butterworth-Heinemann.

Howard, T. J., Culley, S. J., & Dekoninck, E. (2008). Describing the creative design process by the integration of engineering design and cognitive psychology literature. *Design Studies, 29,* 160–180.

Hsiehand, Y.-M., & Kao, K.-C. (2012). Joys of community ensemble playing: The case of the happy roll elastic ensemble. *International Journal of Community Music, 5*(1), 45–57.

Hunt, J. G., Stelluto, E., & Hooijberg, R. (2004). Toward new-wave organization creativity: Beyond romance and analogy in the relationship between orchestra-conductor leadership and musician creativity. *The Leadership Quarterly, 15*(1), 145–162.

Johnson, D. W., Johnson, R. T., & Smith, K. A. (2014). Cooperative learning: Improving university instruction by basing practice on validated theory. *Journal on Excellence in University Teaching, 25*(3–4), 85–118.

Jones, S., Lefoe, G., Harvey, M., & Ryland, K. (2012). Distributed leadership: A collaborative framework for academics, executives and professionals in higher education. *Journal of Higher Education Policy and Management, 34*(1), 67–78

Katzenbach, J., Gratton, L., Harvard Business Review, & Eisenhardt, M. K. (2013). In *HBR'S 10 must reads on teams.* Harvard Business School Publishing Corporation. Last accessed March 11, 2019 at https://learning.oreilly.com/library/view/hbrs-10-must/9781422191460/Text/04_Copyright_Page.html#page_iv.

Katzenbach, J. R., & Smith, D. K. (2013). *HBR's 10 must reads on teams* (*with featured article The discipline of teams*). Harvard Business School Publishing Corporation. Last accessed March 11, 2019 at https://hbr.org/product/hbrs-10-must-reads-on-teams-with-featured-article-the-discipline-of-teams-by-jon-r-katzenbach-and-douglas-k-smith/11365E-KND-ENG

Keller, P. E., Novembre, G., & Loehr, J. (2016). Musical ensemble performance: Representing self, other and joint outcomes. In S. Obhi & E. S. Cross (Eds.), *Shared representations: Sensorimotor foundations of social life* (pp. 280–310). Cambridge: Cambridge University Press. Last accessed February 10, 2019 at http://www.jdloehr.net/publications/KellerNovembreLoehrInPress.pdf.

Keppell, M., O'Dwyer, C., Lyon, B., & Childs, M. (2010). Transforming distance education curricula through distributive leadership. *Research in Learning Technology, 18*(3), 165–178.

Kerr, N. L., & Tindale, R. S. (2004). Group performance and decisionmaking. *Annual Review of Psychology, 55*(1), 623–655.

Khodyakov, D. (2014). Getting in tune: A qualitative analysis of guest conductor-musicians relationships in symphony orchestras. *Poetics: Journal of Empirical Research on Culture, the Media and the Arts, 44,* 64–83.

Kirrane, M., O'Connor, C., Dunne, A. M., & Moriarty, P. (2017). Intragroup processes and teamwork within a successful chamber choir. *Music Education Research, 19*(4), 357–370.

Koivunen, N. (2003). *Leadership in Symphony Orchestras. Discursive and Aesthetic Practices.* Tampere: Tampere University Press.

Koivunen, N. (2006). Auditive leadership culture: Lessons from symphony orchestras. *Advances in Organization Studies, 16,* 91.

Koivunen, N., & Wennes, G. (2011). Show us the sound! Aesthetic leadership of symphony orchestra conductors. *Leadership, 7*(1), 51–71.

Lewis, L. A. (2012). *The incomplete conductor: Theorizing the conductor's role in orchestral interpretation in the light of shared leadership practices.* Dissertation presented to the faculty of the Royal Holloway College, University of London in partial fulfillment of the requirements for the Ph.D. degree in Music.

Lodders, N., & Meijers, F. (2017). Collective learning, transformational leadership and new forms of careers guidance in universities. *British Journal of Guidance and Counselling, 45*(5), 532–546.

Marotto, M., Roos, J., & Victor, B. (2007). Collective virtuosity in organizations: A study of peak performance in an orchestra. *Journal of Management Studies, 44*(3), 388–413.

Mauceri, J. (2017). *Maestros and their music, the art and alchemy of conducting.* New York: Alfred A. Knopf.

Parasuraman, S., & Nachman, S. A. (1987). Correlation of organization and professional commitment: The case of musicians in symphony orchestras. *Group and Organisation Studies, 12*(3), 287–303.

Piqueé, M. M. (2015). *The Berlin Philharmonic; culture and leadership.* Thesis submitted in partial fulfillment of requirement for the programme in Arts Management, University of the Arts Helsinki, Sibelius Academy, Helsinki, Finland.

Preston-Shoot, M. (2007) *Group dynamics. The construct.* Last accessed March 11, 2019 at http://www.siliconyogi.com/andreas/l/The%20Construct/GroupDynamics.html.

Randel, D. M. (Ed.). (2003). *Harvard dictionary of music* (4th ed.). Cambridge: The Belknap Press of the Harvard University Press.

Rink, J., Gaunt, H., & Williamon, A. (Eds.). (2017). *Musicians in the making: Pathways to creative performance.* New York: Oxford University Press.

Robbins, S. P., & Judge, T. A. (2016). *Essentials of organizational behavior* (13th ed.). New Jersey: Pearson Education.

Sawyer, R. K. (2006). Group creativity: Musical performance and collaboration. *Psychology of Music, 34*(2), 148–165.

Sawyer, R. K. (2007). *Group genius: The creative power of collaboration.* New York: Basic Books.

Saylor Academy. (2012). Last accessed March 11, 2019 at https://saylordotorg.github.io/text_organizational-behavior-v1.1/s13-02-group-dynamics.html.

Schoor, C., Narciss, S., & Körndle, H. (2015). Regulation during cooperative and collaborative learning: A theory-based review of terms and concepts. *Educational Psychologist, 50*(2), 97–119.

Seddon, F., & Biasuttii, M. (2009). A comparison of modes of communication between members of a string quartet and a jazz sextet. *Psychology of Music, 37*(4), 395–415.

Sicca, L. M. (2000). Chamber music and organization theory: Some typical organizational phenomena seen under the microscope. *Studies in Cultures, Organizations and Societies, 6*(2), 145–168.

Spillane, J. P. (2005). Distributed leadership. *The Educational Forum, 69*(2), 143–150.

Spillane, J.P., Halverson, R. & Diamond, J.B. (2001).*'Towards a theory of leadership practice: A distributed perspective', Institute for policy research working article.* Northwestern University.

Spillane, J. P., Halverson, R., & Diamond, J. B. (2004). Towards a theory of leadership practice: A distributed perspective. *Journal of Curriculum Studies, 36*(1), 3–34.

Stacey, R., Griffin, D., & Shaw, P. (2000). *Complexity and management fad or radical challenge to systems thinking?.* London: Routledge.

Swearing, K. (2019). *An investigation and an exploration into the use of group work by lecturers in tertiary institutions across the Caribbean: A mixed methods approach.* Submitted to the Faculty of PSB Paris School of Business in partial fulfillment of the requirements for the degree of Doctorate in Business Administration.

Westby, D. L. (1960). The career experience of the symphony musicians. *Social Forces, 38,* 223–230.

Whitman, D. S., Caleo, S., Carpenter, N., & Horner, M. T. (2012). Fairness at the collective level: A meta-analytic examination of the consequences and boundary conditions of organizational justice climate. *Journal of Applied Psychology, 97*(4), 776–791.

Williamon, A. & Davidson, J. W. (2002). Exploring co-performer communication. *Musicae Scientiae, 6*(1), 53–72.

Wilson, K. J., Brickman, P., & Brame, C. J. (2018). Group work. *CBE-Life Sciences Education, 17*(1), 1–5.

Wise, K., James, M., & Rink, J. (2017). Performers in the practice room. In J. Rink, H. Gaunt, & A. Williamon (Eds.), *Musicians in the making: Pathways to creative performance* (pp. 141–163). Oxford: Oxford University Press.

Young, V. M., & Coleman, A. M. (1979). Some psychological processes in string quartets. *Psychology of Music, 7*(1), 12–18.

Zaccaro, S. J., Ely, K., & Shuffler, M. (2008). The leader's role in group learning. In V. L. Sessa & M. London (Eds.), *Group work learning: Understanding, improving & assessing how groups learning in organizations* (pp. 193–214). New York: Lawrence Erlbaum Associates.

Prof. Phillip Cartwright holds a Ph.D. degree from the Univ of Illinois, Urbana-Champaign. He is an economist with thirty years of academic and executive experience. He is Professor of Economics at PSB Paris School of Business and Visiting Researcher at the Royal College of Music, London. He focuses on artists' skills and attributes as determinants of success, innovation and creativity. He has served as senior researcher at the University of Georgia, Imperial College, and INSEAD. He pursues advanced music studies at Berklee College of Music, and as Founder and CEO, HorizonVU Music LLC, he participates in the music business as a member of The Recording Academy, Audio Engineering Society, Royal Musical Association, American Society of Composers, Authors, and Publishers (ASCAP).

Dr. Kadeshah Swearing is an Assistant Professor at the University College of the Cayman Islands. She holds a Doctorate in Business Administration (DBA), a Masters in Economics with concentrations in Financial Economics and Game Theory, Post-Graduate Certificates in Credit Risk, Portfolio Management and Market Risk and Middle Office Management, a Bachelors Degree in Mathematics and she is a trained teacher of Mathematics and Computer Studies and Complimentary Literature. She is a Chartered Economist (ChE) and currently pursuing the Chartered Financial Analyst (CFA) designation where she has cleared level 1. Dr. Swearing has approximately fifteen years of teaching experience. Prior tot UCCI, she lectured at both the University of Technology, Jamaica and The University of the West Indies (Mona).

Chapter 21
Facilitating Communication in Adaptive Planning Processes for Inclusive Innovation: Discussing an Integrative Approach

Bernardo Alayza and Domingo Gonzalez

Abstract In this chapter, it is argued that linear approaches have influenced innovation strategies for local development, having limitations for addressing long-standing socio-economic problems in unequal and exclusionary societies such as Peru. However, the facilitation of a broader communication perspective could contribute to configure more inclusive innovation processes for a number of reasons: it can contribute to understand the nature of inclusive innovation processes; it can serve to recognise entry points for inclusive innovation as well as to reorient innovation processes towards inclusivity. This discussion allows a rethinking of the current innovation strategies for local development and proposes a set of recommendations to enable communication with adaptive planning processes for a broad-based inclusive development strategy in the Peruvian context.

21.1 Introduction

The important contribution of innovation to economic growth, development and social welfare is recognised throughout the world. Over the past decade, the Peruvian government has been implementing a set of innovation policies, programs and proyects as part of a national strategy to address long-standing socio-economic problems (Ismodes 2006; Villaran 2010; OECD 2011; Bazán et al. 2014; Kuramoto 2014; European Commission 2014; Ismodes and Manrique 2016).

Innovation strategies for local development in Peru are mainly focused on the promotion of innovation through linear communication understandings that are mostly based on a technology transfer approach (see CONCYTEC 2016). Despite some positive advances in the identification, recognition and support of local technological innovations, these initiatives have limitations in addressing socio-economic problems

B. Alayza (✉) · D. Gonzalez
Pontifical Catholic University of Peru (PUCP), Lima, Peru
e-mail: b.alayza@pucp.edu.pe

© Springer Nature Switzerland AG 2020   351
N. Pfeffermann (ed.), *New Leadership in Strategy and Communication*,
https://doi.org/10.1007/978-3-030-19681-3_21

such as unequal access to public services and the exclusion of rural communities and socially vulnerable groups[1] (Bazán et al. 2014).

Johnson and Andersen (2012) argue that more inclusive development strategies will generate more inclusive innovation processes, in which excluded groups may have the opportunity to shape their future by interacting with other relevant stakeholder groups (p. 10). Inclusive innovation is defined, then, as a process in which historically excluded groups may have the opportunity to participate in innovation processes oriented to their local development (IDRC 2011; Cozzens and Sutz 2014; Bazán et al. 2014; Dutrénit and Sutz 2014; Heeks et al. 2014; Joseph 2014; Schillo and Robinson 2017).

Broader strategies for inclusive innovation that question linear technological assumptions have contributed to the creation of new local networks, new forms knowledge and new technologies oriented to transform exclusionary societies (Fressoli et al. 2014; Smith et al. 2014). In this context, the facilitation of communication become central not only to disseminate innovations but also to create environments for configuring inclusive innovation processes (Alayza 2017).

While changes into unequal, complex and uncertain societies require broad-based socio and technical strategies (Geels 2004; De Melo 2014), adaptive planning processes may permit to design well-strategized communication interventions that can give some insights for configuring innovation for inclusive development.

This chapter presents a review of the conceptualisations of communication in innovation in order to discuss the role of communication for inclusive innovation. This allows proposing broad-based strategies, in which adaptive planning processes could instrumentalise communication in order to understand, recognise and reorient actions towards to this aims.

21.2   Interpretation of Innovation for Local Development

Innovation is defined as a process through which new products, services, processes, organisational methods and practices are used by people in a given context (BID 2010; World Bank 2010). Strategies for promoting innovation in Latin America and principally in Peru are mainly focused on a technology transfer approach (e.g. CONCYTEC 2016), having multiples limitations for reaching local development.[2]

Since its early use in the 1970s, technology transfer has been interpreted as the action of transmitting ideas, information, knowledge and technology between

---

[1]Excluded groups are defined as groups or subgroups that lives under an unequal distribution of a valued good or basic service such as health, food, water and/or energy. For example, people who live in isolated rural villages in Peru or people who live in peri-urban zones of Peru that are lacking a basic service, live with less than 3.5 American dollars and have an informal job could be considered as excluded.

[2]Local development is defined as the interactive process of expansion of capabilities and collective liberties in a territory for the reduction of disparities and the participative role of the excluded populations in their development (Sen 2000).

academia to the business and government sector (Rogers 2002). This logic has also been extended to diverse programs and projects promoted by governments and NGOs to support agricultural, food and water in order to improve local practices with the adoption of new technologies (Vanclay and Leach 2011; Coutts and Roberts 2011).

Technology transfer strategies have been influenced by the linear notion of innovation that mostly conceptualises innovation as a process of discovery associated with scientific inventions in isolated places such as laboratories, and assume that the benefits of transference of technologies would transform society as a whole (see Latour 1987). The ability of technology transfer as one of the main strategies to promote innovation for local development has been debated by several scholars, proposing more participative and constructive strategy that need to align and configure different socio and technical aspects for the generation, sustain and expansion of innovation in a territory (Smits and Kuhlmann 2004; Wieczorek et al. 2012; Thomas et al. 2012; Fressoli et al. 2013; Leeuwis and Aarts 2016; Harman 2018).

Broader notions of innovation sustain that innovation emerges from a systemic process generated by the multiple interactions of technological, social and institutional aspects (Smits 2002). In this systemic and co-evolutionary view, businesses, universities, governments and users participate in the creation of innovative process according to the demands and needs of a local and global marketplace (Freeman 1995; Lundvall 1992; Nelson 1993). This perspective also suggests that several components within innovation processes co-evolve, and in every interaction, there is an alignment of diverse processes in which social and technological changes are produced at almost the same time (Geels 2002, 2004).

Adopting a similar argument under a socio-technical framework, Thomas (2009) explains in a philosophical way that society is technologically configured as technologies are socially constructed (p. 15). Thus, innovation is not only ruled by economic, social or technological separate actions, but is configured by societal and technical components that influence each other through every interaction (Bijker et al. 1987).

The socio-technical perspective and the systemic and co-evolutionary notion of innovation have conceptual similarities, but most importantly, they challenge linear technological conceptions that motivate particularly technology transfer strategies for innovation; and propose a broader understanding of innovation in which the relationships and interactions (i.e., communication) are the pillars of innovation processes.

21.3 Interpretations of Communication in Innovation

Communication is highly relevant and influential in innovation studies (See Rogers 2003; Leeuwis 2004a, b; Zerfass and Huck 2007; Hülsmann and Pfeffermann 2011; Pfeffermann et al. 2013; Leeuwis and Aarts 2016; Pfeffermann and Gould 2017). In Peru, communication in innovation has been mostly interpreted as a process of transference in strategies that promote local development (See CONCYTEC 2016). Nevertheless, there are broader conceptualisations that give more coherent communication views for implementing broad-based strategies in order to support innovation effectively. Over time, the notion of communication has evolved during the last century to broader conceptualisations. Although there is not a comprehensive chronology in the literature showing how ideas of communication have been interpreted in innovation studies (Mattelart and Mattelart 2003), there are three interconnected views of communication that can help to understand how communication has been interpreted and implemented in innovation strategies for local development.

21.3.1　The One-Way Idea of Communication in Innovation

The one-way idea of communication is a linear understanding which is the most basic and instrumentalist conception of communication. It recognises that communication is produced by senders and receivers who exchange messages through several communicational channels, in which the "noise" or the "interference" may cause failure in the communication between individuals. This idea emphasises the message emitted by a sender, which can be seen as a linear understanding of communication (see Berlo 1960).

The linear understanding of communication has similar principles to the linear ideas of innovation. Thus, linear concepts of innovation interpret communication with a separation between the "scientific" and "real" world, in which communication may connect those separate spheres (see Latour 1987). From this perspective, communication is seen as the intermediary between science and society, or as the intermediary between the "the natural world" (i.e., laboratories or scientific spheres) and the "social world" (i.e., the society). Here, existing information is transferred from one side to the other, principally from scientists to the population (Latour 1987). In other words, it is assumed that existing messages (e.g. technologies) are created in a natural world (separate from society) and their mere transference may generate an impact in the social world.

For example, in rural communities of Peru, one of the long-held assumptions in the implementation of innovation strategies is that there are individuals who know or possess knowledge (i.e., scientists, experts or scholars) and there are other individuals who do not possess such knowledge and must need to acquire that knowledge (i.e., the rural population). Thus, this notion emphasises the transmission from the individuals who are considered that "know" and relegate the individuals who are the receivers, which are considered "vulgar" (which constitutes the origins of the word "divulgation") or lacking skills or knowledge (Alfaro 2006).

Experiences of technology transfer processes in Latin America and Peru oriented to local development show that the dissemination of pre-conceived ideas and technologies has not achieved the desired results concerning local development (see Box 1). These strategies have been promoting with a little engagement of the users, generating in most of the cases some contradictory effects (Escobal et al. 2012; Herrera 2011; Fressoli et al. 2013; Harman 2018). Such processes have mostly been unsuccessful because they did not promote any room for negotiation (Fressoli et al.

2013). Furthermore, most of them have failed because they have replicated linear and top-down orientations, in where the potential beneficiaries are considered passive actors (Harman 2018).

The linear interpretation of communication in the field of innovation in Latin America has centred on spreading technologies, knowledge or information rather than on creating interactions between diverse groups to facilitate or improve social practices. This reduces communication in the transmission process where the possibility of a modification in the message is not recognised, and therefore, not integrated into the process; reducing the important interactions that are beneficial for the generation and expansion of local innovation processes (Alayza 2017).

> **Box 1:** *The Technology Transfer Fair for Development: The Experience of an Innovation Policy Instrument in Peru*
> Local fairs are cultural, economic and social events that promote the traditional values and customs of the regions, cities, towns or villages. In Peru, it is a widespread practice for fairs to be organised in the productive, livestock and agricultural sectors.
>
> Placing the local fair custom in the framework of national innovation policies, local authorities the academia, and the local business sector organised the Technology Transfer Fair for Development (TTFD) as a policy instrument that seeks to create environments and incentives to identify, recognise and support local technological innovations that address socio-economic problems at the regional level.
>
> In Cusco,[3] Peru, since 2008, the TTFD supported more than 330 potential innovations from medium and low-income communities and villages through promotion and display at the fair, and more than 90 local innovators were given monetary and technical assistance in the areas of agribusiness, metalworking and handicrafts. In addition, during the last decade, the TTFD has attracted high levels of attendance from the local population, with attendances of more than 10,000 people, and received significant media coverage at local, regional and national levels. The pedal-operated fodder shredder and the brick production fan are interesting initiatives identified and supported by the TTFD with technical and instrumental business plans that reached unexpected and some contradictory results (see Boxes 2 and 3).

---

[3]Cusco is a region in the southeastern Sierra of Peru located at 3300–4500 m above sea level. It has a population of roughly 1.2 million inhabitants. Cusco has a millenary tradition, having an impressive cultural heritage that came from the Inca culture. However, it is one of the poorest regions in Peru with more than 60% of people living in poverty. Also, more than 80% of the local population is involved in informal activities with very low economic productivity, with a monthly average income of approximately US$85 per capita (INEI 2015).

The complexity of each proposal supported by the TTFD, which implied the participation of heterogeneous actors, high uncertainty to extend those initiatives and the turbulent and unequal context in where they run, were not sufficiently backing and articulated by the actors that organised the TTFD. In fact, it can be affirmed that the influence of the TTFD ideas of technology transfer has generated some effects that can be seen as contradictory effects to their participants.

Taking into consideration that the majority of proposals supported by the TTFD belonged to or were oriented towards low and middle-income groups, the technology transfer approach—an integral part of the name, brand and actions of the TTFD—presented many drawbacks to recognise and strengthen important aspects that can configure inclusive innovation processes.

While new local networks of heterogeneous actors have been built as a consequence of the TTFD, strategies were focused on support only specific problems, mostly technical, rather than on creating spaces for sharing knowledge, strengthen local capabilities, and open new opportunities that could allow to create and extend inclusive innovation processes according to the local need in the region.

*This case illustration is based on Alayza (2017).

21.3.2 The Persuasive Ideas of Communication in Innovation

A better understanding of the linear ideas of communication has led to more persuasive interpretations. The subjective model interprets communication as the relationship between senders and receivers, in which alternative interpretations, previous knowledge, and experiences are part of the process of communication (Dervin 1983). This improved understanding of communication has generated more emphasis on comprehending how the receiver creates messages in order to develop tactics and strategies that promote behavioural changes and the adoption of new practices (Mattelart and Mattelart 2003; Leeuwis 2004a, b; Alfaro 2006).

This persuasive concept of communication has oriented communication actions to generate stimuli, responses and meanings through communicational intervention, media channels and campaigns. This concept of communication assumes that the quality of the stimulus and message will determine the success of the communication (Leeuwis 2004a). Although these ideas of communication have adopted demographic and sociological concepts to understand human relationships, there is always an emphasis on persuasion for disseminating ideas, practices and technologies to change behaviour (Beltran 2005).

In the field of innovation studies, the subjective model of communication has been operationalised in the diffusion of innovation theory. This theory orients the actions of communication to the adoption of technologies, practices and knowledge. Rogers (1962, 2003) proposed the diffusion of innovation theory to promote the adoption of products in the agricultural sector, and these ideas have been extended to diverse spheres of the development world (i.e., development projects, programs and movements). From that perspective, the practice of communication has been oriented as the way to disseminate products and also as the way to convince and persuade potential users to adopt a product or technology.

The diffusion of innovation theory has been reviewed and improved since its creation in the 1960s. New concepts such as social networks, decision-making mechanisms and relationships with the environment have been added to discuss communication more broadly. Despite changes and contributions in the field, one of the main weaknesses of this conception is that it continues to assume that innovations (i.e., ideas, knowledge, messages and technologies) have in themselves a positive association for the adopters, and does not deal with issues such as the way power is exercised behind innovation processes that may be negative for deprived and poor groups (Mattelart and Mattelart 2003).

This practice of communication is a significant evolution in comparison to the linear model of communication. However, it has limited effects because efforts have been focused mainly on singular changes through diffusion (e.g., the introduction of a new technology or practice) rather than a constructive and collective way for change that must have more relevance deal with unequal societies.

Box 2: *Why Technology Transfer Approach Influenced the Not Expansion of the Pedal-Operated Fodder Shredder into Other Peruvian Villages?*
The majority of people who live in rural villages in Peru are small farmers dedicated to agriculture and livestock-raising. Both practices are complementary because the residues of the crop production, with suitable treatment and cutting, can serve to feed the local cattle (including cows, sheep and llamas). This generates better production and, in turn, better incomes. However, small farmers waste around 40% of fodder material because of inefficient practices, reducing the possibility of obtaining better prices for selling their livestock.

Understanding this problem, a professor at a local university (i.e. the innovator) worked with a group of small farmers to create technology in the form of a pedal-operated fodder shredder with the aim to reduce the amount of waste. The machine cuts fodder material into usable pieces without wasting too much material. This technology operates using human energy, and it is adaptable to farmers' conditions at an altitude of over 3900 m above sea level. In comparison with similar products, the fodder shredder has several features that suit the local conditions. It is portable (it has wheels), it operates with mechanical energy (it has pedals to generate energy), it is relatively cheap (it costs roughly five times less than conventional machines), it is easy to repair, and the materials to make it can be sourced locally.

The adaptation of this new technology in combination with social and technical aspects in some rural villages generated an inclusive innovation process for the following three reasons. First, it was created as a new technology oriented to meet a local necessity of small farmers who live under the poverty line (i.e., people who earn approximately US$3.50 per day). Second, the process involved the generation of a new network (comprising the innovator, the university, small farmers, local authorities and municipalities) which enabled the participation of different actors in diverse stages in the innovation process such as the R&D process, the implementation and use, and the expansion of new practices and technologies. Third, the new institutional arrangements and new actions facilitated an improvement in the management of livestock-raising, creating new opportunities.

Due to the success of the process, the demand for the fodder shredder increased considerably in surrounding villages. This led to the participation of the innovator in the TTFD to scaling up their innovation. As a strategy for expansion, the TTFD encouraged the innovator to conduct a technology transfer process to a local enterprise. After this process, the expansion of the inclusive innovation process was interrupted, and later on, this initiative stopped.

Based on the situation described for the fodder shredder, two aspects can be highlighted for the thwarted expansion of the technology's inclusive innovation processes to other villages. First, the strategy of generating relationships and agreements changed in favour of a technical strategy fostered by the TTFD. The previous interactive R&D process adopted by the innovator not only contributed to the creation of new technology, but it also allowed the innovator to gain an in-depth understanding of local practices and needs and thereby to develop a better strategy for adapting and expanding the technology. The strategy was based on the mobilisation of diverse actors who were related to the problem, not only generating interest in the technology but also establishing relationships with the actors involved. Relationships were forged based on respect for local traditions, creating the foundation for an interactive social learning process between the innovator and the potential users.

The local enterprise that later took responsibility for producing the technology did not replicate those actions. It used some communicational tools to show the advantages of the technology, but its main emphasis was on disseminating the product rather than mobilising diverse actors to create new opportunities for local farmers. Second, the technology transfer process to pass the technical information to the local enterprise was not adequate. The technology transfer actions undertaken by the innovator did not convey all the complexity that would be involved in facilitating and extending the inclusive innovation processes.

The technology transfer from the innovator to the local enterprise was limited to the technical aspects, taking for granted the other actions that were vital for the adaptation of the technology in rural villages. Yet the passage of time proved the strategies of the local enterprise to be limited considering the expectations of the rural farmers and in comparison with the previous example set by the innovator in forming relationships with other actors.

The case of the fodder shredder suggests that, rather than facilitating an inclusive innovation process, the technology transfer strategy promoted by the TTFD undermined the expansion of this inclusive innovation process. By comparison, a people-centred, inclusive technology adaptation and application approach would have been adequate to forge a local strategy in the rural communities that had a need for the innovation.

This case illustration is based on Alayza (2017).

21.3.3 The Construction Notion of Communication in Innovation

A broader interpretation of communication is the social construction model, which proposes that communication must be understood in the context of the interrelationships of the protagonists that requires time and involves a constant process of renegotiation (Hajer and Laws 2006). This concept of communication has its roots in a constructivist understanding in which the protagonists of communication construct meaning in their multiple interactions (Te Molder and Potter 2005).

This idea of communication takes into consideration the multiple interpretations of relationships that allow individuals to generate dialogue focusing on shared experiences, facilitating new ways of understanding themselves, other people and their reality (Leeuwis and Aarts 2011). This permits to interpret communication as the form to create new discourses and actions towards a social, political, cultural and humanistic change (Alfaro 2006; Dutta 2012).

Understanding that innovation implies change in the status quo that occurs by the constant co-evolution of technological and societal aspects in relation to each other in a territory (Smits 2002), communication is central to configuring innovation because it can facilitate the alignment of discourses and actions for the construction of new agreements, networks, and capabilities that enables innovation (Leeuwis and Aarts 2011, 2016).

While there is a lack of opportunities to deal with local structural problems-mostly for excluded groups in Latin America and Peru- the facilitation of communication can open new opportunities, in which multiple actors can participate in the configuration of more inclusive innovation processes for local development.

21.4 Communication for Inclusive Innovation

The debate about how to generate innovation that allows new opportunities in unequal structures challenge traditional frameworks by proposing more participative and broader concepts create broad-based inclusive strategies. Reflecting on the Latin American context, barriers exist that prevent local needs being met through fair redistribution and lack of opportunities. This situation limits citizens' access to essential services and infrastructure and reduces their ability to fully exercise their rights (Dutrénit and Sutz 2014).

Innovation could be understood as inclusive because of a number of intercorrelated factors. First, new technologies, ideas, processes, goods and/or services focus on overcoming local problems of deprived and excluded groups. The nature of these processes encourages the participation of excluded groups in the different stages of the innovation process (Fressoli et al. 2014; Smith et al. 2014); generating social learning processes and new opportunities for negotiate change (Dutrénit and Sutz 2014; Papaioannou 2014). Then, the combination of these factors encourages the further generation of new opportunities for social and technical transformation, especially within excluded groups (Johnson and Andersen 2012; Foster and Heeks 2013; Chataway et al. 2014; Dutrénit and Sutz 2014).

In this regards, communication can enable more inclusive innovation processes due to multiple actor interactions that allow the alignment of discourses and actions towards their local development (Alayza 2017). Thus, communication as a strategy that seeks inclusive innovation can provide a deeper understanding of how stakeholders frame problems, make negotiations, take decisions, construct networks, shape behaviours, exercise power and generate processes of inclusion and exclusion (Thomas et al. 2012; Heeks et al. 2014; Leeuwis and Aarts 2016; Schillo and Robinson 2017).

Inclusive innovation for local development can be forged with planning processes motivating the engagements of actors to reconfigure a better future (Andersen and Andersen 2014, 2017). Nevertheless, change generated by innovation is not rigid; therefore, change cannot be simply and linear planned (Leeuwis and Aarts 2011). This is why it is pertinent to discuss adaptive approaches that enable more participative planning processes to configure inclusive innovation for local development.

21.5 Adaptive Planning Approach

Adaptive planning is focused on generating flexible plans that allow reconstructing models of reality for a suitable and realistic social and technical change. From a constructivist perspective, adaptive planning is oriented on re-building a better future, which is possible when multiple reconfigurations and readjustments occur at the social and the technical level, involving the participation of various parties that dialogue, negotiate and complement diverse perspectives based on their models of reality.

The work of Emery and Trist (1973), Trist (1976a, b, c), Ackoff (1974) and Ozbekhan (1973) in the early 70s, contributed to shifting ideas of rigid planning processes from seeing a more participative and adaptive perspective that implies changes at a social and technical level.

The philosophy of adaptive planning partly on the assumption that changes need to deal with elements such as uncertainty, complex problems, and the participation of distinct protagonist (Emery and Trist 1965; González 1997; Burns et al. 1983; De Melo 2014).

In adaptive planning processes, the active participation of multiple actors can facilitate the constant evaluation and reorientation of actions to achieve changes at organisational and inter-organisational level (De Melo 1985). Nevertheless, progressive's processes for re-evaluation, decision-making, learning made by the actors' interactions in the planning process are even more relevant than reaching specific goals because it can lay the foundations for a suitable and consistent change (Babüroglu and Ravn 1992).

Thus, adapting planning processes can be understood as a continuous social learning, characterised by flexibility, dynamic adaptation, and by the constant evaluation of the carried out actions, which allows making more compatible with the interactions between different actors that seeking innovation (De Melo 2014).

In this regards, when multiple actors work together on resolving complex problems or undertaking innovative initiatives for social change, non-synoptic (i.e. that does not follow a schematic and rigid course) and adaptive planning processes may motivate to accommodate agreements between different kind of actors in order to configure innovation according to local capabilities and opportunities in a territory (González and De Melo 2004).

While adapting planning actions require the active participation of the protagonist of change, communication for inclusive innovation may give some insights in order to open new possibilities for change in unequal structures, contributing to understanding the roots of the problems for innovation, identifying key players for configuring innovation and opening spaces for dialogue in order to reorient actions toward a desirable social and technical transformation.

Box 3: *How the Interruption of an Adaptive Planning Process Stopped the Expansion of the Brick-Making Business?*

The artisanal manufacture of bricks constitutes a significant problem in Cusco, Peru because it generates high levels of contamination in the environment and is detrimental to human health. The artisanal brick-making business is an informal activity in Peru because it is not covered by government regulations. The majority of people involved in the small-scale brick-making business in Cusco are low-income families, and even entire villages, who live in precarious conditions in peri-urban zones in the region.

Artisanal bricks are produced outdoors using traditional ovens that burn the clay in order to create small units of building material for local commercialisation. The brick-making production generates high contamination from the materials used as fuel including wood, plastics, rubber tires, textiles and even rubbish. The high emission of carbon monoxide affects surrounding villages and the brick manufacturers themselves. The methods and technologies used, combined with the high altitude of Cusco, make it difficult for the ovens to reach sufficient temperatures to produce bricks. This situation obliges brick manufacturers to use a significant amount of material to generate energy. It also requires them to be exposed for long hours to noxious smoke, damaging their health considerably.

Understanding this problem, a local innovator, with the support of a research centre from a local university, an international NGO and the active participation of a brick producers' village, created a ventilation system to improve the combustion in the artisanal ovens for more efficient brick production. The use of the fan for artisanal brick production improves the energy efficiency, accelerating the time required to reach the optimal temperatures to fire clay, reducing the working time for the brick manufacturers, and most importantly, reducing the emission of carbon monoxide by 70%. In comparison with other similar products, the fan technology can operate at high altitudes (over 3500 m above sea level) and under low temperatures, making it suitable for Cusco conditions.

The constant interrelations between the actors through adaptive planning processes that were agreed according to each interest, expectation and resources permitted to configure an inclusive innovation process, in which a new technology in combination with societal (social, economic, institutional and political aspects) was adapted to the context of small villages in Cusco. This was possible because it was built a new brick production network comprise by different groups such as the innovator, a local university, village's brick producers, and local authorities. This brick production network was undertaken adaptive actions that facilitated the participation of various actors in different stages of the innovation process such as the R&D process, the implementation, the dissemination, the generation of new local business and the creation of other complementary actions such as roundtables and open events for dialogue. Finally, it was implemented new policies, new institutional arrangements and new methods to fire the clay led to cleaner and more efficient brick-making production in some poor villages.

After the expansion of the inclusive innovation process in some villages in Cusco, the innovator participated in the TTFD and received an award. As part of the award, the organisers of the TTFD brought in some technology-based companies from overseas to participate in the technology transfer process for some local innovations including the brick production fan. The involvement of the overseas companies served to train the local innovator; on the other hand, it encouraged these companies to bring their technological offerings into the Cusco market, undermining the future expansion of the inclusive innovation process for the brick production fan into other villages with similar character-istics.

Analysis of this case indicates two main reasons why the inclusive inno-vation process for the brick production fan was not expanded. First, commer-cial perceptions prevailed over a participative and adaptive process that has undermined the extension of inclusive innovation processes. The commercial opportunities taken up by foreign enterprises undermined the mobilisation of new actors regarding this local problem. This meant that the previous adaptive planning actions of building networks with excluded brick producers and other relevant actors were replaced for individual actions. Although more technolo-gies were inserted into the local market, which could be seen as a positive in terms of improving the technological offerings in the region, this stopped diverse actions that were configured as an inclusive innovation process in local villages such as dialogue roundtables, new training processes and new policy regulations. This also reduced the opportunities to replicate adaptive planning actions oriented to improve brick production in other small-scale brick pro-ducer villages. Second, the authorities did not realise all the implications of supporting local inclusive innovation processes. While their intention was to improve the local brick production through the transfer of technologies from a foreign enterprise to local innovators, they did not see that this may backfire and create unfair competition for local innovators. In this regard, the lack of protection mechanisms makes local innovators vulnerable to powerful com-petitors.

In light of this example, it can be suggested that actions promoted by the TTFD focusing on transfer technology approach under a commercial view undermined previous adaptive plans that were configured an inclusive innova-tion process in brick producer villages in Cusco, Peru.

This case illustration is based on Alayza (2017).

21.6 Facilitating Communication in Adaptive Planning Processes for Inclusive Innovation

Adaptive planning processes can generate innovation oriented towards local devel-opment (De Melo 2014). The current socio-economic and the lack of opportunities caused by unequal structures in Latin America and Peru, challenging to find new ways to support and promote inclusive innovation processes for local development. This situation allows discussing an integrative approach that encompasses communication in adaptive planning processes that can help to understand the nature of innovation, recognise entry point for innovation and re-orient actions towards inclusivity (see Table 21.1).

Table 21.1 Integrative approach

	Inclusive innovation	
	Communication	Adaptive planning
Approach	– Construct meaning in interaction	– Active adaptation for a coherent change
Orientation	– Allow understanding of themselves, other people and their reality	– Better futures based on the participation of their protagonist
Characteristics	– Interpretative – Complex – Contextual	– Adaptive – Flexible – Complex – Uncertainty
Functions	– Dialogue – Construct network – Social learning – Negotiation – Diffusion	– Involvement of their protagonist – Social learning – Constant re-evaluation of objectives – Redefinition of paths
Integrative approach	– Understand the nature of inclusive innovation processes – Recognise entry points for inclusive innovation – Reorient innovation processes towards inclusivity	

21.6.1 Communication in Adaptive Planning for Understanding the Nature of Inclusive Innovation

Taking into consideration the nature of innovation which has high levels of complexity and uncertainty (Smits 2002; Smits and Kuhlmann 2004), adaptive planning processes can help to motivate the inclusion of actors in the solution of complex problems in the society (De Melo 2003), in which this active involvement could be crucial for dispelling uncertainty in innovation processes (Powell and Grodal 2009).

In this regards, the role of communication is central because it can enable a better comprehension of the nature of local innovation processes due to it can open spaces for dialogue around a real problem (Dutta 2012). For example, events such as local fairs (see Box 1) that concentrate diverse actors can be conceptualised as a communication space not only for promoting the diffusion of ready-made technologies but also for fostering communication for inclusive innovation, in which various actors with different roles, responsibilities, aims, necessities, capabilities and opportunities may find room to manoeuvre to shape viable options for change in their territory.

Furthermore, given that adaptive planning is considered as a continuous process of social learning (De Melo 2014), promote this kind of spaces for dialogue could contribute to building interactive learning spaces (Lundvall et al. 2009; Johnson and Andersen 2012) that allow generating innovation according to different needs. Thus, rather than the instrumentalist forms of disseminating new ideas or technologies such as communication is commonly understood in innovation, adaptive planning processes can help to facilitate communication oriented to comprehend what kind of new technologies or solutions could be adapted according to social, cultural and economic backgrounds. This may allow avoiding strategies that are regarded as appropriate in advance for building, in a participative way, strategies that embrace diverse technological, social and institutional aspects, permitting to shape strategies in a more effective and participative ways towards more inclusive development.

21.6.2 Communication in Adaptive Planning for Recognising Entry Points for Inclusive Innovation

Adaptive planning proposes the introduction of incremental changes with real implications in society, in which actors are gradually articulate with others, and in this articulation, it can be recognised multiple options for innovation (Almeida and De Melo 2017).

Given that the current strategies to promote innovation in Peru present various constraints, dialogical communication may help to identify, recognise and support local inclusive innovation processes taking into consideration local capabilities, resources and opportunities. For example, the formation of local networks that have that enabled dialogical communication to strengthen local capabilities in excluded groups, opening new opportunities for innovation through more participative decision-making and knowledge-sharing processes (See Boxes 2 and 3).

Thus, an alternative for integrating excluded groups is promoting the formation of new networks focused on resolving local problems with the participation of different kind of actors. This can influence the generation of social and political decision-making concerning what kind of innovation process should be undertaken or supported according to the local reality (Thomas 2012). Building new networks according to current problems may permit orienting actions towards some specific goals in terms of seeking innovation for inclusive development, but, even in cases innovations fail, the ties and learning processes generated can contribute to the configuration of other innovation processes (Lundvall et al. 2009; Borrás 2011; Wieczorek et al. 2012; Arocena and Sutz 2014).

21.6.3 Communication in Adaptive Planning for Reorienting Innovation Processes Towards Inclusivity

Considering that local innovation strategies in Peru have limitations regarding enhanced social inclusion, it seems to be necessary to design flexible planning processes in which multiple actors that can orient actions towards more inclusive innovation processes.

González (1997) states that participation in collaborative arrangements made by adaptive planning actions motivates active process of innovation. These arrangements permit to built networks not only for sharing knowledge but also for increasing the efficiency in processes for change (De Melo 2014). In this context, promoting communication may help to find entry points for identifying and give voice several kinds of actors (Dutta 2012), which can influence the generation of social and political decision-making concerning the direction of innovation processes. This is why that, strategic and inclusive local networks should be oriented to enable new agreements and new forms of organisation that open possibilities for creating new lines of production according to local characteristics (Thomas et al. 2012). For example according to the cases illustrated in the Boxes 2 and 3, the formation of strategic networks around innovations oriented to resolve local needs, not only contributed to improving the quality of life of excluded groups but also created new lines of production in the processes of design, construction and expansion of local innovations. However, unequal power relations embedded in technology transference strategies undermined the extension of those inclusive innovation processes.

In this regards, power relations have to be taken into consideration in innovation processes, especially when these processes are oriented towards supporting excluded groups. Power is the foundation of the different ways of organisations in societies, in which different forms of communication constitute critical sources of power and counter-power for social transformations (Castells 2010). Since ways of thinking can be turned into ways of doing, communication may constitute the way of creating coordination among diverse actors and encourage the generation of changes through the formation of counter-power networks in societies (Castells 2011).

Inclusive innovation is not a mainstream discourse, and this is why counter-power networks should have an "explicitly normative agenda, which seeks to mobilise distinctly political processes, such as claims to social justice, and often questions organisational and economic assumptions in conventional innovation policies" (Fressoli et al. 2012, p. 2).

Thus, adaptive planning processes must be oriented to create a new agenda with the participation of multiple networks that can support different stages of innovation processes, strengthening local capabilities and opening new opportunities for accommodating a broad base inclusive development strategy.

21.7   Conclusion and Recommendations

Discussing on the literature about communication, inclusive innovation and adapting planning processes in the Peruvian context, it can be affirmed that there are conceptual relations between those concepts as well as there is relevance for using together to accommodate a broad-based inclusive strategy for dealing with exclusionary structures. In this scenario, the role of communication under a social construction conception is crucial because it can permit to adapt actor's views towards more doable and coherent innovation processes in diverse contexts.

While adapting planning processes can help to construct multiple options for change, communication can be instrumentalised for helping to understand the nature of inclusive innovation processes, recognising entry point for innovation and re-orienting actions towards inclusive innovation for local development.

This also opens new possible roles and contributions for communication specialists, who can act as cross-linker between diverse actors and networks for facilitating an adaptive social and technical change. Such communication specialists can be vitally instrumental in the conceptualisation, facilitation and expansion of inclusive innovation processes.

Regarding the roles of communication specialists, the following are some recommendations that must be considered in order to facilitate adaptive planning processes for configuring inclusive innovation:

Enable communication as the creation of meaning in the multiple interactions of the involved actors. This can help to promote the creation of dialogue rather than just the dissemination of preconceived ideas or technologies, enabling more creative, coordinated and relevant actions according to every context. Communication can be operationalised by creating roundtables, agendas, events to discuss how to create implement, extended, monitoring and/or evaluate innovation processes in priority areas such as health, food, water and/or energy.

Build new networks towards inclusive innovation. Communication facilitates integrative actions in order to interconnect multiple actors' interactions. In this regards, communication specialist can motivate the engagement of local organisation such as universities, government agencies, and innovative enterprises to find feasible solutions and to resolve some bottlenecks that can appear in the trajectory of the inclusive innovation process at different levels and due to various aspects, mostly related to human relations.

Orchestrate adaptive planning processes in order to create innovation processes towards inclusivity. Communicational efforts can strengthen the networks relations with a view to sustaining them and permitting to plan new initiatives that are seeking inclusive innovation. In that sense, a communication specialist can facilitate that relevant actors and networks can work together in flexible action plans helping to discover entry points for making inter-stakeholder agreements, strengthening local networks, systematising and monitoring learning processes and visualising relevant achievements of those interactions.

Promoting an inclusive and participative approach. Through adaptive planning processes, it can find strategic ways to include excluded groups, not only with conventional technology transference or market strategies but also with creative and collaborative mechanisms that could be oriented to improve local capabilities and generate new opportunities at different levels. Collective actions based on resolving local needs can make innovative solutions but also can create new lines of production related to the topic areas (e.g. health, food, water, energy) that may give a broader, inclusive and sustainable socio and technical change.

Bibliography

Ackoff, R. (1974). *Redesigning the future: A system approach to societal problems.* New York: Willey-Interscience.

Akrich, M., Callon, M., Latour, B., & Monaghan, A. (2002). The key to success in innovation Part I: The art of interessement. *International Journal of Innovation Management, 06*(02), 187–206. http://dx.doi.org/10.1142/s1363919602000550.

Alayza, B. (2017). *Communication for inclusive innovation: Rethinking science, technology and innovation policies for development in Peru* (M.Phil thesis). School of Communication and Arts, The University of Queensland. https://doi.org/10.14264/uql.2017.893.

Alayza, B., & Ismodes, E. (2011). Estrategias de comunicación para aumentar el éxito en proyectos de fomento a la innovación tecnológica a nivel descentralizado: experiencia del proyecto RAMP-PERÚ [Strategies of communication to improve the success of innovation projects at national level: Experience of RAMP-PERU project]. *Anales del XIV Congreso Latino Iberoamericano de Gestión Tecnológica "Innovación para el crecimiento sostenible en el marco del Bicentenario".* ALTEC, Lima.

Alfaro, R. (2006). *Otra Brújula: Innovaciones en comunicación y desarrollo [Innovation in communication and development]* (1st ed.). Lima, Perú: Asociación de Comunicadores Sociales CALANDRIA.

Almeida, M. F. L. D., & De Melo, M. A. C. D. (2017). Sociotechnical regimes, technological innovation and corporate sustainability: From principles to action. *Technology Analysis & Strategic Management, 29*(4), 395–413.

Andersen, A. D., & Andersen, P. D. (2014). Innovation system foresight. *Technological Forecasting and Social Change, 88,* 276–286.

Andersen, A. D., & Andersen, P. D. (2017). Foresighting for inclusive development. *Technological Forecasting and Social Change, 119,* 227–236.

Arocena, R., & Sutz, J. (2014). Innovation and democratisation of knowledge as a contribution to inclusive development. In G. Dutrénit & J. Sutz (Eds.), *National innovation systems, social inclusion and development: The Latin American experience* (pp. 15–33). Cheltenham: Edward Elgar Publishing Ltd.

Babüroglu, O. N., & Ravn, I. (1992). Normative action research. *Organization Studies, 13*(1), 019–34.

Bazán, M., Sagasti, F., & Cárdenas, R. (2014). National system of innovation for inclusive development: Achievements and challenges in Peru. In *National innovation systems, social inclusion and development: The Latin American experience.* Cheltenham, UK: Edward Elgar Publishing. http://dx.doi.org.ezproxy.library.uq.edu.au/10.4337/9781782548683.00011.

Beltran, R. (2005). La comunicación para el desarrollo en Latinoamérica: un recuento de medio siglo [Communication for development in Latin America: A recount of half a century]. *III Congreso panamericano de la comunicación Panel 3: Problemática De La Comunicación Para El Desarrollo, 54.*

Berlo, D. K. (1960) *The process of communication: An introduction to theory and practice.* New York: Holt, Rinehart & Winston.

BID. (2010). *La necesidad de Innovar. El camino hacia el progreso de América Latina y el Caribe.* Paper prepared for the EU-LAC Summit of Heads of State and Government. Madrid: Inter-American and Development Bank (IDB). Retrieved from: http://idbdocs.iadb.org/wsdocs/getdocument.aspx?docnum=35168065.

Bijker, W., Hughes, T., & Pinch, T. (Eds.). (1987). *The social construction of technological systems. New directions in sociology and history of technology.* Cambridge, MA: MIT Press.

Borrás, S. (2011). Policy learning and organizational capacities in innovation policies. *Science and Public Policy, 38*(9), 725–734. https://doi.org/10.3152/030234211x1307002163332.

Burns, T. (1980). *Network agents and community governance* (Tesis de Doctorado). University of Pennsylvania, Filadelfia.

Burns, T., Cohen, B., De Melo, M. A. C., Hawk, D., Keidel, R., Pava, C.Y., et al. (1983). *Where planning fails: A perspective of adaptive incremental learning in complex organizational settings*. Filadélfia, Mimeo: University of Pennsylvania.

Castells, M. (2010). *The information age* (1st ed.). Oxford, England: Wiley-Blackwell.

Castells, M. (2011). A network theory of power. *International Journal of Communication, 5*, 773–787. http://dx.doi.org/1932–8036/20110773.

Chataway, J., Hanlin, R., & Kaplinsky, R. (2014). Inclusive innovation: An architecture for policy development. *Innovation and Development, 4*(1), 33–54. https://doi.org/10.1080/2157930x.2013.876800.

CONCYTEC. (2016). *Política nacional para el desarrollo de la ciencia, tecnología e innovación tecnológica - CTI* [National policy for the development of science, technology and technological innovation—ST&I] (p. 124). Lima, Peru: CONSEJO NACIONAL DE CIENCIA, TECNOLOGÍA E INNOVACIÓN TECNOLÓGICA.

Coutts, J., & Roberts, K. (2011). Theories and approaches of extension: Review of extension in capacity building. In J. Jennings, R. Packham, & D. Woodside (Eds.), *Shaping change: Natural resource management, agriculture and the role of extension* (pp. 22–31). Australia: Australasia-Pacific Extension Network (APEN).

Cozzens, S., & Sutz, J. (2014). Innovation in informal settings: Reflections and proposals for a research agenda. *Innovation and Development, 4*(1), 5–31. https://doi.org/10.1080/2157930x.2013.876803.

De Melo, M. A. C. (2014). Sustainable innovation and the dimensions of complexity. *READINGS BOOK, 122.*

De Melo, M. A. C. (1985). Action research and planning process: A learning perspective. In *Proceedings of the V International Congress of Industrial Engineering*, Florianopolis, Brazil.

De Melo, M. A. C. (2003). Innovatory planning: Methods and instruments. In *Readings Book of X Annual Conference of GBATA—Global Business and Technology Association*, Budapest.

Dervin, B. (1983). Information as a user construct: the relevance of perceived information needs to synthesis and interpretation. In S.A. Ward & L.J. Reed (Eds.), Knowledge, structure and use: Implications for synthesis and interpretation, pp. 153–83. Philadelphia: Temple University Press.

Dutrénit, G., & Sutz, J. (2014). *National innovation systems, social inclusion and development: The Latin American experience*. Cheltenham: Edward Elgar Publishing Ltd.

Dutta, M. J. (2012). *Communicating social change structure, culture, and agency*. New York: Routledge/Taylor & Francis.

Emery, F. E., & Trist, E. L. (1973). *Towards a social ecology, contextual appreciation of the future in the present*. London: Plenum Press.

Escobal, J., Ponce, C., Pajuelo, R., & Espinoza, M. (2012). Estudio comparativo de intervenciones para el desarrollo rural en la sierra sur del Perú [A comparative study of rural development interventions in the south highlands of Peru]. Lima: GRADE, c2012.

European Commission. (2014). Study on strengthening regional innovation systems in Peru: Policy lessons. (p. 86). Cusco-Tacna: Germán Granda (Director), María Angélica Ropert (Consultant), Lenia Planas (Consultant).

Foster, C., & Heeks, R. (2013). Conceptualising inclusive innovation: Modifying systems of innovation frameworks to understand diffusion of new technology to low-income consumers. *The European Journal of Development Research, 25*(3), 333–355. https://doi.org/10.1057/ejdr.2013.7.

Freeman, C. (1995). The national system of innovation in historical perspective. *Cambridge Journal of Economics, 19*(1), 5–24.

Fressoli, M., Arond, E., Abrol, D., Smith, A., Ely, A., & Dias, R. (2014). When grassroots innovation movements encounter mainstream institutions: Implications for models of inclusive innovation. *Innovation and Development, 4*(2), 277–292. https://doi.org/10.1080/2157930x.2014.921354.

Fressoli, M., Garrido, S., Picabea, F., Lalouf, A., & Fenog, V. (2013). Cuando las "transferencias" tecnológicas "fracasan". Aprendizajes y limitaciones en la construcción de Tecnologías para la

Inclusión Socia [When technological "transfers" fail. Learning and limitations in the construction of technologies for social inclusion] l. *Universitas Humanística, (76)*, 73–96.

Fressoli, M., Smith, A., & Thomas, H. (2012). *Grassroots innovation movements: Enduring dilemmas as sources of knowledge production.* Retrieved August 20, 2015, from https://grfscp.files.wordpress.com/2012/05/grf-2012-rio-smith-et-al.pdf.

Geels, F. (2004). From sectoral systems of innovation to socio-technical systems: Insights about dynamics and change from sociology and institutional theory. *Research Policy, 33,* 897–920.

Geels, F. W. (2002). Technological transitions as evolutionary reconfiguration processes: A multi-level perspective and a case-study. *Research Policy, 31*(8–9), 1257–1274.

González, M., & De Melo (2004). Planificación interoganizacional y desarrollo emprendedor: un estudio de caso. In *Anais da III Conferencia Internacional de Pesquisa em Emprendedorismo na América Latina (CIPEAL)*, Río de Janeiro (pp. 11–13).

González, M. D. (1997). *Planejamento dos Pólos Tecnológicos: un enfoque adaptativo* (Tese de Doutorado em Engenharia de Produção). Pontifícia Universidade Católica do Rio de Janeiro, Brasil.

Hajer, M. A., & Laws, D. (2006). Ordering through discourse. In M. Moran, M. Rein, & R. E. Goodin (Eds.), *The Oxford handbook of public policy* (p. 249266). Oxford: Oxford University Press.

Harman, U. (2018). *Inclusive innovation in rural communities: Three case studies of rural electrification in Peru* (Ph.D. thesis). School of Agriculture & Food Sciences, The University of Queensland. https://doi.org/10.14264/uql.2018.149.

Heeks, R., Foster, C., & Nugroho, Y. (2014). New models of inclusive innovation for development. *Innovation and Development, 4*(2), 175–185. https://doi.org/10.1080/2157930x.2014.928982.

Herrera, A. (2011). La recuperación de tecnologías indígenas. Arqueología, tecnología y desarrollo en los Andes. [The recovery of Indigenous technologies. Archaeology, Technology and Development in the Andes]. Lima: IEP.

Hülsmann, M., & Pfeffermann, N. (Eds.). (2011). *Strategies and communications for innovations: An integrative management view for companies and networks.* Berlin: Springer Science & Business Media.

IDRC. (2011). *Innovation for inclusive development, program prospectus for 2011–2016.* Accessed: December 1, 2015. http://www.slideshare.net/uniid-sea/october-2011-innovation-for-inclusivedevelopment-program-prospectus-for-20112016.

INEI. (2015). *Evolución de la Pobreza Monetaria 2009 – 2014. Informe Técnico* [Evolution of monetary poverty 2009–2014. Technical report]. Lima: Instituto Nacional de Estadística e Informática.

Ismodes, E. (2006). *Países sin futuro*: qué puede hacer la Universidad? [Countries without future. What can the university do?] (1st ed.). Lima: Fondo Editorial de la Pontificia Universidad Católica del Perú.

Ismodes, E., & Manrique, K. (2016). *Estudio de Caracterización del Sistema de Innovación del Perú* [Characterization study of the Peruvian innovation system] (p. 192). Lima.

Johnson, B., & Andersen, A. D. (2012). *Learning, innovation and inclusive development. New perspectives on economic development strategy and development aid.* Globelics. Thematic Report 2011/12. Aalborg University Press.

Joseph, K. J. (2014). Exploring exclusion in innovation systems. *Innovation and Development, 4*(1), 73–90.

Kuramoto, J. R. (2014). Inclusive innovation against all odds: The case of Peru. In G. Dutrénit & G. Crespi (Eds.), *Science, technology and innovation policies for development: The Latin American experience* (pp. 109–131). Cham: Springer. https://doi.org/10.1007/978-3-319-04108-7.

Latour, B. (1987). *Science in action: How to follow scientists and engineers through society.* Cambridge: Harvard University Press.

Leeuwis, C. (2004a). *Communication for rural innovation: Rethinking agricultural extension* (with contributions by A. van den Ban). Oxford: Blackwell Science.

Leeuwis, C. (2004b). Changing views of innovation and the role of science. The 'socio-technical root-system' as a tool for identifying relevant cross-disciplinary research questions. In *Pre Proceedings of the Sixth European IFSA Symposium* (Vol. 2, pp. 773–782).

Leeuwis, C., & Aarts, N. (2011). Rethinking communication in innovation processes: Creating space for change in complex systems. *The Journal of Agricultural Education and Extension, 17*(1), 21–36.

Leeuwis, C., & Aarts, N. (2016). Communication as intermediation for socio-technical innovation. *Journal of Science Communication, 15*(6), 1–12.

Lundvall, B. Å. (1992). *National systems of innovation: Towards a theory of innovation and interactive learning*. London: Pinter.

Lundvall, B. Å., Joseph, K., & Chaminade, C. (2009). *Handbook on innovation systems and developing countries: Building domestic capabilities in a global context*. Cheltenham, UK: Edward Elgar Publishing.

Mattelart, A., & Mattelart, M. (2003). *Theories of communication: A short introduction*. Thousand Oaks, Calif; London: Sage Publications.

Nelson, R. (1993). *National innovation systems: A comparative analysis*. New York, NY: Oxford University Press.

OECD. (2011). *Organisation for Economic Co-operation and Development (OECD) reviews of innovation policy: Peru 2011*. Paris: OECD Publishing.

Papaioannou, T. (2014). How inclusive can innovation and development be in the twenty-first century? *Innovation and Development, 4*(2), 187–202.

Pfeffermann, N., & Gould, J. (2017). *Strategy and communication for innovation: Integrative perspectives on innovation in the digital economy* (3rd ed., 2017 ed.). Cham: Springer International Publishing.

Pfeffermann, N., Minshall, T., & Mortara, L. (Eds.). (2013). *Strategy and communication for innovation*. Berlin: Springer.

Powel W. W., & Grodal, S. (2009). Networks of innovators. *The Oxford handbook of innovation* (p. 78).

Rogers, E. M. (1962). *Diffusion of innovations*. New York: Free Press.

Rogers, E. M. (2002). The nature of technology transfer. *Science Communication, 23*(3), 323–341.

Rogers, E. M. (2003). *Diffusion of innovations*. New York: Free Press.

Schillo, R. S., & Robinson, R. M. (2017). Inclusive innovation in developed countries: The who, what, why, and how. *Technology Innovation Management Review, 7*(7).

Sen, A. (2000). *Social exclusion: Concept, application, and scrutiny*. Social Development Papers No. 1. Manila: Asian Development Bank.

Smith, A., Fressoli, M., & Thomas, H. (2014). Grassroots innovation movements: Challenges and contributions. *Journal of Cleaner Production, 63*, 114–124. https://doi.org/10.1016/j.jclepro.2012.12.025.

Smits, R. (2002). Innovation studies in the 21st century: Questions from a user's perspective. *Technological Forecasting and Social Change, 69*(9), 861–883.

Smits, R., & Kuhlmann, S. (2004). The rise of systemic instruments in innovation policy. *International Journal of Foresight and Innovation Policy, 1*(1/2), 4. https://doi.org/10.1504/ijfip.2004.004621.

Te Molder, H., & Potter, J. (2005). *Conversation and cognition*. Cambridge: Cambridge University Press.

Thomas, H. (2009). *Tecnologías para la inclusión social y políticas públicas en América Latina*, 32. [Technologies for social inclusion and public policies in Latin America]. Grupo de Estudios Sociales de la Tecnología y la Innovación IESCT/UNQ CONICE. Retrieved from http://www.redtisa.org/Hernan-Thomas-Tecnologias-para-la-inclusion-social-y-politicas-publicas-en-America-Latina.pdf.

Thomas, H. (2012), Tecnologías para la inclusión social en América Latina. De las tecnologías apropiadas a los sistemas tecnológicos sociales. Problemas conceptuales y soluciones estratégicas

[From technologies appropriate to social technological systems. Conceptual problems and strategic solutions]. In H. Thomas, M. Fressoli, & G. Santos (Eds.), *Tecnología, desarrollo y democracia. Nueve estudios sobre dinámicas socio-técnicas de exclusión/inclusión social* (pp. 25–78). Buenos Aires: Ministerio de Ciencia, Tecnología e Innovación Productiva.

Thomas, H., Fressoli, M., & Becerra, L. (2012). Science and technology policy and social ex/inclusion: Analyzing opportunities and constraints in Brazil and Argentina. *Science and Public Policy, 39*(5), 579–591. https://doi.org/10.1093/scipol/scs065.

Trist, E. (1976a). *A concept of organizational ecology.* Pennsylvania: Management and Behavioral Science Center, Wharton School, University of Pennsylvania, Mimeo.

Trist, E. (1976b). Action research and adaptive planning. In *Experimenting with organization life: The action research approach, Clark, A. (org.).* London: Plenum Press.

Trist, E. (1976c). *Some concepts of planning.* Pennsylvania: Management and Behavioral Science Center, Wharton School, University of Pennsylvania, Mimeo.

Vanclay, F., & Leach, G. (2011). Enabling change in rural and regional Australia. In J. Jennings, R. Packham, & D. Woodside (Eds.), *Shaping change: Natural resource management, agriculture and the role of extension* (pp. 62–71). Australia: Australasia-Pacific Extension Network (APEN).

Villaran, F. (2010). *Emergencia de la Ciencia, la Tecnología y la Innovación (CTI) en el Perú.* Organización de Estados Iberoamericanos [Emergency in science, technology and innovation in Peru. Organization of Iberoamerican states]. Lima. Retrieved from: http://www.oei.es/salactsi/EmergenciaDeCtiEnPeru.pdf.

Wieczorek, A., Hekkert, M., & Smits, R. (2012). *Contemporary innovation policy and instruments: Challenges and implications.* Innovation Studies Utrecht (ISU) Working Paper Series, 45. Retrieved from http://www.geo.uu.nl/isu/pdf/isu0912.pdf.

World Bank. (2010). Innovation policy. A guide for developing countries. Washington, DC. Retrieved from: http://www.innovation.lv/ino2/publications/Innovation_policy_World_Bank.pdf

Zerfass, A., & Huck, S. (2007). Innovation, communication, and leadership: New developments in strategic communication. *International Journal of Strategic Communication, 1*(2), 107–122.

Bernardo Alayza is a doctoral candidate in Innovation Management at Pontifical Catholic University of Peru (PUCP). Master Philosophy at The University of Queensland of Australia (UQ), Master in Technology and Innovation Management from PUCP and Bachelor's Degree in Communication majoring in Development Communication. Lecturer in the Faculty of Communication and Arts, and in the Postgraduate School at PUCP in topics such as communication, innovation and research methodologies. His area of work and research is focused on communication for inclusive innovation, facilitating coordination and cross-sector collaboration to implement broad-based innovation for development strategies and policies for Peru and Latin America. Bernardo's contributions in this field were recognised by international awards, and he has received scholarships to complete postgraduate studies in Peru and Australia.

Domingo Gonzalez is Professor of the Engineering Department of the Pontifical Catholic University of Peru (PUCP). Industrial Engineer from PUCP, Master and Doctor in Industrial Engineering from Pontifical Catholic University of Rio de Janeiro. From 2001 to 2008, he was the Director of the Center for Development and Innovation of PUCP. He was president and founder of Peruvian Association of Incubators (PERUINCUBA). He was former Head of Engineering Department of PUCP and president of the Latin American Association for Technology Management (ALTEC). His current research interests are innovation management, interorganizational planning, innovation and entrepreneurial systems.

Chapter 22
Corporate Values and People Attributes. Practical Leadership Applications in the Asset Management Industry

Stefan Hofrichter

Abstract In this chapter, the author explains how Allianz Global Investors' Corporate Values (Excellence, Passion, Integrity, Respect) and related People Attributes (Entrepreneurship, Trust, Collaborative Leadership, Customer and Market Excellence) can be applied in the day-to-day management of an Economics and Strategy research team and how the Values and People Attributes are interrelated. The team manager's role is very much too lead by example. An appropriate organizational framework set up by the company is required in addition.

22.1 Introduction: It's not only About the Content

Economists like me in the financial industry are trained to see the world through economic models, fitted to data and tested against actual developments in the capital markets. The purpose is generating insights on the world, notably the financial markets, deriving superior investment conclusions, and helping internal and external clients make the right decisions.

In other words, at first sight, an economist's job seems to be only about generating and communicating content and investment performance. But while this is essential for every asset management firm, it is not enough. To us at Allianz Global Investors, creating value for our clients goes beyond delivering pure investment returns. It is about having a trusted, stable and long-term relationship that generates value for our clients every step of the way a relationship that is mutually beneficial for our clients and for our firm.

This also has profound implications for every team leader. Notably, when working with an international team of investment professionals (in my case economists and investment strategists based around the world), I also have to think about *how* to achieve high-quality research and money-making investment ideas for our clients on a *sustainable* basis.

S. Hofrichter (✉)
Allianz Global Investors, Frankfurt, Germany
e-mail: stefan.hofrichter@allianzgi.com

© Springer Nature Switzerland AG 2020
N. Pfeffermann (ed.), *New Leadership in Strategy and Communication*,
https://doi.org/10.1007/978-3-030-19681-3_22

To that end, Allianz Global Investors has developed a set of Values and a set of People Attributes that we expect to see in our employees. I will discuss corporate Values and the People Attributes in general terms in the first part of this paper. In the second part, I will discuss some practical applications.

22.2 Corporate Values and People Attributes—A Real Life Example

The four Corporate Values that we at Allianz Global Investors live every day—or are at least trying to are-*Excellence, Passion, Integrity* and *Respect.* We are committed to delivering *excellence* for our clients, their advisors, our employees, our parent company and within our industry. This means we seek to inspire both our clients and our colleagues with a *passion* for what we do. Furthermore, we put our clients' needs first, behaving in an open and transparent way and treating people fairly, which means acting *respectfully* and with *integrity.* We respect difference and diversity, and we reward individual performance as well as teamwork.

Our values are complemented by our "People Attributes"—i.e. those characteristics that describe the ambitions of our colleagues and expected behaviours and set-ups in a more concrete way: *Entrepreneurship, Trust, Collaborative Leadership,* and *Customer and Market Excellence.* Graphic 22.1 describes our people attributes in more detail.

22.3 A Practical Application of Corporate Values and People Attributes in the Asset Management Industry

What do these concepts imply in real life? How can they be applied in the day-to-day management of investment professionals in an asset management firm, notably in the management of a buy-side economic and strategy research team. First, let's review Allianz Global Investors' corporate Values by having a closer look at the corresponding People Attributes.

22.3.1 Entrepreneurship

Entrepreneurship is, first and foremost, about *taking risk, being innovative and taking responsibility for own decisions*. It is our firm belief at Allianz Global Investors that a working environment that gives investment professionals as much leeway as possible—provided that they "play by the rules" and collaborate within and across

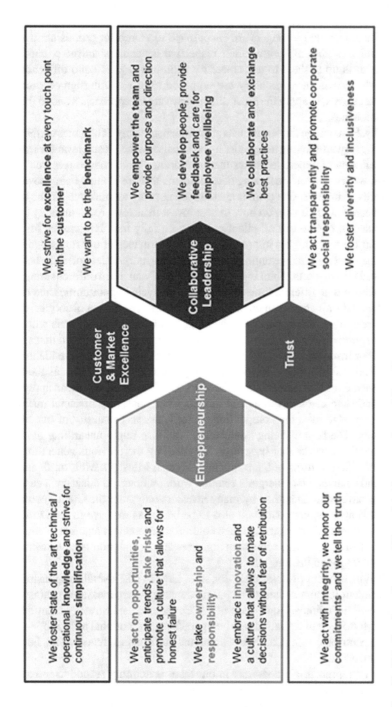

Graphic 22.1 People attributes at Allianz global investors. *Source* AllianzGI

departments in order to generate synergies—is the most motivating environment, and one that leads to the best results possible.

For instance, our fund managers are encouraged to search for promising investment opportunities not only within their respective benchmark universe, but also outside of it. It is up to them to use creativity, thorough research and diligence to come up with investment ideas. They are supposed to invest with high conviction against market trends, to apply the most suitable investment approaches and to freely test new approaches.

Personally, I encourage the economists and strategists on my Economics and Strategy team to be forwardlooking and to identify on their own the most relevant research topics in their area of expertise, rather than just doing "maintenance research". In other words, we try to avoid analyzing developments that are already being covered by the majority of brokers. Our team is encouraged not only to think outside the box, but also to decide on their own on how to best invest their resources—notably their time—provided of course that all client requests are fully met. It is crucial for the success of our team and highly important for the performance of our funds—meaning our clients' assets—that we anticipate investment opportunities and risks before they are priced into markets. Thinking "outside the box" can mean different things to different colleagues at different times. It can mean analyzing opportunities and risks in financial markets for an investment horizon that is different (longer or shorter) than the conventional investment horizon of the majority of market participants with the goal of identifying market inefficiencies that can be exploited by our fund managers and clients. For instance, very early during the Great Financial Crisis, we identified "Financial Repression", that is low or even negative real interest rates, as the key investment theme for capital markets in the years to come. It can also mean applying innovative and state-of-the-art tools and models when analyzing financial markets that give us an edge over the competition. Indeed, the vast majority of our tools are proprietary. The idea of being "innovative" can also imply unearthing models and economic frameworks that have been forgotten by the previous generation of investors but which, it turns out, are timeless. Hyman Minsky's work on financial instability and Charles Kindleberger's seminal work on financial crisis are a case in point as they have been unearthed by many investors only after the Great Financial Crisis. All this also requires our team also to be brave, as we have to dare to look at markets differently than others and take contrarian investment recommendations. The upside is that this is, more often than not, a rewarding and promising investment proposition—at least in the long run.

Individual investment professionals, for sure, have to take overall responsibility for the investment recommendations taken: fund managers are measured against the performance of their funds. In the case of the Economics and Strategy team that I lead, our portfolio recommendations are being tracked in a "notional portfolio"—and these results contribute significantly to our annual performance reviews and, hence, to our compensation.

Being an entrepreneurial economist in our team is not only related to research, but it also includes marketing activities—at least in a broad sense. For instance, my colleagues and I actively promote articles in financial newspapers and on social media; we contribute to scientific journals; we are editors of and contributors to finance books; and we originate and chair client conferences. As a manager, I am more than happy to support these activities as they also help promote our firm's brand.

An environment that encourages and promotes risk-taking has to, almost without saying, allow for honest failure. This is particularly true for us investors who work in the asset management industry: none of us has perfect foresight and even the best among us make decisions or give recommendations that turn out to be wrong around 40% of the time.

Calling for entrepreneurship is often easier said than done, as John Maynard Keynes put it so wisely in his book "The General Theory of Employment, Interest, and Money": "Worldly wisdom teaches that it is better for reputation to fail conventionally than to succeed unconventionally". It is the role of a team leader in the asset management industry to help achieve exactly the opposite: to encourage team members to take risks, to be original and to be creative and not to follow the crowd.

22.3.2 Trust

The attribute *Trust* includes, in our firm's interpretation, the aspects *Integrity and Honesty*, *Diversity and Inclusion* and *Social Responsibility*. Trust is an important—if not vital—commodity for our industry. One could argue that we trade in trust—tangible or not—more than anything else. It is therefore imperative that we understand how to first earn and build trust with our clients, and then how to continue to maintain this trusted status through turbulent market environments. For Allianz Global Investors, trust is one of the most distinguishable and differentiating factors for positioning our firm as a market leader deserving of long-term, sustainable client relationships.

Integrity and Honesty

For a money manager, integrity is an asset in itself. This includes, first and foremost, trust from our clients, which has to be earned every day by acting with integrity on their behalf when fulfilling our fiduciary duties. This sounds obvious, but there are many areas in which mistakes could be made unintentionally by employees. The interests of clients as well as employees are best protected when there are clear guidelines in place that help avoid conflicts of interest, prohibit market manipulation and secure data privacy. We have clear policies in place covering customer complaints, bribery, gifts and entertainment, money laundering, fraud, donations and political contributions, employee social networking and personal account dealing. The role of a manager here is to lead by example—to make sure that employees are fully aware of the internal rules and, of course, follow them in spirit as well as by the letter.

Honest communication is not only important when dealing with the outside world. It is also crucial internally when interacting with internal clients or peers. This is particularly true when one's investment process is built on an open and active exchange of information and thoughts. In the Economics and Strategy team, we uphold high standards for the quality and transparency of our research. Credibility, reliability and trust are key: good reputation often takes years to build, but can be wiped out in an instant if your work lacks integrity and honesty.

Diversity and Inclusion

Diversity and inclusion are another important aspect of trust. The traditional argument in favor of having a diverse and inclusive workforce, especially as an asset manager, is that we want to embrace diversity of thought in our investment process and mitigate potential risks from "groupthink". An inclusive workplace helps us attract, develop and retain our best talent which is a necessary, albeit not sufficient, condition for us to better serve our global client base. Moreover, our goal is to create a working environment characterized by mutual respect. In that context, having a heterogeneous workforce with respect to age, gender, socio-economic background, education, marital partnership status, race, religion etc. has practical and tangible benefits for our business.

Let's think of diversity also in terms of diversification in a portfolio context: As an economist in the financial industry, I find it particularly appealing to relate the topic to the findings from Modern Portfolio Theory, as Andrew Haldane, the Bank of England's Chief Economist, has done so nicely in his Financial Times op-ed from July 22, 2018. Let me explore this in more detail.

For the sake of the argument, and following along the lines that Andrew Haldane did, just think of the staff of an asset management firm—all employees, a department or a team—as a portfolio. This portfolio may be a fund management team that generates returns (be it absolute or in excess of a benchmark) with some volatility around it. This portfolio could also be my Economics and Strategy team, which gives portfolio recommendations to our internal clients—fund managers and research analysts—and external clients. Let's describe the expected returns and the volatility (standard deviation) of the expected returns with the value of 5% respectively. Let us also assume that the team can hire new team members. Three options for these hires are available.

Let's start with Option 1—a team with an expected return of 10 and a volatility of 6. In other words, option 1 is expected to generate higher returns than the existing team, but the returns are expected to be more volatile as well. Option 1 is supposed to be of a similar breed as the existing team—e.g. it has a similar background and investment process, which implies that integrating them may be quite smooth. Let's also say that both the existing team and Option 1 are "fundamental investors", meaning they decide on fundamental economic considerations (which is not to say that the fundamentalists don't apply a rigorous, structured and datadriven approach). Consequently, as both investment approaches are similar, the results of both teams are likely to be highly correlated: let's say, the correlation coefficient of the two team's results is 0.75.

The stats for Option 2 are different: a lower expected return than Option 1, say 8, but with the same volatility of 6. However, the results of Option 2 and of the existing team are totally uncorrelated: the correlation is 0. This may be explained by a totally different investment approach, e.g. Option 2 may apply a purely quantitative investment style, based on mathematical and statistical models, rather than looking at fundamental consideration.

Finally, there is Option 3, with an expected return of 8, a volatility of 8, but with a correlation coefficient of −0.75. In other words, this team does a good job on average, even though results are more volatile than that of the existing team as well as of Option 1 and 2. However, their results move countercyclical to that of the existing team: Option 3 tends to be good when the existing team underperforms and vice versa. How can this be explained? Option 3 may apply a totally different investment approach and may invest with a different investment horizon than the existing team. For instance, Option 3 may follow a "contrarian" investment style, i.e. they prefer to buy when the majority in the marketplace including the existing team sells. As contrarian investors, they also tend to have a much longer time horizon. Over the long run, though, Option 3, just like the existing team, tends to generate superior returns.

Which team should be picked, assuming that the existing team will make up ¾ of the combined team?

If you only consider expected returns, you would clearly go for Option 1, as the expected return of the combined team is 6¼ when adding Option 2 or 3 respectively. However, this higher expected return comes at a price: when hiring the "fundamental investors", the volatility of the combined team is 5%, it is 4% when adding the "quants" but it is only 2.6% when hiring the "contrarians". Hence, if you adjust the expected returns for the expected volatility of the overall outcome, the best choice is to hire the "contrarians" (Option 3) even though their expected returns are lower and the volatility of their returns is higher than for the other two options. It is the negative correlation with the existing team that very much explains this outcome. As Option 3 tends to deliver superior returns, when the existing team does not (and vice versa), the results of the combined entity are expected to be quite smoothed-out and the risk djusted returns are relative stable. On the other side, hiring the "fundamental investors" (Option 1) turns out to be the wrong decision, since their approach is roughly in line with the existing team. Hiring the "quants" leads to a second-best solution in this example.

This illustration clearly highlights the advantage of diversity, notably in the investment process. The implications are obvious, and not only for the asset management industry: try to build a heterogeneous team, that combines, for instance, the experience and wisdom of seniors with the state-of-the-art technical know-how and understanding of new trends by younger team members. A team, that blends members with different academic backgrounds and professional experience and a good mix of introvert number crunchers as well as outgoing characters who rely on intuition and who are good in communicating with clients. In general, the best advice is to build a team that benefits from the diversity of backgrounds, skills, and experiences.

However, while diversification seems to be straightforward at first sight, it is not that clear-cut when you think about it twice. Which kind of diversity should you aim for? If you go for the "wrong" diversity in your team structure, you may not end up with the intended results.

Let's illustrate the challenge with an example: assume you are the head of an investment team and could choose between (a) a team that consists of men and women of different ages and nationalities from all over the world, all with a finance degree from the same business school plus a CFA designation; or (b) a team of only European "pale males" (or Asian women, or any other group of homogeneous origin) of similar age but totally different academic backgrounds—let's say, a team of economists, MBAs, mathematicians, rocket scientists, biologists, historians and lawyers.

So which team is more diverse? The answer is not obvious. As an economist, I am inclined to say "it depends". Assuming that all members of the second group have appropriate work experience, training and passion for managing money, their (likely) different approach and methodology in solving problems and making investment decisions may lead to superior results compared to the first group, which is quite heterogeneous in terms of non-professional criteria but very homogenous in their academic training and, hence, possibly their mindset.

This is not to say that having diversity with respect to non-professional criteria is not desirable nor that it can't lead to better results. Quite the opposite: first, nobody should be discriminated against—discrimination reveals a lack of respect for others, not to mention the fact that it is illegal. Hiring people of different genders, ethnicities, religious beliefs etc. is a straightforward way of fighting discrimination. Second, offering equal opportunities may very well—all else equal—enhance productivity both on a micro level and in terms of the broader economy. People of different social backgrounds, genders, ethnicities, etc. have had different experiences in life and, hence, can bring different perspectives to the table when tackling professional challenges. In that sense, too, the risk of groupthink may be reduced. Having different experiences, however, is totally different from saying that people of different genders or ethnicities think differently. Saying that is clearly an expression of prejudice.

At the same time, diversity goes beyond just having a mixed group of people in a team. Notably, when active internal discussions among a team is a key part of generating ideas—which is typically the case in the asset management industry, particularly within Allianz Global Investors—the team must work together while challenging each other with respect. The overall team result is much more than just the sum of the parts.

In that sense, the team leader's role goes way beyond just intervening in order to achieve consistency of portfolio structures or macro and market forecasts. He or she has to make sure that experts in one field are being challenged—in a respectful way—by experts in related fields. For instance, a US economist will surely have deep insights into the future of path of US monetary policy and what it possibly means for US sovereign bond yields. However, insights from economists specializing in the eurozone and Japan who can comment on the ECB's and Bank of Japan's monetary policy, respectively will surely be beneficial for the US economist's assessment of the Fed's monetary policy. Moreover, a global fixed-income expert can contribute a market perspective to the outlook for US monetary policy and its implication for bonds. In the same vein, a cross-asset specialist will add a different perspective to the discussion, taking into account feedback loops from the global equity, real estate and currency markets on sovereign bonds. In this context it is important to point out that being a generalist—e.g. analyzing an asset across regions or carrying out cross-market analysis—is a specialization in itself that requires specific skills. Unfortunately, asset class or regional specialists can all too often perceive their views to be superior to the ones expressed by generalists. This ignores the fact that—at least in my experience—looking at a specific question from different angles is most likely to generate better and more robust conclusions.

The arguments for diversity and inclusion, therefore, seem to be straightforward. Does this imply that becoming more diverse and inclusive is easy to do? No, it is not. But the difficulty often has little to do with an organization's lack of willingness and more with the institutional and legal framework in which a company operates. For example, a firm's success at having more women not only in the workforce but in leading positions depends also to a significant degree on the company's ability to balance private and professional life. I will discuss this aspect in more detail under *Collaborative Leadership.*

Moreover, conscious and unconscious biases when hiring or promoting people also plays a role here. For instance, how many managers don't accept a male team leader going on paternity leave? Not so long ago taking a break from work as a young father was perceived to ruin your career and in some cases it still does. I personally heard about this in 2012 from employees working for a major US investment bank, when I told them that taking two months off as a father following the birth of your child has become quite common at Allianz Global Investors.

Likewise, are there still employers or managers out there in the 21st century who view being a young woman to be an impediment for climbing up the career ladder? Or does it matter for hiring or promoting candidates if they originate from a "developing" country rather than from the developed West—and does it matter if the candidates have totally different lifestyles? I hope not, but not too long ago it still did.

However, it is much more likely that candidates with similar views as the hiring manager, e.g. with a similar general assessment on financial markets and or the investment process, have an advantage. Research demonstrates that as human beings, we have an inbuilt propensity to gravitate towards similarity (of thought, background, appearance, etc.)—rather than embracing difference. When managers at a modern workplace receive guidelines on how to avoid these kinds of unconscious biases, as we do at our firm, it's easier to be more diverse and inclusive. Tackling conscious biases is more difficult, admittedly.

Efforts to actively promote diversity have to be well explained and communi-cated—and convincingly executed—in order to be supported by employees. The perception of "positive discrimination"—i.e. hiring and promoting candidates from underrepresented groups when it is not fully justified by their credentials—has to be avoided.

Social Responsibility

At an asset management firm, just like at any other private for-profit organization, meeting the shareholders' return targets is a goal in itself. However, this does not preclude doing this in a socially responsible way. You could argue that fulfilling fiduciary duties and managing clients' money prudently is already a way of addressing social responsibility, and I would agree. However, this is not enough. We at Allianz Global Investors are convinced that investing can and should also embrace environmental, social and corporate governance (*ESG*) considerations. Moreover, we are of the strong opinion that ESG investing does not only add value to the society at large: It also helps to reduce portfolio risks without sacrificing returns, as academic studies are proving. ESG investing can, therefore, enhance risk-adjusted returns to the benefit of our clients. Consequently, ESG is an integral part of our investment decision-making. Moreover, some of our funds are dedicated socially responsible investments (*SRI*), which explicitly invest in assets with high ESG scores. Finally, a small but growing section of our funds are so-called *impact funds*, which explicitly aim at achieving more than just financial returns.

What is key in ESG and SRI investing is to actively engage with companies in order to improve their "ESG behavior". In other words, it is not only about penalizing companies—for example, for environmental issues, child labor practices, discrimination against minorities, etc.—by selling their stocks and bonds. It is also about engaging with companies and trying to convince them to change their behavior. We do this not only to be the "good guy", but to enhance the performance of the securities issued by the respective companies and ultimately to improve the returns of our clients' funds.

Likewise, we are also in regular dialogue with governments, central bankers, international organizations as well as regulators. When sharing our insights with them and advising them on policy issues from the perspective of a financial market participant, we hope and strongly believe that this will lead to superior decision-making on their behalf. Just to give some examples: central bankers have a vested interest in knowing how market participants perceive their decisions and communications, because it helps them get a better gauge of the monetary transmission mechanism of their monetary policy decisions. Policy makers and regulators need advice from investors on how, for instance, they can help promote sustainable finance in their area of jurisdiction. Managers at Allianz Global Investors explicitly encourage our teams to participate in these kinds of meetings, and to become members of advisory boards and contact groups with the public sector.

Another way of demonstrating social responsibility is to be active in transferring know-how, e.g. by contributing to economic discussions in public, informing clients, supporting colleagues in giving guest lectures at universities and schools, working in non-profit organizations or contributing in an honorary capacity in the political field, just to give a couple of examples.

22.3.3 Collaborative Leadership

Collaborative leadership for us at AllianzGI means operating in a holistic manner, empowering the team, developing and caring for employees, and fostering collaboration.

Empowering the Team

Empowering the team is closely related to the need for entrepreneurship: You can't have one without the other. At AllianzGI, when employees are encouraged and expected to take responsibility and search for innovative solutions, they are also entitled to make decisions to get things done. In the same vein, this necessitates trust from the managers in the team members' intuition, skills and motivation. Likewise, employees have to trust their managers that honest failure will not be held against them—and everyone must collectively learn from mistakes and actively seek to develop continuously as a team. Finally, team members have to be aware of the firm's and department's vision in order to march in the right direction.

Development and Care of Employees

Developing employees must always be a cornerstone of any successful organization and development is closely tied to the idea of having a diversified workforce. After all, diversity can't be achieved only by hiring. It has to be followed up by developing employees of different backgrounds from entry level to senior management.

Developing people can take different forms, but sending them to a formal training is only one small aspect. Institutionalized development programs and secondments play an important role as well, but these are usually only open to a limited number of colleagues. The most efficient and practical way of developing employees is to widen the scope of their jobs, to expose them to new tasks to give them new and more responsibilities and to give them the chance to connect within the organization. Also important is giving team members room to "shine" by showing their skills to a broad audience, whether inside or outside the organization. Yet this is not only the manager's job: The organization itself must be open and ready to listen to ideas from everyone, including junior-level employees. For instance, at our firm's most important internal investment meetings, it is not only the most senior team leaders who discuss markets. Rather, we explicitly invite internal investment specialists and younger colleagues to share their insights.

As mentioned previously, diversity can also be negatively affected if employees find it difficult to combine professional and private life appropriately. There are multiple options available to help employers and managers *take care* of their employees.

First, it is about offering flexible work arrangements. Employees at Allianz Global Investors who need to leave early or arrive late for personal reasons—e.g. because the kids need to see the doctor—can easily do so compensating by staying longer in the office on another day or by working remotely. Of course, business needs need to be taken into account, and our clients always come first. Overall, however, this flexibility is highly welcomed and fosters our employees' loyalty to our company.

Second, employers may also offer employees assistance from a professional service provider in organizing care for children and elderly family members. This kind of service is very well received at our firm.

Last but not least, creating an environment in which taking parental leave or a sabbatical is acceptable also helps to break down traditional role models. Undoubtedly, this not only helps employees to recharge their batteries and come back with a fresh mindset, but it also helps to promote diversity in the long run.

Collaboration

In a global asset management firm, active collaboration with colleagues within a department as well as across departments is key. The reasons are obvious: having superior processes and insights into markets and client needs is an important differentiating factor in a highly competitive industry. Investment professionals are specialists—they are either experts in one specific asset class or investment region or are cross-asset specialists (as mentioned above, this is another form of specialization. As a result, everyone needs to rely on other experts' insights in order to get a comprehensive view of financial markets. Team members have to actively exchange views on markets and be willing to share their insights, but they must also be willing to incorporate other team members' views in their own decision making. Otherwise, the team can't reap the benefits of specialization.

To that end, we at AllianzGI use a technology called "Chatter" that permits individuals to share research insights in a very efficient way. It operates much like LinkedIn or Facebook do: the system is a repository for research notes, and it also helps colleagues share views and comment on other colleagues ideas. Of course, sharing written research is a complement to, and not a replacement for, other forms of formal and informal person-to-person sharing for example, in investment meetings, on the trading floor or in the cafeteria.

In the same vein, all of our investment professionals and non-investment professionals (including sales, marketing and communication experts) need to closely collaborate in order to truly act in our clients' best interest. Managers who lead by example can prove that effective collaboration is easy to do and very rewarding. Here are a few examples of how this works on AllianzGI's Global Economics and Strategy team.

When putting together a presentation, team members are supposed to rely on insights from colleagues—picking and choosing others' work as it fits, rather than reinventing everything on their own. Obviously, this is not only more efficient, but it also enables our team to incorporate other specialists' insights as well. In short, it enhances the quality of our output.

When on a business trip to see a particular client, we offer our services to other clients by actively contacting sales colleagues from other sales channels. Additionally, we set up meetings with colleagues in other offices if they happen to be in town for a business trip and we try to find out their needs and offer them advice and insights. My own experience is that colleagues are always more than happy to get this kind of support, whether for client interactions or to help develop the insights they need to get this done. When our offers to help are turned down, which, of course, happens occasionally, it is for pure professional reasons only and does not discourage us from trying again next time. As always, leading by example is a necessary—albeit not sufficient—condition for the team's willingness to collaborate. Individual goals have to be set in such a way that active collaboration is rewarded.

22.3.4 Market and Customer Excellence

If an asset management firm strives to be *perceived* as best in class—the *benchmark* it has to *be* best in class. That is why at Allianz Global Investors, we emphasize hiring and developing talent. Moreover, we recognize that taking care of our employees will help us retain them. We have to empower them and encourage them to make entrepreneurial decisions—i.e. to take risks and feel responsible. We are convinced that this is a precondition for achieving top-quality results.

However, encouraging an entrepreneurial spirit is not just the manager's job: each individual employee must also be responsible for trying to achieve his or her best results. This takes curiosity, the willingness to learn every day, and the drive to improve hard and soft skills not only early in his or her career, but also at a later stage.

This is critical in the asset management industry, where market circumstances are changing every day. In order to be able to deal with this change, employees need to have a deep and broad understanding of capital markets, of financial market and economic history and of behavioral finance. A university degree alone is not sufficient. Academic knowledge has to be complemented and augmented by lifelong learning which includes getting an understanding of new developments, instruments and theories. However, as I alluded to in the first section on entrepreneurship and innovation, some of the biggest treasures in finance literature can be found in theories and research results that have been forgotten by many market participants.

The lifelong pursuit of knowledge may also lead employees to sit for a post-graduate exam designed for practitioners, e.g. the Chartered Financial Analysts program, which requires a high degree of self-motivation, diligence and discipline. Moreover, learning and growing means leaving the comfort zone of our usual work that we have grown accustomed to, sometimes over many years, and instead voluntarily taking on new responsibilities and challenging ourselves with a new environment, new themes or new colleagues and a new office location following a relocation. This is cumbersome initially. Nevertheless, it is a promising way to develop and *excel* as a person and as a professional.

Improving and building up know-how should not be mixed up with striving for absolute perfection. Perfection in the professional world is rarely possible, as it would usually require an unlimited amount of resources—which are typically not available in the real world. Instead, perfection means to go for the best and/or maximum output given the resources at hand.

Increasing the quality and/or quantity of output can only be achieved by improving productivity. This requires a reduction in complexity in processes or outputs: we have to *simplify* our work. Using new technologies, of course, is one aspect. But simplification goes way beyond that. It requires us to reflect constantly on what kind of output is needed and how to produce it—or how to produce something similar that serves the same purpose with fewer resources. In the investment world, simplification means thinking hard about what kind of information is required for achieving superior investment returns—and what is not.

A couple of examples illustrate this point. We know from the field of behavioral finance that the amount of information required for efficient investment decision-making is rather small. What matters more is having the *relevant* pieces of information. Of course, investors are too frequently showered with non-relevant details about all kinds of macro and market developments. Identifying that which is relevant is not always easy: Deep insights, know-how and experience are needed to be able to differentiate between noise and essentials. To illustrate this challenge, consider that in the asset management industry, one frequently hears requests for a detailed quarterly breakdown of GDP estimates, point forecasts for earnings growth, or year-end forecasts for equity, bond and currency markets. The issue is that these data sets are, to a large extent, irrelevant for superior decisionmaking. Moreover, point forecasts that create the illusion of being precise can also confuse the audience—plus they are frequently inaccurate and, consequently, damaging to the credibility of the author. That is why the Economics and Strategy team prefers to research those factors that we deem to be true drivers of financial markets. Instead of making point forecasts, we focus on getting the direction of market driving factors right: Are high-frequency economic data accelerating or decelerating? Are structural developments in an economy improving or deteriorating? Are financial-stability risks building or declining? Are asset markets over- or undervalued? Are investors overreacting to the news flow? Which factors are relevant at which point in time? To be sure, all this work involves working quantitatively and running a lot of data. The key difference to the "old school" approach is that we really try to concentrate on data that matter so we can come up with simplified outputs. In other words, we "reduce information to the max".

Getting this message across is sometimes tricky, since our internal and external clients can still be distracted by data that hardly matter for good decision making. We therefore have to try and convince our clients, in a respectful way, that our research style is more promising than the traditional one.

22.3.5 People Attributes' Interdependence and Relation to Corporate Values

At Allianz Global Investors, the People Attributes explained previously don't stand in isolation: they are interdependent and reinforce each other. These attributes are meant to enhance results and to help individuals improve their hard and soft skills. This can happen directly by empowering employees to take risks and to try innovative solutions, developing their skills on the job or through dedicated training. Yet it can also happen indirectly by increasing diversity in a team, which facilitates learning about different views and discovering new ways to find solutions. In the same vein, engaging with companies and policy makers allows us to look at things from different angles, to gain deeper insights and to foster communication and persuasion skills.

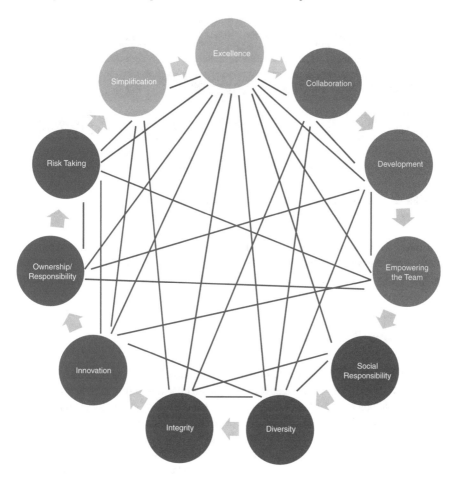

Graphic 22.2   Interdependence of people attributes

There are many more examples for how People Attributes are interrelated. For example, encouraging *diversity* is one aspect of being *social responsible* and living with *integrity*. Moreover, building a more *diverse and inclusive* environment allows colleagues of different backgrounds to develop within the organization. *Innovative* approaches, almost by definition, imply *taking risk*—trying new and less tested ways of solving problems. *Taking responsibility* for one's decisions, i.e. taking risks, necessitates that you are *empowered* by your manager to do so. When *collaborating* with colleagues, it has to be done in a *respectful* way. Collaboration can also lead to *simpler* solutions as one builds on the experience and wisdom of colleagues with different backgrounds.

Graphic 22.2 illustrates the interdependence of our People Attributes.

To what extent do the people attributes complement our corporate values of *Excellence, Passion, Integrity* and *Respect?* Not all of our four values apply to our people

		Excellence	Passion	Integrity	Respect
Entrepreneurship	Risk Taking	X	X		
	Ownership & Responsibility	X		X	X
	Innovation	X		X	
Trust	Integrity			X	X
	Diversity & Inclusion	X		X	X
	Social Responsibility	X		X	X
Collaborative Leadership	Empowering the Team	X	X	X	X
	Development & Care	X		X	X
	Collaboration	X		X	X
Customer & Market Excellence	Knowledge/ Excellence/ Benchmark	X	X		
	Simplification	X	X		

Source: AllianzGI

Graphic 22.3 Corporate values and the importance for people attributes

attributes to the same extent, as Graphic 22.3 illustrates: usually, some values are more important than others for specific people attributes. However, every one of our people attributes can be linked to at least two of our company values.

22.4 Final Remarks

The asset management industry is very much a "people's industry," and there are particular people-oriented challenges for managers whose leadership is built on essential corporate values and corresponding people attributes. One of the most important is giving team members room to work independently and in a creative way so that they can grow—in addition to setting motivating goals and incentives.

Perhaps the most critical thing for a manager to do, however, is to lead by example. Leaders need to live the corporate values and exhibit the People Attributes every day. Of course, nobody is perfect—we are all human beings and we all make mistakes. However, leaders are measured in many ways. Not only do their superiors measure their numbers and performance, but team members assess how the manager lives the firm's values and people attributes every day. Failing to do so can backfire quickly if the leader doesn't live up to the values and the people attributes, you can hardly expect the team to do so. Even worse, morale can fall in sync with declines in the leader's credibility. That's why corporate values and corresponding people attributes should be guiding principles for the leader as much as the team.

References

AllianzGI. Active is: Embracing diversity (in-house brochure).
AllianzGI. Flexible Work Arrangements Principles & Guidelines (in-house brochure).
Elton, E. J., Gruber, M. J. (2009). Modern portfolio theory and investment analysis.

Haldane, A. (2018). Diversity versus merit is a false choice for recruiters. Financial Times, July 22, 2018.

Keynes, J. M. (1936).The general theory of employment, interest and money.

Kindleberger, C. P. (1978).Manias, panics, and crashes.

Minsky, H. (1986). Stabilizing an unstable economy.

Stefan Hofrichter is Allianz Global Investors' Global Economist and head of Macro Research since 2011, based in Frankfurt. Together with his team he is advising clients, in-house investment professionals and sales colleagues on global economic trends and asset allocation. Stefan joined the firm in 1996 as an equity portfolio manager and assumed the role of economist and strategist in 1998. Between 2004 and 2010 he also had responsibility for various multi-asset manadates. In 2016 Stefan became a member of the ESMA (European Securities and Markets Authority) Group of Economic Advisers. He holds a degree in Economics from the University of Konstanz and in Business Administration from the University of Applied Sciences of the Deutsche Bundesbank, Hachenburg and is a CFA Charterholder.

Chapter 23
Innovation in a Turbulent World: The Case for Creative Leadership

Joseph Press, Sandy McLean and Cindy McCauley

Abstract In this chapter, we begin by illustrating how leaders in turbulent times demonstrate vision, understanding, creativity, and agility to impact our fraught, fast-moving age. However, we argue that contemporary leaders must go beyond mere capabilities to seek meaning for users, organizations and society. To respond to these acute conditions, we propose an alternative leadership ontology whose purpose is to co-create the meaning required to engage members of a collective to cultivate the conditions to achieve mutual long-term goals. We examine how Creative Leaders can cultivate Creative Leadership, if they establish direction, alignment and commitment across the collective. Our call to action for learning leaders is to develop meaning-makers and innovators who can tackle the wicked, existential problems, facing humanity.

23.1 Introduction

We live in many worlds, bounded primarily by our identities and perceptions of reality (Chrobot-Mason and Ernst 2010; Goodman 1978). One world can be described as a steady state, where the raw materials are known, and the variables change only neglibly over time (Gagniuc 2017). In the steady state, we believe we can control the outcomes, using knowledge and structure to achieve common goals. We can continue to govern our lives and institutions according to known frameworks and philosophies. This world and the inherent attitudes inform our behaviours and actions. In the context of humanity, we may say that this world is predominantly in our mind (Natelson and Natelson 1975). Akin to Daniel Kahneman's System 2, we perceive a slower pace, enabling the analyses which allows for reason to dominate (Kahneman 2011).

Another world is more turbulent. It has more interdependencies, arising from the complexities of systems, stakeholders, and societal institutions. Described by the

J. Press (✉) · S. McLean · C. McCauley
Centre for Creative Leadership, Greensboro, USA
e-mail: pressj@ccl.org

S. McLean
e-mail: sandy@sandymcleanpr.com

© Springer Nature Switzerland AG 2020   391
N. Pfeffermann (ed.), *New Leadership in Strategy and Communication*,
https://doi.org/10.1007/978-3-030-19681-3_23

U.S. Army War College following the Cold War (Kraaijenbrink 2018), this volatile, uncertain, complex and ambiguous world (VUCA) is a closer reflection of our reality. The majority of us reside in this inherently instable world. Emergent technologies and evolving wicked problems only exacerbate the challenges in the tumultuous world. The frame of mind in a turbulent world is quite different from that of the steady state—we seek rationality amid the complexity to inform behaviour.

The challenge for leadership is to operate within the nexus of these two worlds, and manage the polarities. While we seek to transform the steady-state, we seek control in the turbulent state. However, neither strategy can work in this age of unprecedented complexity and possibility. System 2 can produce poor (sometimes irrational) results, despite being conscious and deliberate (Kahneman 2011). On the other hand, the managerial mindset, finely tuned for a 'plan-and-control' world, cannot cope with the uncertainties of contemporary life. The challenge for humanity is existential. Already back in 2004, Harvard Business School Professor and former Medtronic CEO Bill George wrote "The time is ripe to redefine leadership for the 21st century." (George 2003).

This chapter explores how to encourage a new type of leader, and then go beyond into a new type of leadership. In the VUCA world, leaders must be able to explore complicated situations, to create visions of what could be, to ideate potential solutions, to build prototypes to test, and to communicate them in compelling ways. However, given the complexity and criticality of solving wicked problems, innovation is not sufficient. We contend that creative leadership is the essential differentiator to create shared meaning, value and impact. Understanding the identity of the creative leader and their responsibilities to cultivate creative communities, can provide a foundation for both innovation and transformation. Since we believe that our mission is to develop leaders who can thrive in a turbulent world, ideally creative leadership will establish an agenda to inform our collective responsibility to nurture leaders' growth, celebrate their successes (and failures), and guide their efforts to change our world for the better.

23.2   Creative Leaders Lead with VUCA

enlargethispage-20ptChange and innovation are typical battle cries of leaders who seek to alter existing situations into preferred ones, to paraphrase Simon (1969). Based on over 40 years of research and experience, the Center for Creative Leadership proposes that for successful leadership in turbulent times, it is necessary to begin leading as a Creative Leader. In the spirit of our VUCA world, for the purposes of this chapter we define these attitudes also as VUCA. As Fig. 23.1 describes, in situations that are Volatile, Uncertain, Complex and Ambiguous, leaders should project a **Vision**, an aspirational direction of what the future will look like. They need to co-create **Understanding** amongst stakeholders by engaging in the hard work of converging on meaning. **Creativity** yields representations and experiences that encourage shared

In VUCA	**Be VUCA**
Volatile - The speed of forces and catalysts that spark changes in situations	**Vision-** imagine an aspirational direction of a meaningful future
Uncertain - The lack of predictability or awareness of issues and events, with the potential for surprise	**Understand** – map the ecosystem and empathize across all stakeholders to identify opportunities for impact
Complex- The lack of a cause-and-effect chain to indicate what may come to a situation an organization	**Create** – visually represent intentions to spark the critical conversations that iterate meaningful experiences
Ambiguous - The haziness of reality, the potential for misreads, cause-and-effect confusion	**Agile-** refine direction or pivot towards converging on shared meaning

Fig. 23.1 Leading in VUCA situations with VUCA

understanding and active engagement. As polarities emerge, **Agility** is required to change direction or pivot in an effective manner, which leads to iterative collaboration and social learning.

23.3   Stories of Creative Leaders

23.3.1   Lucid.Berlin's Innovation Prize

To understand how creative leaders innovate and lead in our turbulent world, we recant stories of innovators. The first is about Felix Matschinske and Jan Schiele, and their company Lucid Berlin, whose Children and Armed Conflict app was awarded the Prize for Innovation in Global Security by Geneva Centre for Security Policy (GCSP). Starting with conversations with human rights experts, it became clear that the constantly growing number of thematic Security Council Resolutions were becoming the primary inhibitor to constructive negotiations. A vision emerged to provide Human Rights and Security Council experts in UN Delegations tools to sift through the complex content and apply frameworks of the Council to country specific negotiations. By spending time *understanding* the challenges of those negotiators, the Lucid team was able to draw *insights* to inspire *vision* and *creativity* and develop their award-winning solution. For them the Geneva Centre for Security Policy prize was a recognition of their focus on ***agility*** when developing strategies and solutions that tackle global challenges, such as climate change, human rights, energy transition, inclusive business, impact investment or corporate climate strategy.

In conversation with both Matschinske and Schiele, they acknowledged that in a turbulent world and challenging situations, their motivation was "It just makes sense." Lucid leadership sought to take part in reinventing global cooperation, or stay stuck in old structures and strategies. They added, as part of the "change generation," their team will never stop innovating to provide solutions. All team members are willing to work hard because it makes sense that it is fulfilling. Their dedication, conviction, and resilience, is common and inspiration for creative leaders.

23.3.2   Iterating the Hippo-Roller

The challenge of obtaining clean water provides a second story of creative leaders. In addition to the challenge of finding clean water, another is actually accessing the water, particularly in remote locations. Two inventions were designed around the same time in the mid-1990s to bring clean water to rural people across Africa. Located in South Africa, Pieter Hendrikse and his brother developed the Q-drum— a plastic container with an empty shaft to make it rollable with a rope. However, the patented rotational molding manufacturing process proved to be a significant challenge. Without a producer, Hendrikse could not test and scale his design, limiting its adoption and impact (Cho 2013).

Nearby Hendrikse in South Africa, Pettie Petzer and Johan Jonker launched the Hippo Water Roller Project in 1994. In 2006, 26-year old industrial designer Emily Pilloton founded Project H Design, a non-profit that helps support, create and deliver life-improving design solutions for the four H's: Humanity, Habitat, Health and Happiness. Project H's first initiative was to raise enough money to deliver 75 Hippo Rollers to Kgautswane, South Africa. When Pilloton went to South Africa for a site visit, she realized that the Hippo Roller had several design flaws. She enlisted Engineers Without Borders to propose a redesign (Walker 2009). With Project H advising how to improve the original design, Grant Gibbs assumed responsibility for bringing the project to market. Despite a higher cost than the Q Drum, the Hippo Water Roller achieved broad social impact. It makes it easy to collect water in tough rural conditions, and is user friendly for all ages and sexes. With a total distribution of over 51,000 Hippo Water Rollers, offering a carrying potential 16 billion liters of water, the Hippo Water Roller has reached close to half a million people (Gibbs Interview 2014).

We see the Hippo Water Roller is a more ***creative*** solution, based on a ***vision*** for a product that affords more uses, and enables improvements throughout daily life. The cap is a hygienic filter, enabling both convenience and health. The frame easily converts into a stand to enable economic activity. Turned around, the roller becomes a wheel to be used for transport. Ideas only become innovative ones if they improve daily lives, not just specific needs, and that requires ***understanding*** people's day-today realities, and **agility** to adapt to those circumstances. Not surprisingly, Grant now uses crowdsourcing platforms and connects to the "Empowering People Network" to match funders with good projects, demonstrating how Creative Leaders continually seek ways leverage and scale innovation (Gibbs Interview 2014).

23.3.3 Improving Newborn Health in Mumbai

Creative Leaders also learn by failure. In a New York Times article author Sam Loewenberg documented the innovative pivot of an initiative to improve newborn health in Mumbai (Loewenberg 2013). Started in the mid-2000s, the program was intended to deploy successful practices of rural communities into the slums of Mumbai to reduce infant mortality. The program facilitated discussions with expectant mothers to raise their awareness of the importance of newborn health. They set up 244 women's groups, which met biweekly with social workers to discuss maternal and child health problems, counsel one another and work to develop solutions.

The researchers also tried using innovative approaches to engage the women, such as role-playing and discussion cards. However, in summer of 2012, researchers at Mumbai's City Initiative for Newborn Health published disappointing results of their three-year effort to implement a community-based maternal- and infant-health initiative in the city's slums. After going into approximately 422 sessions, the results unfortunately demonstrated very little impact, measured against any changes in infant mortality. The team realized that the transition from rural settings to a hyper-urban mega-city like Mumbai was too difficult, and even measuring improvement in infant health is much easier in rural areas than in urban ones.

Rather than stop the program, the researchers decided to take this as an opportunity to pivot, to understand and learn from what they went through, to recast the initiative. The newfound **vision** was not on educating, but providing services that would clearly fill the needs of both the expectant mothers and their newborn children, such as immunization, family planning and help navigating Mumbai's Byzantine health and social service systems. This new **understanding** encouraged leadership to re-launch the program in a more **creative**, engaging way. In partnership with the Gates Foundation and a local television network NDTV, they produced a promotional video that took a positive approach by suggesting solutions to the problems affecting women and children. The **agility** leaders of the research team demonstrated by admitting failure enabled the redesign and reboot of the program. The creative leaders who sought to change Newborn Health in Mumbai lead with VUCA to create sustainable innovation with impact.

23.3.4 Learning from Creative Leaders

As Loewenberg notes "… making a difference in the world is hard, often messy." Creative leadership starts with a **vision** to frame a direction to a "wicked problem"—a complex, inter-dependent situation that resists resolution because of incomplete, contradictory, and changing requirements. Setting a visionary direction requires imagining an idea for a different experience or the intent of a new reality to improve a current situation. All the leaders sought a problem, or wicked challenge, that had some important meaning for society. They imagined possibilities, not accepting the way things currently were, similar to the way the Lucid team didn't accept the way negotiators had to fumble through papers and sought to empower their negotiations. Others found a way to empathize with the individuals involved and provide a meaningful experience.

With a vision in hand, creative leaders can seek **understanding** key stakeholders involved in a situation. The consumers, organization, and the broader community need to find experience *meaning*—emotions that fulfil deeper needs of stakeholders. All of the innovators not just came up with ideas, they sought to propose and try. The leaders were able to then visualize and communicate the experience in a compelling way. When meaning delivers impact to society, teams, and people, the leader herself will be engaged to use **creativity**. Representing ideas and intentions of the new reality, using techniques like storyboarding, enable users to visual the experience towards creating shared understanding. The resultant feedback, particularly critique, sparks iteration. As Professor Roberto Verganti writes in Harvard Business Review (2016), critique sparks collaborative learning with users to encourage reflection and convergence of meaning.

These situations of iterative collaboration that yield innovation by social learning are a direct result of **agility**. The leader's ability to change direction or pivot in an efficient and effective manner is essential to *adapt*ing the vision and realign meaning to ensure broad feasibility. When they would try, they may fail, but then they would **learn and adapt** by incrementally improving their innovative ideas to the realities of the turbulent world. This is a fundamentally different way of leading—in the past we wouldn't admit failure, nor admit that we can learn. Changing because something didn't work out was the sign of a weak leader. However, in the turbulent world, change is the way to bring people together and create diverse communities. Formulating and communicating new proposals encourages coalition building. Realigning plans to reflect evolving situation and new preferences ultimately leads to revitalized personal meaning with purpose, essential for leading in turbulent times.

23.4 Creative Leadership in VUCA

23.4.1 The Search for Meaning

However, the success of these, and other leaders, goes beyond their VUCA capabilities. If it was just a matter of inculcating competencies, or encouraging attitudes, we would have an abundance of creative leaders and a smaller set of wicked problems in the world. Viktor Frankl, in his seminal work *Man's Search for Meaning* (1946), proposed a fundamental truth of our existence, that humans "…can only live by looking to the future." Our VUCA leaders all had a purpose, a "… call of a potential meaning waiting to be fulfilled…" (Frankl 1946). This intrinsic motivation to realize the potential impact of their innovation was a key source of resilience to overcome the challenges the leaders in our stories faced. When leaders also seek to impact a grand challenge—like clean, accessible water or security—leaders find themselves in a place where they significantly amplify their engagement and can actualize their identity as a creative leader.

The importance of purpose, or meaning, has increased significantly across all dimensions of society. According to the 2018 Global Impact Investing Network annual survey, five-year repeat respondents increased the amount of capital invested by 27% over five years to a total of USD 35.5 billion into 11,136 deals during 2017. The rise in impact investment demonstrates that the business community has awakened to purpose and profit. In an open letter to CEOs published in January 2019, Larry Fink, CEO of BlackRock, wrote "Profits are in no way inconsistent with purpose—in fact, profits and purpose are inextricably linked…Profits are essential if a company is to effectively serve all of its stakeholders over time—not only shareholders, but also employees, customers, and communities."

In a world overcrowded with ideas, achieving shared meaning with multiple stakeholders is no simple task (Verganti 2017). As Wenger (1998) writes "… human engagement in the world is first and foremost a process of negotiating meaning." The leader who seeks convergence of meaning, in the form of an innovation with purpose, finds herself in the middle of strenuous tripartite polarity. Today's leaders must achieve organizational objectives and customer expectations, while simultaneously achieving some form of societal impact. Although the ubiquity of human-centered design, in its most common form called "design-thinking", may contribute to fulfilling the needs of users and decision-makers, it does not address the challenge of convergence, which is critical success criteria for leading in a VUCA environment, and is therefore insufficient to realize innovation with sustainable impact.

To realize a meaningful and motivational purpose, creative leaders must engage in the act of *making common sense*. In their book of the same title, Paulus and Drath (1994) propose to view leadership as a social meaning-making process that occurs in groups of people who are engaged in an activity together—commonly referred to as a community of practice. According to Paulus and Drath, "[M]eaning-making consists of creation, nurturance, and evolution (or revolution) of cognitive and emotional frameworks. When making such frameworks happens in a community of practice (people in an enterprise with a shared purpose or history), then we can say that leadership happening."

23.4.2 Direction, Alignment, Commitment: A New Ontology of Leadership

Traditionally, Leadership is said to exist when there are leaders, followers, and shared goals. Dubbed the "tripod ontology" by Warren Bennis, in his introduction to a special American Psychologist issue on leadership, he wrote: "In its simplest form [leadership] is a tripod—a leader or leaders, followers, and a common goal they want to achieve." (Bennis 2007). In this paradigm, leaders and followers are in turn recognized by their respective roles in an asymmetrical influence relationship. However, in a VUCA world, where leadership is called to become increasingly peer-like and collaborative, the tripod imposes limitations on leadership theory and practice.

To respond to new conditions, The Center for Creative Leadership developed an alternative ontology based on the view that leadership encompasses a full range of human activity (including leaders influencing followers) whose purpose is to bring members of a collective into the conditions required for the achievement of their mutual long-term goals (McCauley 2019). These conditions are **Direction** (agreement on the collective's overall goals), **Alignment** (the coordination of work in the collective), and **Commitment** (mutual responsibility for the collective), or **DAC**. In other words, leadership is said to exist when direction, alignment, and commitment are present among people with shared work. As illustrated in Fig. 23.2, DAC enables Creative Leadership in a VUCA world by encouraging:

- **Direction**—The exploration and emergence of new perspectives to co-create the meaning of the collective and shape its overall goals.

Fig. 23.2 How the search for meaning stimulates creative leadership

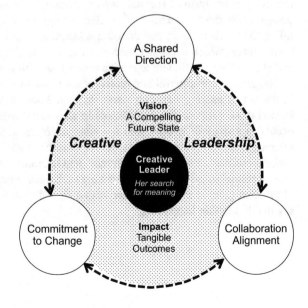

- **Alignment**—The on-going mutual adjustment among system-responsible people to coordinate the work within the collective.
- **Commitment**—The desire to be engaged in developing a community with a strong sense of mutual responsibility for the collective.

DAC is a relational framework, which is socially constructed. It can be purpose-driven, creating meaning for a shared outcome. This emerging practice of Creative Leadership requires the development of a theory of how leadership cultures develop to deal with increasingly complex challenges.

23.4.3   Creative Leadership in India

To explore the integration of such practices aimed at developing individuals who simultaneously seek changing organizations and communities, we will examine the case of RB (Reckitt Benckiser). India is one the world's fastest growing economies. While no one can doubt its increasing influence on the world stage, it also faces the uncomfortable truth that many people still struggle with life on the poverty line. The paradox is that India is a technological leader and home to titans of industry, yet millions still live in the most basic conditions with inadequate sanitation and healthcare.

Prime Minister Narendra Modi has not shied away from this reality. In October 2014, he launched Swachh Bharat Mission, the "Clean India" Mission. At that time, virtually every second person in India had to defecate in the open, more people had a mobile phone than a toilet and every third person was drinking unsafe water. This could be no longer be tolerated. Modi set a target date of 2 October 2019 for achieving the Swachh Bharat Mission (SBM), the 150th anniversary of Gandhi's birth. Reckitt Benckiser (RB), one of the world's largest health and hygiene companies, wanted to help make the Clean India Mission a success. RB believes that ensuring healthy lives and happy homes for everyone means assuring everyday hygiene for each person. RB found a natural fit with the Indian government's *Swachh Bharat Mission* and decided to align with the call for a clean and hygienic India through the its *Banega Swachh India* campaign which was aimed to drive positive behavioural change towards hygiene and sanitation habits across the country by providing innovative solutions.

Patty O'Hayer, RB's Global Head of Communications and Government Affairs told us major public health issues are nothing new, and we often look to governments to solve them. According to O'Hayer, RB has learned that these issues are more successfully resolved when the government partners with other stakeholders such as businesses, media and NGOs. RB's Banega Swachh Mission is one of those successes.

O'Hayer described how RB approached the government and proposed that the company become a partner in Prime Minister Modi's Clean India initiative. Working with RB headquarters, Ravi Bhatnagar, who joined RB in 2015, put together a partnership framework and developed the strategy and tactics. According to O'Hayer, timing was also a key element of success. RB got in on the beginning of the government initiative, which gave the project a clear mandate and helped attract important partners. Securing the support of the two biggest India media companies—NDTV and Dainik Jagran—boosted Banega Swacch's reach to millions of people across the country. Banega Swachh is thus an excellent example of RB's purpose-driven ethos.

23.4.4 DAC at Banega Swacch

RB focused on fundamental issues of cleanliness and scaled them to a mass level. The company's leaders at UK headquarters and in India drafted a five-year vision which helped set the **direction** and committed resources through dedicated specialists. RB's staff in India employed their creativity to come up with innovative tactics to tackle their ambition. In every step, employees, partners and specialists were **aligned** with Banega Swachh's direction and approach.

RB targeted communities with 85 thousand families to get directly involved and push for behaviour change by introducing a three-year Hygiene Curriculum in 125,000 schools. Understanding cultural habits and beliefs as well as the role education could play in changing behavior was key to development of the curriculum and winning celebrity support. Opinion leaders and influencers from diverse fields including television and films, sports, and politics were recruited to support the Banega Swachh India campaign and advanced its credence.

23.4.5 Engaging Partners and Employees

RB has always believed that transformational change can be successfully achieved with shared responsibility. The success of such a huge initiative as Banega Swachh India depended on finding the right partners to lend their expertise in shaping it. They had to buy into the societal meaning and purpose of Banega Swachh, and more broadly, Prime Minister Modi's Clean Indian Mission. This was the **commitment** phase of the programme.

Aiming towards mass outreach across the country and to provide better living to as many people as possible, RB partnered with two of the biggest Indian media companies—NDTV and Dainik Jagran. RB was joined by influential and popular actor, producer and TV host Amitabh Bachchan as campaign ambassador.

RB also partnered with the Central and Local governments along with the Ministry of Drinking Water and Sanitation and the Ministry of Urban Development. Reputed non-profit organizations and international aid agencies such as United States Agency for International Development (USAID), Bill and Melinda Gates Foundation, Aga Khan Foundation, XSEED, Butterfly Fields, Jagran Pahel, NDTV and the Swades foundation, FICCI (India Sanitation Coalition) gave valuable inputs and support.

Additionally, the entire RB India workforce pledged over 20,000 h to clean-up drives and organized neighbourhood clean-ups.

23.4.6   RB's Impact

RB's VUCA-driven innovation, framed by DAC, was focused on new ways to build capacity and spread behaviour change to large populations. The impact has been phenomenal because of concerted efforts from people across the country who shared RB's purpose. Two hundred villages are now open defecation free, with 35,000 new toilets being built. Handwashing behaviour has risen to a great extent in the six targeted states and all of it simultaneously contributed towards a healthy growth of the company along with better market shares.

By spearheading a campaign for change, leaders injected new vigour within RB, one that comes from making a difference. Deploying some of the company's biggest brands in a nationwide drive has had a positive impact on profits, outlook and vision as well as on RB's commitment to core values, strengthening passion to even further drive business with purpose.

23.4.7   Learnings from Creative Leadership

Before embarking upon Banega Swachh, RB leaders in India and at corporate head-quarters, including CEO Rakesh Kapoor, had to collectively *envision* a meaningful **Direction** before they could agree on aspirations and goals. Working with government officials and NGOs, they shared a joint exploration of a wicked problem: poor sanitation, and the life-threatening diseases such as diarrhea and cholera that result.

Quickly RB's team moved to develop shared ***understanding*** across the collective to **Align** resources and coordinate effort. RB India achieved agreement with corporate HQ on how to mobilize and deploy resources, and also worked with local and international partners. This collaboration among diverse participants seeking systemic change yielded coordinated and consistent impact.

RB's team on the ground used ***creative*** tactics and communicated a compelling beneficial outcome for each person involved. They enabled the community to experience the growth and success of their joint efforts, thereby increasing participant engagement and **Commitment** across the community. For example, women were empowered with the knowledge of best hygiene and sanitation practices, and this in turn opened up economic opportunities for them in their communities.

In order to demonstrate personal ***agility*** and encourage others to cross personal, organizational and community boundaries, RB staff working on Banega Swachh upped their individual **direction, alignment and commitment** skills, increasing the impact for ordinary Indian citizens, the organization, and society. One of those was Ravi Bhatnagar, who is now RB's Director External Affairs and Parnerships, Africa,

Middle East and South Asia, who attended CCL's Leadership Development Program (LDP)® in Brussels. Bhatnagar cites the direct positive impact the program has had on his work on Banega Swachh. In 2017, he joined the high level technical team of the public/private Hygiene Index, and became part of the effort to help 100 million Indians lead healthier lives by 2020. That helped him to stretch into a more proactive and collaborative leadership role.

23.5 The Case for Creative Leadership

Unfortunately, in today's turbulent world, humankind doesn't lack opportunities to benefit from the potential of creative leadership. The UN has identified 21 global issues, including Aging, Big Data, Climate Change, Gender Equality, Peace and Security and Refugees (United Nations 2019). Worryingly, the World Economic Forum states: "Humanity has become remarkably adept at understanding how to mitigate countless conventional risks that can be relatively easily isolated and managed with standard risk-management approaches. But we are much less competent when it comes to dealing with complex risks in systems characterized by feedback loops, tipping points and opaque cause-and-effect relationships that can make intervention problematic."

Given there is much opportunity for self-actualization, and increasing appreciation of its financial value, where are all the **individuals** *willing* to step to these opportunities? More importantly, how can we develop **individuals** who are *able* to step up to these opportunities? In our experience, **co-**creating new meaning, for organizations, customers, and communities, can accelerate developing innovation leaders. Within this triangulated space, leaders find within themselves with the passion to look into the future, the inspiration to create meaningful directions, and the resilience to struggle against the challenges of bringing shared value to the world. In this space they must overcome many boundaries– between individual identities, and between those identities and collective (Chrobot-Mason et al. 2009). Given the breath and depth of boundaries, creative leadership is dependent on social integration as the driver for transformation (Pasmore 2015).

When leaders are given an opportunity to create propositions that bring together the needs of those three pillars, they are enabled and empowered to have significant impact for the daily lives of those they serve. As their vision and mission—how to get there—becomes apparent through co-creative, "designerly" activities of iteration, critique, and dialogue, their passion and resilience to make an impact can increase exponentially (Cross 1982).

Searching for meaning is an engaging way to accelerate the development of creative leaders, because the search is, at its core, is an act of self-actualization (Maslow 1943). By positioning leaders and their development within this space, we see significant personal, organizational, and societal impact. By co-creating shared meaning, engagement evolves in an implicit and complicit way that is essential to bring meaning to the world. By providing leaders the opportunity to create a vision that gives them passion, intrinsic motivation to solve the problems of society, meet the strategic objectives of the organization, and contribute to the benefit of all individuals in the collective community.

23.5.1 Our Responsibility to Cultivate Creative Leadership

From the personal to planetary, change is upon us. Increasingly, change is forced upon organizations because their leaders lack courage and imagination to envision an alternative set of futures ahead of the curve. When the disruptive forces strike, including technology and competition, the unprepared are left scrambling to reinvent their organisations, in hopes of reclaiming their position in the marketplace.

When people seek innovation together, they seek to make sense, or meaning, of an existing situation to inform a preferred situation. The process of participating in collective making meaning—the striving to make situations make sense for the community—makes leaders out of people. As Bennis (1991) states: "A leader creates meaning. You start with vision. You build trust. And you create meaning."

As learning leaders, we propose it is our responsibly to frame the spaces where leaders can explore, define and negotiate meaning. By setting out this space and asking leaders to set a vision to specify what their mission is, with very specific and concrete propositions, we are able to significantly contribute to their development as leaders in general, and specifically innovative leaders. As a learning leader, all of us should not be hesitant about challenging participants with finding that meaning. As Frankl writes, 'We should not be hesitant about challenging man with a potential meaning for him to fulfil.' Unfortunately, too often we shy away from this meaningful mandate. Encouraging creative leadership is a shared responsibility. If we are all able to lead creatively, then we will certainly impact the VUCA in our world. We will also create experiences and the communities to thrive with shared meaning.

References

Bennis, W. (1991). Creative leadership. In *Executive Excellence*. pp 5–6 as quoted in Drath, W., & Palus, C. (1994). pp. 10.

Bennis, W. G. (2007). The challenges of leadership in the modern world: An introduction to the special issue. *American Psychologist, 62*(1), 2–5.

Cho, K. (2013, February 27). Social entrepreneurship research: Q Drum. Retrieved at: http://risddesekcho.blogspot.com/2013/02/social-entrepreneurship-research-q-drum.html.

Chrobot-Mason, D., & Ernst, C. (2010). *Boundary spanning leadership: Six practices for solving problems, driving innovation, and transforming organizations*. New York, NY: McGraw-Hill Education.

Cross, N. (1982). Designerly ways of knowing. *Design Studies, 3*(4), 221–227.

Chrobot-Mason, D., Ruderman, M. N., Weber, T. J., & Ernst, C. (2009). The challenge of leading on unstable ground: Triggers that activate social identity faultlines. *Human Relations 62*(11), 1763–1794.

Drath, W., & Palus, C. (1994). *Making common sense*. Greensboro, NC: Center for Creative Leadership.

Fink, L. (2019, January). CEO letter. Retrieved at https://www.blackrock.com/corporate/investor-relations/larry-fink-ceo-letter.

Frankl, V. First published in 1946 in German entitled *Ein Psycholog erlebt das Konzentrationslager,* and in 1959, *Man's search for meaning*, Boston, MA: Beacon Press.

Gagniuc, Paul A. (2017). *Markov chains: From theory to implementation and experimentation* (pp. 46–59). USA, NJ: Wiley.

George, B. (2003). *Authentic leadership: Rediscovering the secrets to creating lasting value*. San Francisco, CA: Jossey-Bass.

Global Impact Investing Network. (2018). Annual impact investor survey 2018, Eighth Edition, Retrieved at https://thegiin.org/assets/2018_GIIN_Annual_Impact_Investor_Survey_webfile.pdf.

Gibbs, G. (2014) *Interview with Grant Gibbs on August 04, 2014, founder of the Hippo Water Roller Project*. Retrieved at https://www.empowering-people-network.siemens-stiftung.org/en/service/press/article-interview/interview-with-grant-gibbs-founder-of-the-hippo-water-roller-project/.

Goodman, N. (1978). *Ways of worldmaking* (Vol. 51). Hackett Publishing.

Kahneman, D. (2011). *Thinking, fast and slow*. London: Macmillan.

Kraaijenbrink, J. (2018, December). What does VUCA really mean? *Forbes*, Retrieved from https://www.forbes.com/sites/jeroenkraaijenbrink/2018/12/19/what-does-vuca-really-mean/#63e443e317d6.

Loewenberg, S. (2013, February 1). Learning from failure. *New York Times*, Retrieved from https://www.nytimes.com.

Maslow, A. H. (1943). A theory of human motivation, originally Published in *Psychological Review*, 50, 370–396.

McCauley, C. (2019, March). From ontology to praxis: The enactment of collective leadership development. *Presentation at the Kravis-de Roulet Leadership Conference*, Claremont McKenna College, Claremont, CA.

Natelson S., & Natelson E. A. (1975). Maintenance of the steady state in the human. In *Principles of Applied Clinical Chemistry Chemical Background and Medical Applications*. Springer, Boston, MA.

Pasmore, B. (2015). *Leading continuous change: Navigating churn in the real world*. Oakland, CA: Berrett-Koehler.

Simon, H. A. (1969). *The sciences of the artificial*. Cambridge, MA: MIT Press.

United Nations. (2019). *Global issues overview*. Retrieved at https://www.un.org/en/sections/issues-depth/global-issues-overview/.

Verganti, R. (2016, January-February). The innovative power of criticism. *Harvard Business Review*. Retrieved from https://hbr.org/2016/01/the-innovative-power-of-criticism.

Verganti, R. (2017). *Overcrowded—Designing meaningful products in a world awash with ideas*. Cambridge, MA: MIT Press.

Walker, A. (2009, July 14). Balancing tradeoffs: The evolution of the hippo roller. *FastCompany* https://www.fastcompany.com/1309505/balancing-tradeoffs-evolution-hippo-roller.

Wenger, E. (1998). *Communities of practice: Learning, meaning, and identity*. Cambridge, MA: Cambridge University Press.

World Economic Forum. (2019). *The global risks report*. Retrieved at https://www.weforum.org/reports/the-global-risks-report-2019.

Joseph Press is the Global Innovator and Strategic Advisor of the Center for Creative Leadership. He is the Scientific Lead of IDeaLs, a global platform co-founded by CCL and Politecnico di Milano to research innovation, design and leadership. He capped his 15-year career with Deloitte as the Director of Deloitte Digital Switzerland. Prior to consulting, he was an architect and design researcher. He completed his B.S. in Managerial Economics at Carnegie-Mellon, Master in Architectural Studies and Ph.D. in Design Technology at MIT. He is an adjunct professor at School of Management & School of Design, Politecnico de Milano. He has taught and lectured at IMD, Parsons The New School For Design, University of St. Gallen, and Bezalel Design Academy.

Sandy McLean a communications and international affairs expert, specializes in media relations, digital communications, and content development for healthcare, renewable energy and executive education publications, among others. She began her career working for a US president, and then for a Member of Congress and law firms as a regulatory analyst. She was hired by an international communications agency in 1988 and worked in their Washington, Paris, Brussels and London offices before setting up her own consultancy. For the past 15 years, she has worked in partnership with the Center for Creative Leadership as a writer and editor.

Cindy McCauley with over 30 years of experience at the Center for Creative Leadership, Senior Fellow Cindy McCauley has contributed to many aspects of CCL's work: research, publication, product development, program evaluation, coaching, and management. As a result of her research and applied work, she is an advocate for using on-the-job experience as a central leader development strategy, for seeing leadership as a collective endeavor, and for integrating constructive-developmental theories of human growth into leader development practice.

Chapter 24
Cognitive Dissonance in Leadership Trainings

Winfried Müller

Various kinds of biases are based on the longstanding theory of cognitive dissonance.

Cohan (2002, p. 283).

Abstract The chapter aims to introduce the importance of cognitive dissonance, a well-established psychological theory. The reduction of the dissonant feeling within an individual has severe negative effects on the organisation. Critical leadership situations, following wrong decisions are not solved but reinterpreted in manipulated, white-washed and ignorant manners. Never the less, academic research and practical leadership training has given little attention to this important relationship. According to author opinion, training and coaching should include the mechanism of cognitive dissonance. Moreover, proposals are provided for ways how to reduce dissonant feelings consistent with the scope of leadership principles.

24.1 Introduction: Psychological Aspects of Leadership

During the last years, leadership and management research has increasingly tightened up to other disciplines namely theories of social psychology (Hinojosa et al. 2017). Exceeding the limited audience of academic literature, influential behavioural concepts got popular by publications of Nobel-prize winners such as Daniel Kahneman, Robert Shiller or Richard Thaler (Posner 2010; Akerlof and Shiller 2009).

This development is predominantly driven by the understanding that the classical view on individual behaviour, being a purely rational decision machine, a homo economicus, has practical limitations (Deutsche Bank Research 2010). In the contrary, individuals permanently behave differently than expected by utility theory and

W. Müller (✉)
BMW Bank GmbH, Munich, Germany
e-mail: Winfried.K.R.Mueller@web.de

© Springer Nature Switzerland AG 2020
N. Pfeffermann (ed.), *New Leadership in Strategy and Communication*,
https://doi.org/10.1007/978-3-030-19681-3_24

the rational actor paradigm has been insufficient to explain and understand human actions (Evans 2013).

Heuristics or cognitive biases are the names of this widely unconscious psychological behaviour. Biases tend to speed up decisions in daily life and reduce cognitive work at the cost of potential results in actions different from initial intentions (Bettinghaus et al. 2014; Evans 2013). No decision is free of such intuitive simplifications and mental shortcuts, even under the environment of ostensibly rational organisational systems. Major drivers behind heuristics and biases are impulses coming from the feeling of cognitive dissonance (Wiswede 2012; Cohan 2002).

Cognitive dissonance is frequently involved when significant failures in companies occur (Amar 2012; Posner 2010). Thus, when reflecting on leadership behaviour during the last financial crisis, relying on new regulatory rules and tighter compliance processes alone might fall short for lessons learned.

This chapter aims to argue that both, academic literature and organisational reality has widely overlooked the powerful impact of cognitive dissonance and the related need for managing the phenomenon when it comes to education and inclusion in leadership trainings.

Within the next section, a brief introduction of cognitive dissonance theory is provided. Section 24.3 reflects on the reasons for giving attention to cognitive dissonance in leadership training. Proposals of how to include aspects of cognitive dissonance in leadership training are the content of Sect. 24.4 followed by the conclusion of this chapter (Sect. 24.5).

24.2   The Importance of Cognitive Dissonance for Leadership Trainings

The American psychologist Leon Festinger (1919–1989) has termed the theory of cognitive dissonance in 1957 (Pickren 2014). Festinger's research can be located at two mutually enriching sub-disciplines of experimental psychology: cognitive psychology, the way of human's "thinking about thinking" (Richards 2004, p. 88), and social psychology, dealing with "motives of people living in social relations" (Richards 2004, p. 162).

Festinger summarized his theory based on few fundamental propositional arguments as follows:

> The holding of two or more inconsistent cognitions arouses the state of cognitive dissonance, which is experienced as uncomfortable tension. This tension has drive-like properties and must be reduced. (Festinger 1957, p. 18)

The term cognition comes from the Latin word 'cognitio' and means 'taking notice of something'. Concerning Festingers' theory, it can be paraphrased with 'going into a huddle with oneself', which explains the threat of not living up to own standards.

Cognitive dissonance occurs if perceptions on how something should be, what is right, which standards and values to apply, contradict with a perceived adverse reality, failure or individual mistake. Therefore and following Festingers original formulation, the theory is applicable to situations where decisions have turned out unfavourable or wrong, when being confronted with information discrepant to expectations (Festinger 1957; Cooper 2007). For instance the clash between the perception of being a successful leader and the occurrence of a professional error, leads—besides extrinsic oriented feelings of fear of punishment, loss of reputation etc.—to the uncomfortable inner feeling of cognitive dissonance. The upper part of Fig. 24.1 shows the arousal process of cognitive dissonance.

It is easy to understand that the negative feeling triggers a desire to eliminate cognitive dissonance. Reduction can occur in three different ways: Firstly, by applying a trivialisation mode. Secondly, by a change of perception and justification and thirdly

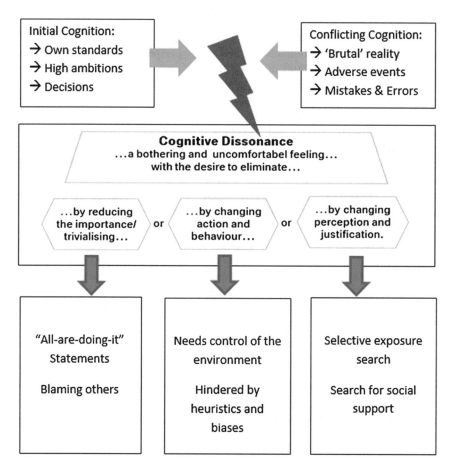

Fig. 24.1 The appearance and ways of reduction of cognitive dissonance following Festinger (1957) and Devine et al. (1999)

by change of action and behaviour (Cooper 2007). All three ways carry the risk of adverse consequences on organisations and leadership behaviour as shown in the lower part of Fig. 24.1.

Firstly, trivialisation, as a way of mental ring-fencing activities ("All others are doing it!") is ignoring the complexity of failures by simplification. A good but dissuasive example is the statement of top executives of Bear Stearns, Lehmann, AIG, and Citigroup during the official investigation of the American Financial Crisis Inquiry Commission (Financial Crisis Inquiry Commission 2011, p. 457) blaming external circumstances:

> We were solvent before the liquidity run started. Someone (unnamed) spread bad information and started an unjustified liquidity run. Had that unjustified liquidity run not happened, given enough time we would have recovered and returned to a position of strength. Therefore, the firm failed because we ran out of time, and it's not my fault.

Secondly, reducing cognitive dissonance by change of perception and justification is similarly not a viable solution from a leadership perspective. Instead of facing the conflict and the adverse result, questing for reasons, why original decisions have been right starts ('selective exposure search', Frey 1986) hindering organisational learning and development. Searching for social support describes the behaviour of a person that is seeking for good feedback instead of true feedback by limiting contact to sources of confirmation instead of critical voices. Both mechanisms aim to legitimate ex-post why a decision make sense in a broader and logical view, by refusing, misperceiving information or seeking for justifying support (Harmon-Jones and Mills 1999).

Cognitive dissonance can finally be reduced is by change of action and behaviour. At first glance, learning from mistakes and developing new ways of acting seems to be the natural solution for responsible leaders, as it stands in contrast to sticking to old decisions too long. However, the necessary control of the environment is often impossible, namely not all decisions can ex-post be made undone. Moreover, research suggests that attitude changes are regularly hindered by internal mechanisms and biases as well (Bettinghaus et al. 2014; Evans 2013; Kahneman 2012). However, despite limitations, change of action and behaviour would be the preferred way of reducing cognitive dissonance.

In summary, all three ways of reduction of cognitive dissonance have the potential to affect the surrounding organisation causing financial, reputational or even human harm. Ignoring or trivialising real reasons for failures, losing crucial time to initiate counter measures by a lack of openness to critical facts and feedback or applying mental heuristic short cuts are manifestations of a phenomena which stands in contrast to basic principles of leadership as f.i. transparency, flexibility and mindful planning and acting.

24.3 Leadership Trainings Neglect Cognitive Dissonance Aspects

In the section before, we have learned that cognitive dissonance is a normal phenomenon in everyone's life that heavily affects organisations as well. Thus, over the last years, the theory has increasingly gained attention in management literature (see Fig. 24.2).

Interestingly, the potential of influencing cognitive dissonance as an object of leadership remains widely untouched in literature so far (Hinojosa et al. 2017). Especially ways, how training could support individual leadership performance remained weak.

This is in line with experiences of the author. During several in-depth interviews, the author has made with senior leaders in the European banking industry, all participants knew and described painful inner feelings and provided various business examples of negative impacts. However, none has ever experienced training or coaching on the impact of the feeling of cognitive dissonance. Even so, the interviews do not provide statistical representativeness, it is fair to formulate the proposition that activities and events where leaders could find space to discuss, to digest and ultimately to manage are at best rare.

As the reasons for dealing with cognitive dissonance are of many kinds, this opens up room for academic research and potential for practical managerial action in organisations.

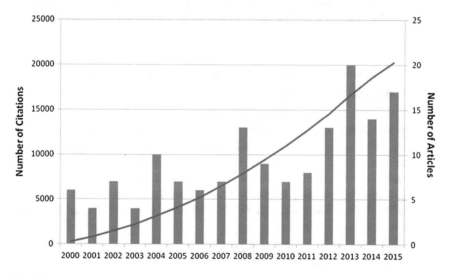

Fig. 24.2 Number of citations (left y-axis) and articles (right y-axis) indicating growing attention on cognitive dissonance theory in management journals between 2000 and 2015 (Hinojosa et al. 2017)

Formally, leadership has a strong behavioural aspect, where deciding is at the heart of leadership behaviour (Mintzberg 2009). This qualifies the activity of deciding as a key quality of leaders and thus exposed to all adverse effects of cognitive dissonance.

Moreover, failure of organisations are often gloomy consequence of a positivistic and almighty view of top managers towards an understanding of the illusion of knowledge and thus the illusion of control. Such unreflect illusions are built over time in order to protect against uncomfortable feelings of inner dissonances, inter alia triggered by cognitive dissonance. Dealing with own limitations could improve leader skills and decision quality in times of disruptive tendencies in many industries.

Unfortunately, professional and personal experiences seem to serve only as limited support in avoiding negative consequences in leadership caused by cognitive dissonance. For instance Tetlock (2000, p. 293) described, long-lasting conservative leaders with "strong preferences for cognitive closure were most likely…to defend simple heuristic driven errors such as over-attribution and overconfidence". Earlier relations and experiences in life and at work influence how later relations are created and managed resulting in a need for examination of cognitive dissonance within leadership training.

Despite the clear and often-confirmed negative effects on the person and the environment, there are only a few links in literature between leadership training and cognitive dissonance. Early research in the aftermath of the Enron failure calls for "emotionally intelligent" staff and leaders (Cohan 2002, p. 298) in an environment, where it is easier for employees to pass on information to superiors quickly and without fears. Recently, the influence of biases is being acknowledged and the importance of de-biasing trainings are stressed (Haerle et al. 2016) but details remain open.

24.4   Inclusion of Cognitive Dissonance Management in Leadership Trainings

The untouched area described in the section before leads to the question of how cognitive dissonance could be included in leadership training.

Addressing organisational and structural activities to reduce the negative impact of cognitive dissonance, like in areas of corporate governance, target setting or decision-making, would exceed the space of this contribution. Therefore, five key aspects are proposed for further discussion focusing on necessary content of leadership training activities.

1. Openness to manage cognitive dissonance
2. Approach towards decisions
3. Dealing with mistakes and learning from errors
4. Importance of resilience
5. Influencing organisational behaviour towards cognitive dissonance.

Dealing with cognitive dissonance from a leadership aspect requires accepting the need for inner examination. In a self-reflected manner, very individual rules of self-steering within the own working life are assessed. This search process has the character of a first-person-research into very personal habits, rituals and paths. While leadership training and coaching can support the reflection work, it is ultimately up to the leader to get involved in this inner 'cognitio', which requires besides self-awareness and self-discipline the willingness of seeking confrontation with some disliked parts of the personality with the aim of allowing the inner conflicts and opening up for managing the feeling of cognitive dissonance.

The second aspect is dealing with the explicit behavioural situation of deciding under uncertainty. Huang and Pearce (2015: 646) describe an individual decision process towards a financial decision in ostensive words: "What do you do, when one aspect looks great and the other awful…you feel conflicted, confused, totally directionless at first…and somehow you acknowledge that chaos; accept it and it starts to feed you some clues".

Mintzberg located decisions as activities in the area between art, applied science and handcraft (Mintzberg 2009) and stated that leaders "operate under conditions of great uncertainty. They often need to make decisions that are not well grounded in any rational analysis of well-defined alternatives but rather on intuitive use of unreliable knowledge" (Jaeger et al. 2001, p. 225), thus linking back to the aspects and parameter crucial for the emergence of cognitive dissonance.

For struggling through such decision processes, some de-biasing thoughts may help to locate the individual position at a point where post-decisional cognitive dissonance either is avoided or can be reduced.

- Be aware of your limitations! Decision-making does neither have the status of pure academia nor of a profession in its traditional sense. Thus, a decision normally only represents the best available solution perceived subjectively under the given circumstances and time (Simon 1997).
- Think probabilistic! What is the probability of a favourable outcome based on your decision? (Schultze and Pfeiffer 2012). This especially helps in diagnosing critical twilight zones (Posner 2010).
- Search for the opposite! What is the value of the most non-conforming information and how likely is it to happen? What would be from your view the most counterfactual scenarios? (Kelly and Milkman 2013; Schultze and Pfeiffer 2012).
- Judge 'political' decisions! Especially of not (only) based on economic preference, make all considerations transparent and try to measure it—even if only by a 'low—medium—high' structure.
- Scarcity of neutral reference points: Who is providing you with the decision proposals? Where do information really come from? What are potential own interests or conflicts of interests of the sources? (Dummond 2014; Sleesman et al. 2012).
- Getting devil's advocates: Especially group decisions carry the risk of overly compliant participants acting in an illusion of unanimity and a culture of mental self-censoring leading to groupthink (Janis 1982). Nominate at least one participant acting as rational Devil's Advocate (an idea also supported by Haerle et al. (2016) and Hoffmann et al. (2017) or develop people that can provide critical feedback in a brutally honest way directly to you. Do not shot the messenger.

– Listen to your inner voice! What are the hardwired assumptions on which the
  decision is taken? Summarise to yourself why you are going to take the decision.
  Reflect internally if these reasons are comprehensive and true. Feeling conflicting
  cognitions already in the moment of the decision is a precursor of strong cognitive
  dissonance to appear if the decision has proven not being right.
– Limitations of decisions: Under which conditions would you reconsider the decision made? When would changing the own standpoint getting important? Avoiding
  giving up too soon or sticking too long on a decision is referring to the implicit
  but crucial question of escalation of commitment leading to cognitive dissonance
  (Drummond 2014).

As a third aspect, even best prepared decisions can turn into a failure. Supporting
the reduction of cognitive dissonance by fostering an attitude of dealing with mistakes and learning from errors links the topic of cognitive dissonance with research
in corporate learning culture and error culture. Admittance of own failures, even if
culturally undesired in leadership, might even foster the own position towards others
as malicious envy is reduced (Huang et al. 2018). A strong concept in this respect is
a way of judgement of results, where not the lucky outcome, but the capable way is
preferred (Drummond 2014). While this might be difficult to be applied in organisations as official incentive system, as a mental model it limits the negative impact
of cognitive dissonance. Clustering mistakes as either acceptable or not acceptable
based on focusing on the quality and thoroughness of the decision process much more
than on the result (Kelly and Milkman 2013) reduces the intrinsic negative feeling
to those cases only, where mistakes are linked to erroneous decision preparation.

Fourthly, inner stability towards negative incidents and own mistakes is a crucial
driver of avoiding cognitive dissonance. Mental resilience training does already exist
(Pickren 2014), however needs to be enriched by the aspect of cognitive dissonance.
Turning mistakes into valuable experiences is an essential learning process for personal growth and expertise development of balanced and self-reflective leaders. The
basic concept is to accept responsibility for an incident but avoiding the arousal of
cognitive dissonance. In terms of the theory, this can occur if the conflict between
two cognitions is reduced. One of the cognitions carries information about the risk
event. This cognition can hardly be denied without manipulating reality. The other
cognition contains the individual claim towards the role. The higher the level of self-
esteem, the more the arousal of cognitive dissonance is unavoidable. If, however,
self-confidence is embedded in a stable intrinsic understanding of the self-esteem,
the quality level an individual sets to him-/herself within the social environment
(Pickren 2014) and is matched it with a possible reality in order to avoid expected
painful losses.

Thus, resilient behaviour is the conscious definition of a balanced claim towards
one's role in the organisation in a more realistic way. Parts of that claim of self-
awareness consist of an inner stance that is in line with outside behaviour. This
will diminish inner pressure, based on the claim of doing what can be done within
one's capabilities and resources and being able to admit errors openly. Elimination
of cognitive dissonance following severe incidents is very difficult if not impossible.

Even though complete recovery is blocked, finding of algorithms on how to cope with the bad memories in order to get inner distance and to accept a situation can still influence and improve the way future decisions are approached and taken.

Finally, influencing organisational behaviour towards cognitive dissonance requests towards leadership to allow communication and exchange of experiences within the team and the organisation. Originally, learning effects from cognitive dissonance are internalised at the level of individuals. Avoiding that persons remain alone with their actor-centred perspective of cognitive dissonance, organisational learning allow the sharing of experiences and encourage team-centred learning. Sharing requires processes, context and forums: weekly exchange in a more informal manner, when people have shared their experiences and speak about what has happened, why it has happened and how they can learn from each other. By witnessing the reactions of others when bringing up issues, they will gain self-confidence. For instance, in banking, latest regulatory requirements regarding strengthening risk culture might be a chance to organisations towards the described view. Exceeding the fulfilment of well-intended regulations, room for free internal discussions on a shared learning platform is seen as a necessary social process behind formal regulatory processes.

24.5 Conclusion

The powerful impact of cognitive dissonance accompanies all individuals though their lives. For leaders, the arousal of dissonance carries the risk to reflect the unpleasant inner feeling towards the outside, resulting in adverse actions with negative consequences for others and the organisation.

Despite vast literature in the area of cognitive dissonance research, proposals of how to reduce the impact remain vague. Thus, from a theoretical, methodological and practical aspect, the key aspects presented are undoubtedly open for discussion and development. It is not a matter of claiming fix processes and contents, but rather defining common ground for future work and debate.

References

Akerlof, G. A., & Shiller, R. J. (2009). *Animal spirits—How human psychology drives the economy, and why it matters for global capitalism*. Princeton: University Press.

Amar, P. (2012). *Psychology du manager*. Paris: Dunot.

Bettinghaus, B., Goldberg, S., & Lindquist, S. (2014). Avoiding auditor bias and making better decisions. *The Journal of Accounting & Finance, May/June, 2014*, 39–44.

Cohan, J. A. (2002). "I didn't know" and "I was only doing my job": Has corporate governance careened out of control: A case study on Enron's information myopia. *Journal of Business Ethics, 40*, 275–299.

Cooper, J. (2007). *Cognitive dissonance: 50 years of a classic theory*. London: Sage.

Deutsche Bank Research. (2010). *Homo Oeconomicus oder doch eher Homer Simpson?* www. dbresearch.de.

Devine, P. G., Tauer, J. M., Barron, K. E., Elliot, A. J., & Vance, K. M. (1999). Moving beyond attitude change in the study of dissonance-related processes. In E. Harmon-Jones & J. Mills (Eds.), *Cognitive dissonance: Progress on a pivotal theory in social psychology* (pp. 297–323). Washington (DC): American Psychological Association.

Drummond, H. (2014). Escalation of commitment: When to stay the course? *The Academy of Management Perspectives*, 284, 430–446.

Evans, D. (2013). *Risk intelligence: How to live with uncertainty*. London: Atlantic books.

Festinger, L. (1957). *Theory of cognitive dissonance*. Bern: Hans Huber. Republished 2012.

Financial Crisis Inquiry Commission. (2011). *Crisis inquiry report: final report of the national commission on the causes of the financial and economic crisis in the United States*. Official Government Edition, U.S. Government Printing Office, Retrieved on 26.12.2015 from http://fcic. law.stanford.edu/report/.

Frey, D. (1986). Recent research on selective exposure to information. In L. Berkowitz (Ed.), Advances in experimental social psychology (pp. 41–80). New York: Academic Press.

Haerle, P., Havas, A., Kremer, A., Rona, D., & Samandari, H. (2016). *The future of bank risk management*. McKinsey Working Papers on Risk. Retrieved on 26.09.2017 from http://www. mckinsey.com/business-functions/risk/our-insights/the-future-of-bank-risk-management?cid= other-eml-alt-mip-mck-oth-1602.

Harmon-Jones, E., & Mills, J. (1999). An introduction to cognitive dissonance theory and an overview of current perspectives on the theory. In E. Harmon-Jones & J. Mills (Eds.), *Cognitive dissonance: Progress on a pivotal theory in social psychology* (pp. 3–21). Washington (DC): American Psychological Association.

Hinojosa, A. S., Gardner, W. L., Walker, H. J., Cogliser, C., & Gullifor, D. (2017). A review of cognitive dissonance theory in management research: Opportunities for further development. *Journal of Management, 43*(1), 170–199.

Hoffmann, N., Huber, M., & Smith, M. (2017). *An analytics approach to debiasing asset-management decisions*. McKinsey & Company Financial Services. Retrieved on 07.01.2018 from https://www.mckinsey.com/industries/financial-services/our-insights/an-analytics-approach-to-debiasing-asset-management-decisions.

Huang, K., Wood Brooks, A., Buell, R. W., Hall, B., & Huang, L. (2018). *Mitigating malicious envy: Why successful individuals should reveal their failures*. Working Paper 18-080. Retrieved on 10.02.2019 https://www.hbs.edu/faculty/Publication%20Files/18-080_56688b05-34cd-47ef-adeb-aa7050b93452.pdf.

Huang, L., & Pearce, J. L. (2015). Managing the unknowable: The effectiveness of early stage investor gut feel in entrepreneurial investment decisions. *Administrative Science Quarterly, 60*(4), 634–670.

Jaeger, C. C., Renn, O., Eugene, A. R., & Webler, T. (2001). *Risk, uncertainty, and rational action*. London: Earthscan.

Janis, I. L. (1982). *Groupthink*. Boston: Houghton Mifflin.

Kahnemann, D. (2012). *Thinking, fast and slow*. London: Penguin.

Kelly, T. F., & Milkman, K. L. (2013). Escalation of commitment. In E. H. Kessler (Ed.), *Encyclopedia of management theory* (Vol. 1, pp. 257–259). Thousand Oaks.

Mintzberg, H. (2009). *Managen*. San Francisco: Berrett-Koehler.

Pickren, W. E. (2014). *The psychology book*. New York: Sterling Publishing.

Posner, K. A. (2010). *Stalking the Black Swan: Research and decision making in a world of extreme volatility*. New York: Columbia University Press.

Richards, G. (2004). *Putting psychology in its place: A critical historical overview*. London: Routledge.

Schultze, T., & Pfeiffer, F. (2012). Biased information processing in the escalation paradigm: Information search and information evaluation as potential mediators of escalating commitment. *Journal of Applied Psychology, 97*(1), 16–32.

Simon, H. A. (1997). *Administrative behaviour: A study of decision-making processes in administrative organisations*. New York: Free Press.

Sleesman, D. J., Conlon, D. E., McNamara, G., & Miles, J. E. (2012). Cleaning up the big muddy: A meta-analytic review of the determinants of escalation of commitment. *Academy of Management Journal, 55*(3), 541–562.

Tetlock, P. E. (2000). Cognitive biases and organizational correctives: Do both disease and cure depend on the politics of the beholder? *Administrative Science Quarterly, 45,* 293–326.

Wiswede, G. (2012). *Einfuehrung in die Wirtschaftspsychologie*. Munich: Max Reinhardt Verlag.

Winfried Müller is CFO of BMW Bank GmbH in Munich. Prior to working for BMW Group, Dr. Müller held various international positions in different areas of banking, finance and risk management. He has worked and lived in China and Italy for several years. He holds a Phd from University of Gloucestershire and an M.A. from ESB European School of Business.

Chapter 25
How to Make A.I. Transformation More Likely to Succeed

Howard Yu and Jialu Shan

Abstract This chapter focuses on artificial intelligence (AI), drawing from examples of how companies invest in AI, why is has been challenging for established companies to truly unleash the full potential of AI as their core strategy and why the management team must think beyond the process of innovation as well as consider an alternative budgeting approach and capital structure to fuel the critical work surrounding AI.

25.1   Introduction

To get the most out of artificial intelligence (A.I.), companies need to think beyond having data, infrastructure, and off-the-shelf analytics, and redesign their investment processes.

"What are we learning about artificial intelligence in financial services?" asked Ms. Lael Brainard, one of the seven members of the Board of Governors of the US Federal Reserve. "My focus today is the branch of artificial intelligence known as machine learning, which is the basis of many recent advances and commercial applications," the governor told her audience in Philadelphia, Pennsylvania. "Due to an early commitment to open-source principles, A.I. algorithms from some of the largest companies are available to even nascent startups… So it is no surprise that many financial services firms are devoting so much money, attention, and time to developing and using A.I. approaches."

JPMorgan Chase is reportedly devoting some USD 10.8 billion to its tech budget in 2018. Europe's largest bank, HSBC, is spending USD 15 billion on new technology. And the biggest spender of all, Bank of America, has set an annual global budget of nearly USD 16 billion for technology and operations. That figure is at least USD 3 billion more than Intel, Microsoft or Apple spent on research and development in 2018. As Andrew S. Grove, the long-time chief executive and chairman of Intel

H. Yu (✉) · J. Shan
IMD Business School, Lausanne, Switzerland
e-mail: howard@howardyu.org

J. Shan
e-mail: jialu.shan@imd.org

© Springer Nature Switzerland AG 2020
N. Pfeffermann (ed.), *New Leadership in Strategy and Communication*,
https://doi.org/10.1007/978-3-030-19681-3_25

Corporation, told a Stanford researcher in 1991, "Don't ask managers, 'What is your strategy?' Look at what they do! Because people will pretend." Whether they are pretending or not, the resource allocation patterns suggest that banks are now effectively IT companies.

What Grove saw as the actual strategy of a firm is the cumulative effect of day-to-day prioritizations or decisions made by middle managers, engineers, salespeople, and financial staff—decisions that are made despite, or in the absence of, intentions. And that's where the problem lies. Money for new investments accounted for only 27% of bank spending on information technology in 2017. According to Celent, a research and consulting company based in Boston, the rest—close to 73% of spending—was allocated to system maintenance. Of the nearly USD 10 billion JPMorgan Chase dished out for IT in 2016, only USD 600 million was in fact devoted to fintech solutions and projects for mobile or online banking, although CEO Jamie Dimon warned shareholders in his letter that "Silicon Valley is coming."

This knowing-doing gap is no simple pretension by senior leadership. Financial institutes I've spoken with have (1) all organized employee seminars inviting motivational speakers to talk about innovation; (2) established corporate venture funds to invest in innovative startups; (3) practiced open innovation, posted challenges online, and run tournaments with external inventors; (4) organized "design thinking" workshops for employees to re-think customer solutions outside the mainstream; and (5) installed *Lean* startup methodologies that allow employees to fail fast in order to succeed early. So widespread is the innovation process, and yet, managers continue to face unyielding organizations whose core business is being encroached on by Google and Amazon, if not Tencent or Alibaba or some other digital upstarts. "Tell me one thing that I should do but haven't done," hissed an executive the moment I mentioned Google Venture. It seems that no matter how hard these in-house innovation experts try, their companies simply won't budge. The ships are not just big; they cannot turn. Why?

Too many innovation experts are focusing solely on the nuts and bolts of everyday implementation: gathering data, tweaking formulas, iterating algorithms in experiments and different combinations, prototyping products, and experimenting with business models. They often forget that the underlying technologies—A.I. in this case—never stay constant. Seizing a window of opportunity is not necessarily about being the first, but about getting it right first. In this instance, that means getting it right for banking clients. Doing so takes courage and determination, as well as vast resources and deep talent.

But the banking industry isn't where Silicon Valley comes first—the auto industry is.

25.1.1   How Likely Is It that Your Industry Will Be Disrupted by the Valley?

No automaker today would speak to investors without mentioning "future mobility." BMW is "a supplier of individual premium mobility with innovative mobility services." General Motors aims to "deliver on its vision of an all-electric, emissions-free future." Toyota possesses the "passion to lead the way to the future of mobility and an enhanced, integrated lifestyle." And Daimler, maker of Mercedes, sees the future as "connected, autonomous, and smart." In contrast to the personally owned, gasoline-powered, human-driven vehicles that dominated the last century, automakers know they're transitioning to mobility services based on driverless electric vehicles paid for by the trip, by the mile, by monthly subscription, or a combination of all three. In the past, mobility was created by individual cars automakers sold; in the future, mobility will be produced by service companies operating various kinds of self-driving vehicles in fleets over time. At the BMW Museum, anyone can witness the gravity of this vision, articulated by its chairman of the board firsthand.

Walking up the spiral ramp of a rotunda inside the BMW Museum, one sees flashes of pictures about BMW history that display in variable sequences, slipping in and out of view like mirages. At the very top of the museum is a "themed area" of about 30 stations demonstrating an emissions-free, autonomously driven future. These are not only a vision, but a real project, begun in earnest in the autumn of 2007 by then-CEO Norbert Reithofer and his chief strategist Friedrich Eichiner. The two men tasked engineer Ulrich Kranz, who had revived the Mini brand in 2001, to "rethink mobility." The task force soon grew to 30 members and moved into a garage-like factory hall inside BMW's main complex.

"I had the freedom to assemble a team the way I wanted. The project was not tied to one of the company's brands, so it could tackle any problem," Kranz said in an interview with *Automotive News Europe* in 2013. "The job was to position BMW for the future—and that was in all fields: from materials to production, from technologies to new vehicle architectures."

And so Kranz and his team decided to explore uncharted territory that included "the development of sustainable mobility concepts, new sales channels, and marketing concepts, along with acquiring new customers." The starting point for "Project i" was, in other words, a blank sheet of paper.

"We traveled to a total of 20 mega-cities, including Los Angeles, Mexico City, London, Tokyo, and Shanghai. We met people who live in metropolises and who indicated that they had a sustainable lifestyle. We lived with them, traveled with them to work, and asked questions," Kranz recalled. "We wanted to know the products that they would like from a car manufacturer, how their commute to work could be improved, and how they imagined their mobility in the future. As a second step, we asked the mayors and city planners in each metropolis about their infrastructure problems, the regulations for internal combustion engines, and the advantages of electric vehicles."

Once the findings came back, Kranz expanded his team by seeking out "the right employees both internally and externally." The result was BMW's gas-electric i8 sports coupe and all-electric i3 people mover, which shimmered under white lights at BMW World, where the company's top automotive offerings are showcased. The i3 had almost no hood, and the front grille was framed by plastic slits that looked like a pair of Ray-Bans. It came in a fun-looking burnt orange. The front seats were so vertically poised, with the dashboard stretching out, that they exuded a "loft on wheels" vibe. Like the interior, made of recycled carbon fiber and faux-wood paneling, the electric motor of the i3 was geared to urban dwellers in mega-cities who yearned for a calm, relaxing drive.

What made BMW all the more remarkable was its timing. Almost two years before Tesla's Model S was introduced, BMW had presented the battery-powered car as a revolutionary product, and committed to build it and deliver it to showrooms by 2013. When the BMW i3 went on sale, Tesla's Model S had spent just over a year on the US market. The 2014 i3 went on to win a World Green Car award, as did the 2015 model, the i8. In short, BMW was fast and early.

Then something terrible happened—or really nothing happened.

The i3 is now five years old, and the i8 is four. The BMW i brand includes the services DriveNow and ReachNow (car sharing), ParkNow (to find available parking), and ChargeNow (to find charging stations). But, besides being featured in occasional press releases, Project i has given way to other BMW sports cars in prime-time TV advertising spots. There's no news from Project i, except that project members are reportedly leaving. Ulrich Kranz, the former manager, joined former BMW CFO Stefan Krause at Faraday Future, and after a short stay, they started Evelozcity in California, where they recruited another i-model designer, Karl-Thomas Neuman. And Kranz is not alone. Carsten Breitfeld, former i8 development manager, is now CEO of Byton, where he also enlisted a marketing expert and a designer from the BMW team.

How much Project i has cost BMW, we'll never know. But if, according to BMW figures, the carbon-fiber production and the autobody works for the i3 set the company back some half a billion euros, the entire project could easily have cost two to three billion—a sum that would have been enough for the development of two to three series of a conventional VW Golf or Mercedes S-Class. Two to three billion euros is also more than fifteen times the USD 150 million Apple spent to develop the first iPhone, which launched in 2007. With so much bleeding, the new CEO Harald Krüger talked of Project i 2.0, a plan to integrate the BMW i sub-brand back into the parent company, and refocus distribution efforts on "classic" products.

In 2018, BMW USA reported just 7% of its sales were cars with a plug, which included all its hybrid offerings. Meanwhile, Tesla reported booming sales of its Model 3, which has become one of the USA's top 20 most-sold vehicles in the third quarter of 2018. Tesla was ranked fourth in luxury car sales during the same quarter. At the time of writing, Tesla has surpassed BMW and Daimler to become the world's second most valuable automaker in terms of market capitalization, trailing only Toyota.

Did Tesla and other start-up companies steal BMW's idea and run with it? No, it's what's called the *Zeitgeist*, a German word meaning "spirit of the time." When the time is ripe, the ideas are "in the air." Competition invariably emerges, and companies have to improve their ideas to stay ahead. They need to come up with demonstrations that excite potential customers, potential investors and, more importantly, potential distributors.

BMW's shift in its distribution of the i sub-brand echoes what Kodak did. Kodak built the first digital camera back in 1975 and was first to put out a competent product, but then ended up folding its consumer digital and professional divisions back into the legacy consumer film divisions in 2003. Meanwhile, Nikon, Sony, and Canon kept innovating in the subsequent decades, with features like face detection, smile detection, and in-camera red-eye fixes. We all know what eventually happened to Kodak.

25.1.2 How to Become Future-Ready

BMW is by no means a laggard in innovation. At IMD business school in Switzerland, we track how likely a firm is to successfully leap toward a new form of knowledge. For automakers, it's the shift from mechanical engineering, with combustion-engine experts, to electric and programming experts of the same kind as those who build computers, mobile games, and handheld devices. For consumer banking, it's the shift from operating a traditional retail branch with knowledgeable staff who provide investment advice to running data analytics and interacting with consumers the same way an e-commerce retailer would. The pace of change may differ between industries, but the directional shift is undeniable.

The IMD ranking measures companies in each industry sector using hard market data that is publicly available and has objective rules, rather than relying on soft data such as polls or subjective judgments by raters. Polls suffer from the tyranny of hype. Names that get early recognition get greater visibility in the press, which accentuates their popularity, leading to a positive cascade in their favor. Rankings based on polls also overlook fundamental drivers that fuel innovation, such as the health of a company's current business, the diversity of its workforce, the governance structure of the firm, the amount it invests in outdoing competitors, the speed of product launches, and so on. According to an objective composite index like this one, BMW is among the best. Table 25.1 below shows the ranking of the top 55 automakers and component suppliers. The methodology of the ranking is described in the appendix.

But the index also points to the general conservatism of large companies. Most radical ideas fail, and large companies can't tolerate failure. It doesn't matter whether you call BMW's strategy "throw everything at the wall and see what sticks" or a groundbreaking, iterative approach to mobility; if the only way to innovate is "to put a few bright people in a dark room, pour in some money, and hope that something wonderful will happen," Gary Hamel once wrote, "the value added by top management is low indeed."

Table 25.1 .

Company names	Score	Rank
Tesla Inc	100	1
General Motors Company	98.357	2
Volkswagen AG	93.221	3
Ford Motor Co.	82.265	4
Toyota Motor Corporation	82.235	5
Nissan Motor Co., Ltd.	81.442	6
Bayerische Motoren Werke AG	71.473	7
Daimler AG	69.570	8
Peugeot S.A.	63.488	9
Visteon Corporation	59.146	10
Honda Motor	56.223	11
AB Volvo	53.885	12
Renault	47.907	13
Ferrari NV	47.710	14
Robert Bosch GMBH	47.094	15
Fiat Chrysler Automobiles N.V.	43.215	16
Brilliance China Automotive Holdings Limited	42.935	17
Audi AG	42.428	18
Continental AG	41.911	19
Valeo SA	41.208	20
Denso Corporation	38.351	21
Cooper-Standard Holdings INC.	36.989	22
Baic Motor Corporation Limited	35.015	23
Skoda Auto, A.S.	34.876	24
Guangzhou Automobile Group	33.444	25
Yamaha Motor Co., Ltd.	32.383	26
Fuyao Glass Group Industries Co., Ltd.	31.058	27
Hyundai Motor Co., Ltd.	29.133	28
Jaguar Land Rover Limited	28.849	29
Aptiv PLC	28.638	30
Suzuki Motor Corporation	27.926	31
Byd Company Limited	27.702	32
Geely Automobile Holdings Limited	27.568	33
Magna International Inc	27.077	34
Mitsubishi Motors Corporation	24.689	35
Chaowei Power Holdings Limited	24.134	36

(continued)

Table 25.1 (continued)

Company names	Score	Rank
Mazda Motor Corporation	22.551	37
Subaru Corporation	22.213	38
Tata Motors Limited	21.093	39
Beiqi Foton Motor Co., Ltd.	20.672	40
Kia Motors Corporation	17.535	41
Isuzu Motors Limited	17.462	42
TS Tech Co., Ltd.	17.074	43
Haima Automobile Group Co., Ltd.	13.603	44
Paccar Inc	11.671	45
Aisin Seiki Co., Ltd.	11.655	46
Saic Motor Corporation Limited	10.135	47
Mahindra & Mahindra Limited	8.539	48
Harley-Davidson, Inc.	7.375	49
China Faw Group Co., Ltd.	6.358	50
Anhui Jianghuai Automobile Group Corp., Ltd.	5.043	51
Jiangling Motors Corporation, Ltd.	4.127	52
Dongfeng Motor Group Company Limited	2.925	53
Chongqing Changan Automobile Company Limited	0.181	54
Great Wall Motor Company Limited	0	55

But it's not just about cars. The dilemma experienced by German automakers is strikingly similar to the ones facing executives in banking and a host of other industries these days. Just as Detroit is confronted by Silicon Valley, Wall Street can see the future of banking everywhere it looks. Turning to China, it sees Alibaba, whose AliPay has become synonymous with mobile payment, and AntFinancial, Alibaba's finance subsidiary, which is now worth USD 150 billion—more than Goldman Sachs. Looking homeward, it sees that start-ups like Wealthfront, Personal Capital, and Betterment have all launched robo-advisors as industry disruptors. In retail checkout lanes, it sees Square or Clover or Paypal Here taking in credit card payments on behalf of millions of small-time merchants. It sees that the future of banking is not only about Big Data analytics, but also about calling on and bundling a group of financial services that happen in real time and with little human interaction. A smart infrastructure that automatically interacts with customers, continuing to improve its algorithm and adjust its response without human supervision as it handles data gushing in from all around the world at millions of bytes per minute, is tantamount to one giant leap forward for every banking incumbent.

Deep-learning-based programs can already decipher human speech, translate documents, recognize images, predict consumer behavior, identify fraud, and help robots "see." Most computer experts would agree that the most direct application of this sort of machine intelligence is in areas like insurance and consumer lending, where relevant data about borrowers—credit score, income, credit card history—is abundant, and goals such as minimizing default rates can be narrowly defined. This explains why, today, no human eyes are needed to process any credit request below USD 50,000. For these businesses, the question of where and how to deploy A.I. is easy to answer: find out where a lot of route decisions are made, and substitute algorithms for humans.

But data can be expensive to acquire, and investment conventionally involves a trade-off between the benefit of more data and the cost of acquiring it. For many traditional banking incumbents, the path to A.I. is anything but straightforward. Managers are often tasked with considering how many different types of data are needed. How many different sensors are required to collect data for training? How frequently does the data need to be collected? More types, more sensors, and more frequent collection mean higher costs along with the potentially higher benefit. In thinking through this decision, managers are asked to carefully determine what they want to predict, guided by the belief that this particular prediction exercise will tell them what they need to know. This thinking process is similar to the "re-engineering" movement of the 1990s, in which managers were told to step back from their processes and outline the objective they wanted to achieve before re-engineering began. It's a logical process, but it's the wrong one.

Consider the process of shopping at Amazon. Amazon's A.I. is already predicting your next purchase under "Inspired by your browsing history." Experts estimate the A.I.'s success rate at about 5%, which is no small feat considering the millions of items on offer. Now imagine if the accuracy of Amazon's A.I. were to improve in the coming years. At some point, the prediction would be enough to justify Amazon pre-shipping stuff to your home, and you'd simply return the things you didn't want. That is, Amazon would move from a shopping-then-shipping model to shipping then shopping, sending items to customers in anticipation of their wants.

The complication lies in *when* Amazon should introduce this A.I.-driven fulfillment service. With the underlying technology improving, Amazon might choose to launch such a service just a year ahead of the competition, when the A.I. prediction is not yet perfect, and suffer a hit on returns and a dip in profitability. Why? Because launching the service slightly sooner will give Amazon's A.I. more data sooner than the competition, which will mean its performance will improve slightly faster than that of others. Those slightly better predictions will in turn attract more shoppers, and more shoppers will generate more data to train the A.I. faster still, leading to a sort of virtuous cycle, a prohibitive lead against competition.

In fact, this data intelligence is the only first-mover advantage that matters. It grows from a positive feedback loop. The more data that is used, the more valuable the business becomes, since getting relevant data in quantity is always difficult and expensive. This is why Google Maps becomes more accurate as more people use it: the underlying algorithms have more data to work with, so the apps become even more accurate. Google has made two decades' worth of investments to digitize all aspects of its workflow, but not because it has a clear notion of what it wants to predict. It had done so before a clear notion of A.I. fully emerged. This is the groundwork that must be laid before a well-defined strategy for effective A.I. can be established. Any data scientists would agree that data sets become geometrically more valuable when you combine them. Combined data sets often reveal insights and business opportunities that could not have been imagined previously. Facebook's photo tagging expanded the social graph. News Feed enriched it further. The Like button deliver data on emotional triggers. Connect tracked users as they went around the web. The value is not in the photos and links posted by users. The real value resides in metadata—data about data—which describes where the user was when he or she posted, what they were doing, with whom they were doing it, alternatives they considered and more.

Put it differently, when Google introduce Gmail, it built a data set of identity. Combining the two data sets created a geometric increase in value, as future AdWords ads would have more value to the advertiser and, by extension, to Google. The same thing happened again with Google Maps, which enable Google to tie identity and purchase intent to location. Each time Google introduce a new service, but Google would find new use cases for user data made possible by combining data sets. One of the most value use cases that resulted from combining data sets was anticipation of future purchase intent based on detailed history of past behavior. When users get ads for things they were just talking about, the key enabler is behavior prediction based on combined data sets. Hence the importance of metadata. Its application potential only means the conventional budget allocation won't work for banking incumbents seeking to scale their footprints in the age of A.I. They have no choice but to follow a disruptive playbook, but with a twist.

25.1.3 How Understanding Disruption Helps Strategists

In the early 1990s, Professor Clayton Christensen of the Harvard Business School noticed an interesting pattern among companies facing the emergence of a new technology. When technological progress was incremental, even if the increments appeared in rapid succession, powerful incumbents always triumphed. Companies that were endowed with vast resources, extensive networks of suppliers, and a loyal customer base were able to command great advantages over their rivals, as would be expected. This is what made IBM a formidable player in the computing industry and General Motors a bellwether organization in the automotive industry.

And yet, there is a class of technological changes in which the new entrant—despite far fewer resources and no track record—almost always topples existing industry giants. This special class of technological changes, Christensen noted, does not have to be sophisticated or even radical.

Take transistor television as an example. When RCA first discovered transistor technology, the company was already the market leader in color televisions produced with vacuum tubes. It naturally saw little use for transistors beyond novelty, and decided to license the technology to a little-known Japanese firm called Sony.

Sony, of course, could not build a TV out of transistors, but it did manage to produce the first transistor radio. The sound quality was awful, but the radio was affordable for teenagers who were delighted by the freedom to listen to rock music away from the complaints of their parents. Transistor radios took off. Still, the profit margins were so low that RCA had no reason to invest further. It was busy making serious money and investing every R&D dollar on improving vacuum tube color TV.

Sony, meanwhile, was looking for the next big thing. It launched a portable, low-end, black-and-white TV at a rock-bottom price, targeting low-income individuals. Called the "Tummy Television," it was tiny enough to perch on one's stomach—the antithesis of RCA's centerpiece of middle-class living rooms. Why would RCA invest in transistors to make an inferior television for a less-attractive market? It didn't.

The real trouble began when Sony finally pushed the transistor's performance to produce color TVs based entirely on the new technology. Overnight, RCA found itself trying to catch up on a technology that it had ignored for the past three decades, which it had ironically pioneered and licensed out. Christensen called this type of technology—inferior at first but immensely useful later—disruptive, a term that has since been immortalized in the business lexicon of executives, consultants, and academics.

What we see today in the financial industry are new entrants leveraging digital interfaces and A.I. decision-making processes that involve minimal manual work to target an underserved market segment. Their technologies cannot satisfy high-end banking customers *yet*. But like the desktops that displaced minicomputers, or the angioplasty that displaced open-heart surgery, A.I. and digital automation will inevitably improve and, one day, these new solutions will be able to meet a substantial part of the need among big clients. The implication is that there will always be space for manual-intensive, human-centric operations, but that space will shrink substantially in the future.

One logical solution is for banking incumbents to create a separate unit and launch "speed boats" that adhere strictly to the playbook of digital disruptors. These will target an underserved market, and provide security services on a digital platform, with minimal human intervention. Initiatives like this are meant to develop a new set of capabilities—advanced analytics, dynamic product deployment, linking to third parties to fill a sudden surge in market demand—initially targeting a new segment that doesn't interfere with the mainstream business of the current banking operation. Over time, such new businesses will develop crucial capabilities that will mature enough to be transplanted back into the mainstream. This approach prevents the often-heard refrain of IT large-scale transformation: overtime, overbudget, and with underwhelming market results. In a way, it's RCA launching Sony's transistor radio, but keeping ownership of it to get future technologies ready.

And here is one last twist. Scaling up a disruptive business will always be costly. But without which, none of these matter. The late Andy Grove, Intel's legendary CEO pointed this out in his 2010 op-ed for Bloomberg:

> Startups are a wonderful thing, but they cannot by themselves increase tech employment. Equally important is what comes after that mythical moment of creation in the garage, as technology goes from prototype to mass production. This is the phase where companies scale up. They work out design details, figure out how to make things affordably, build factories, and hire people by the thousands. Scaling is hard work but necessary to make innovation matter.

And yet, scaling up disrupotion is where a company is likely to suffer financial loss for years, if not decades, and in the foreseeable future, carry with itself a business that is unlikely to achieve the same level of profitability of its core business. BMW has been profitable for a very long time; Tesla is still operating at a loss today, as is Uber.

That's why from Amazon to Square to Ant Financial, profitability is not the most important metric for managers; user base and market share are. That's also why banking incumbents need to consider an alternate investment structure, allowing third parties, venture capitalists, and even competitors to take an equity stake. Such a structure seems controversial, but is not unprecedented: Alibaba doesn't own all of Ant Financial; Uber now owns a minority share of its Chinese rival Didi after exiting China. This is the same strategy of GM's CEO Mary Barra, and it paid off handsomely in May 2018 when SoftBank announced a USD 2.25 billion investment in Cruise Automation—the self-driving unit of General Motors, headquartered in San Francisco. The investment pushed Cruise, originally purchased by GM for USD 581 million, to USD 11.5 billion. It does take more than vision, belief, passion and experimentation in A.I. to transform a company. It takes a pocket so deep that it requires other people's money to act on that aspiration. It's an unconventional approach taken during an unconventional time.

25.1.4   One Last Flashback…

Adjacent to the Mercedes-Benz museum in Stuttg art, Germany is one of the largest Mercedes dealerships in the world, which I also visited during the autumn of 2018. Its cavernous main hall is preceded by a restaurant, a café, and a shop hawking Mercedes-Benz merchandise. I saw a vertical banner stretching from the ceiling to the floor along the glass panels on one wall. "Ready to change," the banner cheered. "Electric intelligence by Mercedes-Benz." It referred to *Concept EQ*, a brand of electric plug-in models first unveiled in Stockholm on 4 September 2018. I found three EQs on display next to an exhibition kiosk that didn't work, but instead displayed an error alert and tangled cables spilling from the back that had come unglued.

Then an escalator took me to the top floor, where I found visitors gawking at a Mercedes-AMG, known for its "pure performance and sublime sportiness." Here was a vision of a forward-looking sports car with all its driving pleasure fully realized. The risers and the wrap-around LCD walls only accentuated the carbon-fiber composite of the chassis, gleaming in matte black. But I also noticed that the CO_2 emissions rating of this Mercedes-AMG GT 63 S, with its 630 horsepower, was an F.

Appendix

This appendix presents a short description of the calculation behind the "Leap readiness index" for the automotive industry in 2018.

This index includes the top 55 automakers and component suppliers by revenue by the end of 2017. The ranking measures four factors: (1) financial performance, (2) employee diversity, (3) research and development, and (4) early results of innovation efforts. These four main factors are tracked by 17 separate indicators that carry the same weight in the overall consolidated result.

Financial performance	Employee diversity	Research and development	Early results of innovation
• % of internatioanl sales last year • 3Y CAGR turnover • 3Y CAGR mkt cap • 3Y average profit change • P/E ratio last year	• % of women employees • % of women management board members • CEO demography • Headquarter competitiveness	• 3Y CAGR R&D intensity • 3Y average R&D intensity • 3Y CAGR R&D expenses	• Press count on "autonomous vehicles" • Press count on "EVs" • Press count on "connected cars" • Press count on "sharing mobility" • Press count on "corporate venturing"

All of our 17 indicators are hard data, i.e. they are publicly available from company websites, annual reports, press releases, news stories, or corporate responsibility reports.

In order to calculate the "Leap readiness index," we first manually collected historical data for each individual company. We then performed calculations for each indicator (e.g. 3-year compound annual growth rate) before we normalized criteria data by scaling it between 0 and 1.

For "early results of innovation," we identified five key trending topics in the automobile industry. These were autonomous vehicles, electric vehicles (EVs), shared mobility, connected cars, and corporate venturing. We consulted Factiva—a global news database that covers all premium sources—and counted the number of press releases on each trending topic over the past three years (2016–18). We then conducted the same normalization for these five indicators.

Finally, we aggregated indicators to build the overall ranking. For the purpose of comparison, we have ranked each company from 1 (best) to 55 (worst) on a scale of 0 to 100.

Howard Yu is the author of LEAP: How to Thrive in a World Where Everything Can Be Copied (PublicAffairs, 2018), LEGO professor of management and innovation at the IMD Business School in Switzerland, and director of IMD's signature Advanced Management Program (AMP). In 2015, Yu was selected by Poets&Quants as one of "The World's Top 40 Business Professors Under 40," and in 2018 he appeared on the Thinkers50 Radar list of thirty management thinkers "most likely to shape the future of how organizations are managed and led." He has delivered customized training programs for leading organizations including Bosch, Mars, Maersk, Daimler, and Electrolux. His articles have appeared in Forbes, Fortune, Harvard Business Review, The Financial Times, and The New York Times. Yu received his doctoral degree from Harvard Business School. Prior to his beginning his doctorate, he worked in the banking industry in Hong Kong.

Jialu Shan is a Research Associate at The Global Center for Digital Business Transformation—An IMD and Cisco Initiative. Her research areas include digital business transformation, business model innovation and new practices, and corporate governance practices. Jialu has published works on a variety of topics, including business model innovation and digital innovation in the Asian market. She obtained her Ph.D. degree in Economics (management) from the Faculty of Business and Economics at the University of Lausanne (HEC, Lausanne) in 2012. Before joining IMD she worked as a lecturer at the International Hotel School of César Ritz Colleges in Brig, Switzerland.

Printed in the United States
By Bookmasters